ACS SYMPOSIUM SERIES **924**

Polymeric Drug Delivery II

Polymeric Matrices and Drug Particle Engineering

Sönke Svenson, Editor
Dendritic NanoTechnologies, Inc.

**Sponsored by the
ACS Division of Polymeric Materials: Science and Engineering, Inc.**

American Chemical Society, Washington, DC

Library of Congress Cataloging-in-Publication Data

Polymeric drug delivery / Sönke Svenson, editor ; sponsored by the ACS Division of Polymeric Materials: Science and Engineering, Inc.

p. cm.—(ACS symposium series ; v. 923-)

"Developed from a symposium sponsored by the Division of Polymeric Materials: Science and Engineering, Inc., at the 226th National Meeting of the American Chemical Society, New York, September 7–11, 2003"—T.p. verso.

Includes bibliographical references and indexes.

Contents: v. 1. Particulate drug carriers

ISBN 13: 978–0–8412–3918–0 (alk. paper)

1. Polymeric drug delivery systems—Congresses. 2. Polymeric drugs—Congresses.

I. Svenson, Sönke, 1956- II. American Chemical Society. Division of Polymeric Materials: Science and Engineering, Inc. III. American Chemical Society. Meeting (226th : 2003 : New York, N.Y.) IV. Series.

RS201 P65P6412 2005
615′.3—dc22 2005053686

The paper used in this publication meets the minimum requirements of American National Standard for Information Sciences—Permanence of Paper for Printed Library Materials, ANSI Z39.48–1984.

Distributed by Oxford University Press

This volume: ISBN 13: 978–0–8412–3976–0 ISBN 10: 0–8412–3976–2

PRINTED IN THE UNITED STATES OF AMERICA

Polymeric Drug Delivery II

Foreword

The ACS Symposium Series was first published in 1974 to provide a mechanism for publishing symposia quickly in book form. The purpose of the series is to publish timely, comprehensive books developed from ACS sponsored symposia based on current scientific research. Occasionally, books are developed from symposia sponsored by other organizations when the topic is of keen interest to the chemistry audience.

Before agreeing to publish a book, the proposed table of contents is reviewed for appropriate and comprehensive coverage and for interest to the audience. Some papers may be excluded to better focus the book; others may be added to provide comprehensiveness. When appropriate, overview or introductory chapters are added. Drafts of chapters are peer-reviewed prior to final acceptance or rejection, and manuscripts are prepared in camera-ready format.

As a rule, only original research papers and original review papers are included in the volumes. Verbatim reproductions of previously published papers are not accepted.

ACS Books Department

Contents

Engineered Drug Particles

Preface

Achieving the therapeutic drug concentration at the desired location within a body during the required length of time without causing local over-concentrations and their adverse side effects is the main challenge in the development of a successful drug delivery system. Key factors to be considered in this effort are drug bioavailability, biocompatibility (or toxicity for that matter), targeting, and drug-release profile. The bioavailability of class II and III drugs (by the Biopharmaceutical Classification System of Drugs) is challenged by their poor water solubility and poor membrane permeability, respectively. Poor water solubility results in the rejection of about 40% of newly developed drugs by the Pharmaceutical Industry, while about 20% of marketed drugs exhibit suboptimal performance due to this challenge. Strategies to overcome this problem include the use of particular drug carriers such as liposomes, (polymeric) micelles, dendrimers, emulsion droplets, and engineered micro- and nanoparticles. These carriers encapsulate a drug, this way enhancing their water solubility (bioavailability) and reducing their toxicity (enhanced biocompatibility). In addition, carriers can transport drugs to the desired location by either passive targeting (enhanced permeability and retention, EPR, effect) or by active targeting through ligands that interact with receptors that are overexpressed at the surface of tumor cells. Alternatively, problem drugs can be encapsulated into depot matrices, whose release profile, defined by diffusion and matrix degradation, determines the concentration level of the free drug. The third strategy does not employ auxiliaries such as carriers or matrices but focuses on particle engineering. For example, reducing the size of a drug particle will increase its surface area and enhance its solubilization rate. All three strategies are explored in university laboratories as well as by the Pharmaceutical Industry.

Thus the main motivation for organizing the *Polymeric Drug*

Delivery: Science and Application symposium during the 226th National Meeting of the American Chemical Society (ACS) in New York, New York, in September 2003, was to provide a forum for both communities, academia and industry, to discuss progress in these three strategic approaches. Seventy-five well-recognized international experts, equally representing both communities, had been invited to present their research being conducted in the United States (57), the United Kingdom (4), Canada (3), Germany (3), France (3), the Netherlands (3), Italy (1) and Spain (1). Highlights from this symposium are presented in two volumes within the ACS Symposium Series: *Polymeric Drug Delivery I:Particulate Drug Carriers* (ACS Symposium Series 923) and *Polymeric Drug Delivery II:Polymeric Matrices and Drug Particle Engineering* (ACS Symposium Series 924). To provide an even broader overview, two contributions from Japanese researchers have been included in the books.

The 44 chapters selected for publication within the Symposium Series are divided into three main sections. Following an overview, the first section describes in twenty chapters the use of carriers such as liposomes, micelles, dendrimers, emulsion droplets, nanoparticles, and yeast cells in the (targeted) delivery of poorly water-soluble drugs, small organic molecules, macromolecules such as proteins and nucleic acids, and metal ions for molecular imaging purposes. Different routes of application are described, including oral and transepithelial delivery (Volume I).

The second section, with 13 chapters, is devoted to the use of polymeric matrices. The application of polymer solutions, nanogels, hydrogels, and millirods in gene and drug delivery, as well the prediction of drug solubility in polymeric matrices, the *in situ* study of drug release from tablets by Fourier-Transform infrared imaging, and antimicrobial release coatings are presented.

The third section details in nine chapters the use of supercritical fluids, controlled precipitation processes, and application of excipients in particle engineering and size control. The final chapter describes the preparation of fast dissolving tablets, another approach to increase the availability of drugs. These two sections, supplemented by an overview of the topics, are presented in Volume II.

These books are intended for readers in the chemical and pharmaceutical industry and academia who are interested or involved in drug delivery research as well as for advanced students who are interested in this active and rapidly developing research area.

I am using this opportunity to congratulate three presenters at the symposium and coauthors of this book, whose contributions received special recognition during the ACS National Meeting. Dr. Theresa Reineke and her co-worker Yemin Liu (University of Cincinnati), received the Arthur K. Doolittle Award for Fall 2003 for the best paper presented at the ACS Division of Polymeric Materials: Science and Engineering, Inc., (PMSE) meeting for their presentation on "Synthesis and Characterization of Polyhydroxyamides for DNA Delivery". Dr. Brian Johnson (Princeton University), received the 2003 ICI Student Award in Applied Polymer Science for his presentation on "Nanoprecipitation of Organic Actives Using Mixing and Block Copolymer Stabilization".

I deeply appreciate the willingness of the authors to contribute to this important overview of drug delivery technologies. I also greatly appreciate the help of many researchers, who have devoted their time to this project by reviewing these contributions, ensuring clarity and technical accuracy of the manuscripts. I thank the PMSE for the opportunity to hold the symposium as a part of their program and especially Ms. Eileen Ernst, the PMSE Administrative Assistant and "Florence Nightingale" to me, for her invaluable help with placing all abstracts and preprints into the system. I appreciate the patience and support of the ACS Symposium Series acquisitions and production team during the production of these books.

Last but not least I thank the ACS Corporation Associates (CA) as the "gold sponsor"; The Dow Chemical Company as the "silver sponsor"; and Elan NanoSystems; Epic Therapeutics, Inc.; Guilford Pharmaceuticals, Inc.; Johnson and Johnson, Center for Biomaterials and Advanced Technologies; and Thar Technologies, Inc. as "bronze sponsors" for their financial support of the symposium. It would have been impossible to assemble this outstanding group of researchers without these contributions.

Sönke Svenson
Dendritic NanoTechnologies, Inc.
2625 Denison Drive
Mount Pleasant, MI 48858
(989) 774–1179 (telephone)
(989) 774–1194 (fax)
Svenson@dnanotech.com (email)

Polymeric Drug Delivery II

Overview

Chapter 1

Advances in Polymeric Matrices and Drug Particle Engineering

Sönke Svenson

Dendritic Nanotechnologies, Inc., 2625 Denison Drive, Mt. Pleasant, MI 48858

An overview is given presenting various strategies to deliver small molecule and macromolecule drugs from polymeric matrices. In addition, drug particle engineering processes are described, including the application of supercritical fluids, controlled precipitation, and co-crystallization of drugs with polymeric templates, with the goal to facilitate drug solubility in water, and therefore, enhance drug bioavailability.

Introduction

There is a considerable interest in the development of drug delivery systems for small molecule and macromolecule drugs that provide stable and controllable blood levels over time to minimize the number of required administrations, and thereby improve patient compliance, and to prevent temporary drug overconcentration with its often adverse side effects. Besides particular drug delivery routes, two major approaches are followed to achieve this goal. *(1,2)* In the first approach, the drug molecules are embedded into a polymeric matrix or the drug molecules are co-crystallized with a polymeric template. In some cases the drug is chemically linked to the matrix, creating a prodrug delivery system. Drug release is a result of either drug diffusion from the matrix and/or degradation of the polymeric matrix. In case of prodrugs, the chemical link to the matrix has to be hydrolyzed as well, triggered either by a change of the local pH or the presence of enzymes. Polymeric matrices utilized in this approach are microspheres, polymer strands, millirods, nanogels, and more 'classical' hydrogels. Desired routes of application include injection and oral delivery.

The second approach relies on engineering the drug particles through processing protocols. Examples for this approach include (i) drug precipitation in the presence of a supercritical fluid, (ii) spray-freezing of drug particles, (iii) controlled precipitation in the presence of a crystal growth inhibitor, (iv) milling of larger drug particles to the desired particle size, and (v) coating of drug particles with a polymeric layer or coating of the drug molecules onto a crystalline surface.

Several of these systems and their utilization in drug delivery and gene transfection are being highlighted in this overview and some will be discussed in detail in the following chapters of this volume.

Results and Discussion

Injectable Polymeric Matrices

Most traditional drug delivery systems involve either suspending a drug into a biodegradable polymer monolith and surgically implanting it into the body, or encapsulating a drug into polymeric microspheres and injecting them subcutaneously. Injectable drug delivery based on polymer solution platforms consist of a biodegradable polymer dissolved in a biocompatible solvent along with the desired drug, usually in the form of suspended particles. A unique feature is that polymer phase inversion occurs on injection of the depot solution into the water-based physiologic environment. Thus the membrane carrier forms *in vivo*, simultaneous with the onset of the drug release. Consequently, the thermodynamic and mass transfer characteristics associated with the *in vivo* liquid demixing process play important roles in establishing the membrane morphology and drug release kinetics. Fast phase inverting systems and a mathematical model describing their behavior are presented in chapter 2 of this volume.

Biodegradable, injectable polymeric microsphere formulations have been commercialized for small molecules, peptides, and proteins. An important principle in developing stable microsphere formulations is minimizing molecular mobility during processing, after administration, and during storage. For example, a process has been developed and commercialized for encapsulating proteins by first suspending a lyophilized form of the drug in a biodegradable polymer solution, spray-freezing the suspension into liquid nitrogen to form nascent microspheres, extracting the polymer solvent into a polymer non-solvent at temperatures below 0°C, and finally drying the product under vacuum to minimize residual solvents. The mechanism of drug release from microsphere formulations involves (i) hydration of the microspheres, (ii) fast dissolution of drug or protein molecules located at the surface or with fast surface access via pores (burst), and (iii) slow drug dissolution through additional pores and channels created by degradation of the polymer matrix. The relative size of the drug particles to the microsphere is an important parameter in controlling the initial release or burst. Encapsulating sub-micron protein particles results in substantially lower initial release than encapsulating larger particles. Finally, the duration of drug release can be controlled by polymer degradation characteristics. For example, blending of poly(lactide-*co*-glycolide) with an amphiphilic polymer is an important tool in tailoring release profiles to meet clinical needs. *(3)*

Matrices based on Polyesters

Poly(orthoesters) (POE) have been under development since the early 1970's and four distinct families have been synthesized. However, long reaction times, scale-up difficulties, and poor control of the molecular weight have so far lowered the commercial potential of POE, despite interesting properties such as excellent ocular biocompatibility. POE hydrolysis proceeds in three consecutive steps. First, the lactic or glycolic acid segment in the polymer backbone hydrolyzes to generate a polymer fragment containing a carboxylic acid end-group that will lower the pH and catalyze orthoester hydrolysis. A second cleavage produces free α-hydroxy acid, the major catalyst for the hydrolysis of the orthoester linkages. Further hydrolysis proceeds in two steps to first generate the diol, or mixture of diols, used in the synthesis and pentaerythritol dipropionate, followed by ester hydrolysis to produce pentaerythritol and propionic acid. Details about synthesis and properties of this very interesting polymeric matrix are provided in chapter 3 of this volume.

Novel biodegradable poly(phosphoester) copolymers containing both phosphonate and lactide ester linkages in the polymer backbone, which degrade *in vivo* into nontoxic residues, have been synthesized by bulk or solvent polymerization techniques. Paclitaxel, incorporated into poly(phosphoester) microspheres by standard methods, was continuously released *in vitro* and *in vivo* over at least 60 days. The PACLIMER® formulation was tested for safety and toxicity in mice, rats and dogs. A no-adverse effect level for PACLIMER® microsphere administrations could not be determined but tolerability was demonstrated and a maximum tolerated dose was identified. Pilot studies in human ovarian patients have shown good tolerability of the drug, with continuous release of paclitaxel from the microspheres over at least 8 weeks at plasma concentrations far below those causing systemic toxicity.

Poly(phosphoester) matrices are described in more detail in chapter 4 of this volume.

The processes used to prepare injectable microspheres typically involve the use of solvents. A new class of absorbable polyester waxes for drug delivery has been developed that allow preparation of microspheres in high yields using solvent-free processes. These polymers are made by polycondensation of monoglycerides and diacids or anhydrides, resulting in an aliphatic polyester backbone with pendant fatty acid ester groups. The expected hydrolysis products are glycerol, a dicarboxylic acid and a fatty acid, all of which are known to be biocompatible. For example, microspheres containing risperidone pamoate as model drug were prepared from poly(monostearoyl glyceride-*co*-succinate) using the spinning disk process. A single dose intramuscular pharmacokinetic study was performed in Beagle dogs using formulations with 25% and 32% drug. The mean plasma concentration was measured, indicating that the 25% drug formulation did not give burst release and provided 30 days of sustained release at target drug plasma levels (≥10 ng/mL), while the 32% drug formulation gave a small burst followed by over 21 days of sustained release at target levels. *(4)*

Drug molecules chemically linked to poly(ethylene glycol) (PEG prodrugs) have an altered biodistribution compared to the free drug, leading to longer half-lives, enhanced drug levels in the blood (area under the curve, AUC), drug release profiles without or reduced burst, and enhanced drug accumulation in tumors through the Enhanced Permeability and Retention (EPR) effect. For example, optimized delivery was demonstrated for the hydroxyl-containing anticancer drugs paclitaxel and camptothecin, chemically linked through ester bonds to PEG of molecular weights 20-40 kDa. Using appropriate linker molecules, PEG could also be linked to amino-containing drugs such as the anticancer agents daunorubicin and doxorubicin. Branching of PEG at the distal ends produced highly loaded prodrugs. Drug release from the carrier matrix can be controlled by specific release mechanisms (e.g., 1,6-benzyl elimination). *(5)*

Matrices based on Polyanhydrides

A newer class of degradable polymeric matrices is composed of hydrolytically labile anhydride linkages. *(6)* These building blocks are easily modified with vinyl moieties to create a photopolymerizable and crosslinkable system. The mild conditions of photopolymerization are conducive to *in situ* formation of the network. Especially appealing is the ability to tailor the crosslinking density to match the requirements of a particular application, e.g., controlled release rates. For example, 2-hydroxyethylmethacrylate (HEMA) has been copolymerized with [1,3-*bis*(carboxyphenoxy)]propyl dimethacrylate to prepare degradable crosslinked copolymer networks. The kinetics of the copolymerizations as well as mass loss and swelling profiles were elucidated, in addition to the release kinetics of various encapsulated materials, all determined as a function of the crosslinking density. It was found that the mass loss followed a surface degradation mechanism for a network composed of 70 wt% crosslinker, but resembled a bulk eroding system more when the crosslinker content was reduced to 30 wt%. *(7)*

In an alternative approach to crosslinking the polymeric matrix via photo-polymerization, crosslinked amino acid-containing poly(anhydride-*co*-imides) have been synthesized and characterized. These polymers are crosslinked exclusively via anhydride linkages, and therefore, can degrade into water-soluble molecules such as amino acid and natural fatty acid. This alternative crosslinking approach avoids the formation of poly(methacrylic acid) as a hydrolysis by-product, a nondegradable macromolecule with poor bio-compatibility. Preliminary *in vitro* results indicated that these polymers can be used as bioerodible supports in controlled drug delivery applications for up to 70 hours. *(8)*

Matrices based on Polyacetals

The prevalence of carbohydrates in naturally occurring interface structures raises the question of whether a biomaterial built of common acyclic carbo-hydrates would have the features necessary for advanced pharmacological engineering, i.e., inertness *in vivo* (non-bioadhesiveness or 'stealth' properties), biodegradability of the main chain, and low toxicity. These interface carbo-hydrates have common non-signaling substructures, i.e., acetal and ketal groups. Hydrophilic polymers (polyals) consisting of these carbohydrate substructures should, therefore, be highly biocompatible. Acyclic hydrophilic polyals can be prepared via either polymerization of suitable monomers or lateral cleavage of cyclic polyals (e.g., polysaccharides). Polyals of various types have been prepared and characterized *in vitro* and *in vivo* as model components of bioconjugates. For example, matrices assembled from 20 kDa poly(L-lysine) as backbone and 10 kDa polyals as protective graft showed strong correlation of blood half-life in rats with the number of polyal chains per backbone. Matrices modified with 10 and 20 polyal grafts per backbone had half-lives of 9.8 and 25.3 hours, while the half-life of unprotected polylysine was ca. 20 seconds. *(9)*

Nanogel, Hydrogel, and Electrogel Matrices

There is a vital need for efficient small (<100 nm), biocompatible vehicles for the delivery of drugs, proteins (enzymes, insulin, antigen protein), and nucleic acids (DNA plasmids) or the extraction of toxics from body fluids. Appropriately modified nanogels meet this need as they are porous in nature. Nanogels are crosslinked macromolecular networks made from water-dispersible, biocompatible polymers that can be designed to either deliver or rapidly absorb large quantity of desired molecules with a response time in seconds *versus* days for 'classical' hydrogels. Nanogels can be formed via chemical crosslinking or physical self-assembly of polymers. Both approaches are discussed in detail in chapters 5 and 6 of this volume.

Matrices based on triblock copolymers made from poly(L-lactide)-poly(ethylene glycol)-poly(L-lactide) (PLL-PEG-PLL) form elastic hydrogels with potential applications in drug delivery. Rheology studies on these gels, formed with varying lengths of the hydrophobic PLLA blocks, revealed strong dependence of the hydrogel strength on the PLLA block length, thus offering a mechanism to control its mechanical properties. The gel strength is dependent upon the network structure, which in turn governs the degradation behavior of the gels and hence the release rate of bioactive molecules. Equilibrium properties of hydrogels such as drying/swelling isotherm, osmotic pressure and water activity were subject of another study, using poly(ethylene oxide)-poly(propylene oxide) diblock copolymers. It was found that swelling of these hydrogels is a diffusion-limited process, while drying is evaporation-limited. A diffusion model to fit water loss or gain as a function of time has been applied to extract useful parameters such as water diffusion coefficients in these hydrogels. In another study, PEO hydrogels were synthesized directly in water or physiological medium by free radical homopolymerization of telechelic PEO macromonomers. Their ability to serve as semi-permeable, biocompatible membranes for an artificial pancreas was examined. *In vitro* tests confirmed their good biocompatibility. Both glucose and insulin diffused through these hydrogels but the behavior of the latter was more complex. The crosslinking reaction could be extended to include direct encapsulation of biologically active materials such as hepatocytes. Details of these studies are discussed in chapters 7 to 9.

The polymeric constituents of hydrogels can be engineered following patterns found in nature. For example, silk-elastinlike protein polymers are a class of genetically engineered block copolymers composed of repeated silk-like (GAGAGS) and elastin-like (GVGVP) peptide blocks. Some of these polymers self-assemble in water and have been extensively studied for drug delivery and gene transfection applications. The thermal characterization, *in vitro* and *in vivo* delivery of plasmid DNA, and the potential of controlled adenoviral delivery from one of these polymers is being reviewed in chapter 10.

Burst release of drugs from a polymeric matrix, often an undesired side effect of matrix-based delivery, can be utilized for a dual-release approach in anticancer therapy. Biodegradable polymer millirods, composed of poly(D,L-lactide-*co*-glycolide (PLGA) and impregnated with anticancer drugs have been designed to be implanted in tumors after image-guided radiofrequency ablation, an alternative therapy to surgery for patients with unresectable tumors. First, burst release will rapidly raise the drug concentration in the surrounding tissue, followed by sustained release to maintain this drug concentration for an extended period of time. This dual-release approach is described in detail in chapter 11.

One major challenge of drug delivery from polymeric matrices is predicting the drug solubility in the respective matrix to optimize permeation flux and meet the therapeutic dose requirement. Although a critical parameter, there is no easy quantitative measurement method available. Therefore, a drug solvation parameter model has been developed for drug delivery from adhesive matrices that replaces the hard to measure quantity by two easily accessible parameters. Details of this model are described in chapter 12 of this volume.

In most delivery systems, drug release is controlled by the interactions between drug and carrier. Weak interactions often result in initial burst release while strong interactions delay or even prevent complete release of the drug. Controlled storage and delivery of drugs from a hydrogel-based system would therefore represent a significant advance in pharmaceutical delivery. A novel redox-activated hydrogel (electrogel) has been developed that achieves trapping and release of drugs by a single mechanism, the reversible switching of the gel from a reduced hydrophobic (storage) state to an oxidized hydrophilic (release) state. The electrogel, derived from a self-doped electroactive polythiophene, is compatible with small molecule drugs in all charge states and has the ability to deliver drugs in molecular or particulate (crystalline, powder, excipient-formulated) forms. Highly potent but extremely toxic anticancer drugs can thus be chemically isolated, sequestered, and delivered locally to tumor sites using the electrogel. *(10)*

Besides using polymeric matrices for drug delivery applications, these materials, when loaded with silver particles, can be utilized in antimicrobial coatings. Loading a polymer coating with a small amount of silver nanoparticles (10 $\mu g/cm^2$) has shown to completely inhibit growth of the Gram-positive bacterium *Staphylococcus aureus*. This study is presented in chapter 14 of this volume.

Visualization of processes occurring within matrix formulations during drug release such as water uptake, drug diffusion, and polymer dissolution or erosion, are crucial for understanding the mechanism of drug release upon contact with dissolution media. A new experimental approach, Fourier Transform infrared (FTIR) spectroscopic imaging, utilizing the macro attenuated total reflection (ATR) IR methology, has been developed to obtain more detailed insight into these processes. The application of this new method to formulations of ibuprofen in PEG matrices is being described in chapter 13.

Effect of Coating and Processing on Polymer Matrices

For some applications, i.e., dosing to elderly and children having problems with swallowing tablets, or in cases where immediate drug release is essential, the availability of fast disintegrating tablets is very important. The key properties of fast dissolving matrices are fast absorption of water into the matrix core and immediate break-up of the matrix to release its load. The use of a highly water-soluble carbohydrate, D-mannose, as the main constituent of a fast disintegrating tablet is being described in chapter 23.

A major challenge for matrix-based delivery is the distribution of drug molecules within the matrix. The formation of drug aggregates, crystallization and changes of the crystalline morphology will affect the dissolution rate, and therefore, the bioavailability of drugs released from the matrix. To obtain formulations containing molecularly dispersed drugs, melt extrusion is gaining more attention in the pharmaceutical industry. Several formulations of the drug fenofibrate have been prepared by hot melt extrusion, and the absence of crystalline drug material was established using differential scanning calorimetry (DSC) and X-ray diffraction. *(11)*

Drug Particle Engineering Using Supercritical Fluids or Liquefied Gases

Supercritical fluids (SCF), i.e., fluids above their critical pressure and temperature, have been the focus of technical and commercial interest in drug particle engineering. Supercritical solutions of drugs and excipients can be rapidly expanded to precipitate the drugs as particles (Rapid Expansion of Supercritical Solutions, RESS). Several variations to this process have been developed. Drugs and excipients that are insoluble in SCF can be dissolved in organic solvents and expanded by an SCF, triggering precipitation of the drug as fine particles. The organic solvent is completely extracted by the SCF anti-solvent (Gas anti-solvent process, GAS). A specific mode of this method is to atomize/spray the solution inside an SCF vessel. Such processes are termed Supercritical Anti-solvent process (SAS), Precipitation under Compressed Anti-solvent (PCA) or Aerosol Solvent Extraction Systems(ASES). *(12)* Several processes to produce drug particles using SCF are described in chapters 15 to 17 of this volume.

Liquid CO_2 as the anti-solvent has been utilized to prepare formulations containing a drug (e.g., griseofulvin) and a crosslinked polymer by swelling the polymer/drug mixture in an organic solvent and removing the solvent with the anti-solvent. *(13)* In a similar way, fine particles of drugs and excipients have been prepared by spraying the organic solution of both under pressure into a cryogenic liquid such as liquid nitrogen (Spray-Freezing into Liquid, SFL) and collecting the particle through lyophilization, and by spraying an alcohol solution of drug and excipient onto a frozen surface, resulting in ultra-rapid freezing, which prevents particle growth during precipitation. Both methods are described in chapters 20 and 21.

Controlled Precipitation in the Presence of a Crystal Growth Inhibitor

Controlled precipitation can be carried out by adding the solution of a poorly water-soluble drug in a water-miscible solvent into water in the presence of a polymeric crystal growth inhibitor. The anti-solvent water triggers drug crystallization, while the excipient prevents crystal growth. In a similar way small molecules, peptides, proteins, antibodies, DNA and other oligonucleotides have been precipitated in the presence of water-soluble polymers in form of microspheres with controllable size distribution (PROMAXX formulation process). In a variation of these approaches, hydrophobic drugs and amphiphilic diblock copolymers dissolved in an organic solvent have been rapidly mixed with a miscible anti-solvent, using a confined impinging jets mixer (Flash Nano Precipitation Process). These approaches are described in chapters 18, 19 and 22. Another way of controlling the crystalline phase and morphology of drug crystals utilizes organized organic assemblies, i.e., self-assembled or Langmuir monolayers on water, as a template. In a proof of concept study, the polymorphism of calcium carbonate crystals has been successfully controlled. *(14)*

Conclusion

An impressive 'toolbox' has been developed to deliver drug molecules from polymeric matrices or to engineer drug particles and crystals to achieve constant blood level concentrations of these drugs and control their release from a matrix or their dissolution behavior from their respective formulations.

References

1. Svenson, S. (ed.), *Carrier-based Drug Delivery*, ACS Symposium Series, Vol. 879, American Chemical Society, Washington, DC, **2004**.
2. Svenson, S. (ed.), *Polymeric Drug Delivery I – Particulate Drug Carriers*, ACS Symposium Series, Vol. 923, American Chemical Society, Washington, DC, **2006**.
3. Tracy, M.A. Controlling Stability and Release of Drugs from Biodegradable Microspheres. *PMSE Preprints*, 226th National Meeting of the American Chemical Society, New York, NY, September **2003**.
4. Nathan, A.; Borgia, M.; Twaddle, P.; Kataria, R.; Cui, H.; Peeters, J.; Verreck, G.; Monbaliu, J.; Fransen, J.; Borghys, H.; Brewster, M.; Arnold, S.; Rosenblatt, J. Absorbable Polyester Waxes for Drug Delivery. *PMSE Preprints*, 226th National Meeting of the American Chemical Society, New York, NY, September **2003**.

5. Greenwald, R.B.; Choe, Y.H.; McGuire, J.; Conover, C.D. Effective Drug Delivery by PEGylated Drug Conjugates. *Adv. Drug Del. Rev.* **2003**, *55*, 217-250.
6. Anseth, K.S.; Quick, D.J. Polymerizations of Multifunctional Anhydride Monomers to Form Highly Crosslinked Degradable Networks. *Macromol. Rapid Commun.* **2001**, *22*, 564-572.
7. Davies, K.A.; Anseth, K.S. Degradable Poly(2-Hydroxyethyl Methacrylate) Biomaterials: The Influence of Crosslinking Density on Network Properties. *PMSE Preprints*, 226th National Meeting of the American Chemical Society, New York, NY, September **2003**.
8. Cheng, G.; Aponte, M.A.; Ramirez, C.A. Synthesis and Characterization of Crosslinked Amino Acid-containing Polyanhydrides for Controlled Drug Release Applications. *PMSE Preprints*, 226th National Meeting of the American Chemical Society, New York, NY, September **2003**.
9. Papisov, M.; Yurkovetskiy, A.; Yin, M.; Leone, P.; Fishman, A.J.; Hiller, A.; Sayed, S. Hydrophilic Polyals: Biomimetic Biodegradable Stealth Materials for Pharmacology and Bioengineering. *PMSE Preprints*, 226th National Meeting of the American Chemical Society, New York, NY, September **2003**.
10. Zakin, M.R.; Gelb, A.H. Electrogels for Delivery of Chemotherapeutic Agents. *PMSE Preprints*, 226th National Meeting of the American Chemical Society, New York, NY, September **2003**.
11. Rosenberg, J.; Degenhardt, M.; Mägerlein, M.; Fastnacht, K.; Berndl, G.; Breitenbach, J. Melt Extrusion: Assessing the Potential of Poorly Soluble Drugs to Form Glass Solutions. *PMSE Preprints*, 226th National Meeting of the American Chemical Society, New York, NY, September **2003**.
12. Muthukumaran, P.; Chordia, L. Supercritical Fluid Particle Engineering: Microparticles to Nanoparticles. Benefits and Limitations. *PMSE Preprints*, 226th National Meeting of the American Chemical Society, New York, NY, September **2003**.
13. Bresciani, M.; Colombo, I.; Gervasoni, D,; Rabaglia, L. Oral Delivery of Poorly Soluble Drugs: Preparation, Characterization and *In Vivo* Behavior of Drug/Crosslinked Polymer Composite Microparticulates. *PMSE Preprints*, 226th National Meeting of the American Chemical Society, New York, NY, September **2003**.
14. Couzis, A.; Maldarelli, C.; Chi, C.; Green, D.A.; Fan, F. Polymorph Selective Crystallization Using Organized Organic Assemblies as Templates. *PMSE Preprints*, 226th National Meeting of the American Chemical Society, New York, NY, September **2003**.

Polymeric Matrices

Chapter 2

Injectable Drug Delivery Based on Polymer Solutions: Experiment and Modeling

Anthony J. McHugh[1,*], Jessica R. DesNoyer[2,3], and Chandrashekar Raman[2]

[1]Department of Chemical Engineering, Lehigh University, Iacocca Hall, Bethlehem, PA 18015
[2]Department of Chemical and Biomolecular Engineering, University of Illinois, Urbana, IL 61801
[3]Current address: Guidant Corporation, 3200 Lakeside Drive, Santa Clara, CA 95054
[*]Corresponding author: ajm8@Lehigh.edu

An overview of injectable drug delivery systems based on polymer solution platforms is presented. The principal formulation parameters that control the phase inversion and drug release characteristics of the injected solution are discussed. These include polymer and solvent type, solution additives, and the bath-side composition. Results from recent studies are presented and a mathematical model describing the release kinetics from so-called fast phase inverting systems is presented and discussed. Predictions of the model are shown to be in good agreement with release data.

Introduction

The past decade has witnessed an explosion of rDNA-derived biopharmaceutical agents (proteins) approved for human therapeutic indications. Challenges facing the prolonged delivery of these compounds include strict dosing and rigorous stability requirements. Most traditional drug delivery systems involve either suspending a bioactive agent into a biodegradable polymer monolith and surgically implanting it into the body, or encapsulating a bioactive agent in polymer microspheres and injecting them subcutaneously. Injectable drug delivery based on polymer solution platforms has gained wide attention in recent years, particularly for protein-based therapies *(1,2)*. These systems consist of a biodegradable polymer dissolved in a biocompatible solvent along with the desired bioactive agent, usually in the form of suspended particles. A unique feature is that polymer phase inversion occurs on injection of the depot solution into the water-based physiologic environment. Thus the membrane carrier forms *in-vivo*, simultaneous with the release of the encapsulated drug. Consequently, the thermodynamic and mass transfer characteristics associated with the *in-vivo* liquid de-mixing process play important roles in establishing the membrane morphology and drug release kinetics.

Injectable Delivery Process

Figure 1 shows a schematic of solution injection. Fundamental parameters include the water influx rate from the physiologic surroundings and the gelation rate of the depot. The water influx rate is critical since it determines the polymer gelation rate as well as the drug dissolution and transport rates from the depot. The micrographs are depot explants from *in-vivo* release studies of human growth hormone (hGH) from poly(lactic-*co*-glycolic acid) (PLGA) *(2)*. They illustrate the two major classes of phase iversion morphology. The depot on the left formed from a solution of PLGA in a strong, highly water-soluble solvent, N-methlypyrollidinone (NMP). In this case so-called fast phase inversion (FPI) of the polymer solution occurs on time scales of seconds to minutes, leading to an interconnected highly porous network of water-rich pockets embedded in a polymer-rich gel framework. These systems are generally characterized by a burst type release profile, followed by a shutdown *(3,4)*. The depot on the right corresponds to a solution in a weak, hydrophobic solvent, benzyl benzoate. In this case much reduced water uptake on injection leads to much slower phase inversion – on the order of days to weeks – resulting in a uniformly dense polymer-rich phase, relatively free of pores. These systems tend to show limited or no bursting and relatively uniform release kinetics *(2)*. In both cases drug release profiles are determined by the phase inversion dynamics and morphology and not by erosion.

Materials and Methods

The goals of our experimental and modeling studies have been to quantify the role of the depot formulation parameters on protein release characteristics *(2-6)*. Solutions and blends of the following polymeric carrier phases have been studied: PLGA, poly(ε-caprolactone) (PCL), and poly(D, L-lactide) (PLA) in the following solvents: ethyl benzoate, benzyl benzoate, NMP, and triacetin. Protein systems have been lysozyme and hGH. Key variables include the polymer and solvent type, the bath-side composition into which the solution is injected, and the protein composition. Phase inversion dynamics, water uptake, and solution gelation rates are monitored *in-situ* using a diffusion cell *(3)*. Dissolution baths and high performance liquid chromatography (HPLC) are used to monitor the protein release rates.

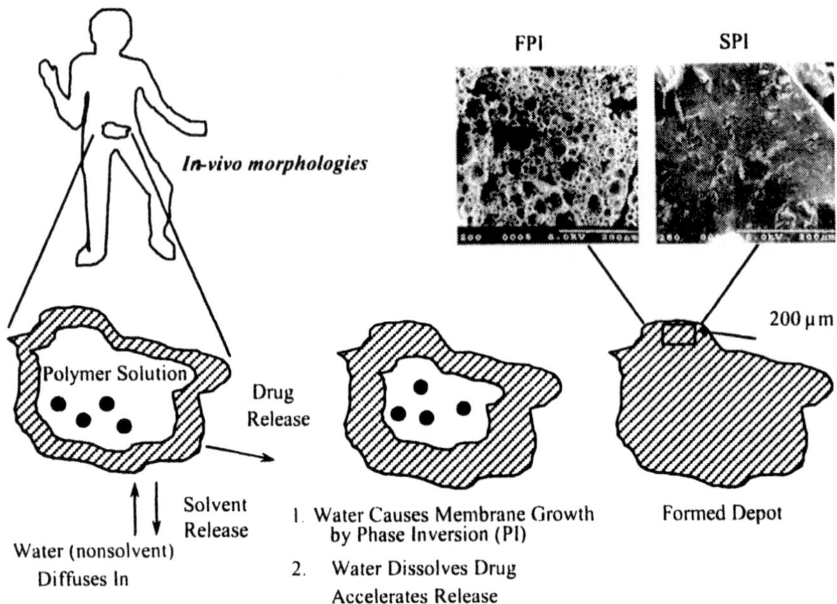

Figure 1. Schematic of injectable solution. Micrographs are depot explants from in-vivo hGH delivery from PLGA solutions and illustrate fast (FPI) and slow (SPI) phase inversion morphologies.

As noted, SPI systems show limited or no burst and exhibit relatively uniform release kinetics over time scales less than depot erosion times. However, since they are highly hydrophobic, protein adsorption to the membrane surface may block the release of the therapeutic agent, hindering their efficacy. Moreover, solution viscosities of these systems are in a range that makes injection difficult unless the system is preheated to 37°C or pre-emulsified *(1)*. While FPI systems typically exhibit burst-like release followed by shutdown, they exhibit lower solution viscosities, making their injection easier. They also provide a more hydrophilic environment that increases biocompatibility. Our recent work has demonstrated that the release characteristics of an FPI system can be modified by using Pluronic additives that preferentially segregate to the liquid-filled pore regions during phase inversion, thereby limiting or eliminating undesirable burst effects *(5)*.

Results and Discussion

Fast Phase Inverting Systems

Figure 2 shows the thermodynamic phase diagram for a fast phase inversion system. The *nose-shaped* binodal curve separates the homogenous region and the region where the solution splits into polymer-rich and polymer-poor phases.

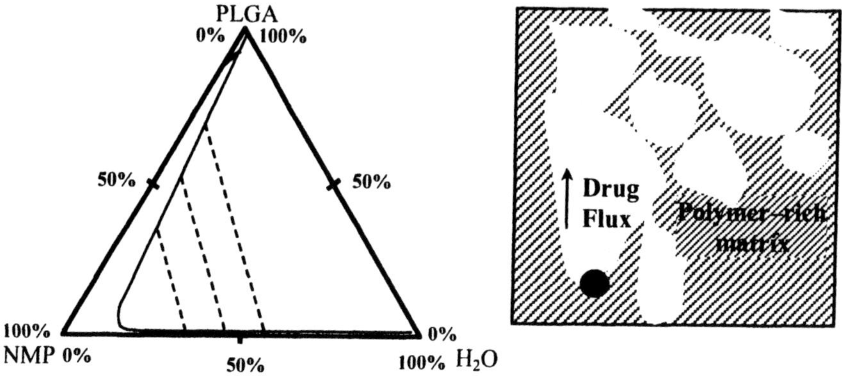

Figure 2. Phase diagram (left) and phase inversion morphology (right) of FPI system based on PLGA/NMP. Dotted lines are tie lines.

Due to the high solvent water solubility, depot compositions are rapidly driven into the two phase region on injection. The steepeness of the tie lines signifies significant enhancement of the polymer concentration in the polymer-rich phase and the formation of a relatively high fraction of water-rich phase. The morphology illustrated on the right corresponds to the FPI micrograph in

Figure 1. In consequence of the fast phase inversion, drug in or near the interconnected water-rich pores rapidly exits the depot, leading to a burst. Ultimately a pseudo-steady state is reached where drug release from the polymer matrix is controlled by the water diffusion to it, and the exchange of the dissolved drug into the pores, followed by diffusive release.

The schematic in Figure 3 illustrates the effect of the Pluronic additive on the phase inversion – drug release process. These systems have lower viscosities and, although they exhibit fast phase inversion, confocal microscopy shows that the Pluronic preferentially accumulates in the interface between the two regions, thus slowing the overall release kinetics *(5)*.

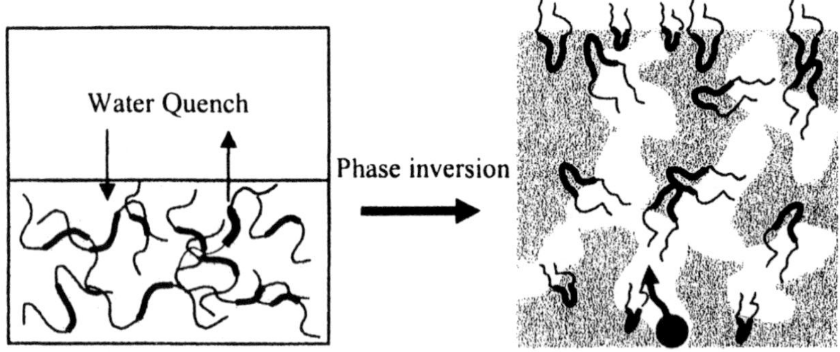

Figure 3. Schematic illustration of Pluronic triblock copolymer additive during phase inversion. Dark lines are hydrophobic blocks. Hydrophilic regions (light lines) segregate to water-rich regions, inhibiting drug diffusion.

Mathematical model of FPI release

Figure 4 shows a schematic of our depot model for FPI systems. Since phase inversion occurs nearly instantaneously compared to the delivery time scales, the two-phase morphology can be considered as fixed throughout the release history. The depot is visualized as consisting of uniformly distributed cylinders of polymer-rich and water-rich phases. The former consists of polymer, solvent, dissolved drug and dispersed drug particles (assumed to be spherical) that dissolve on contact with water. In reality the polymer-rich phase does not have a uniform geometry. However, examination of typical micrographs *(5)* suggests that the long, finger-like structures commonly seen in fast phase inverting systems, can be approximated as cylinders. The water-rich phase contains water, solvent and dissolved drug and is assumed to be continuously interconnected with no isolated droplet regions. The dynamics of the model can be summarized as follows: Water enters the polymer-rich phase from the water-rich phase and dissolves dispersed drug particles. The dissolved drug and solvent diffuse radially through the polymer-rich phase to the water-rich phase. Once in the water-rich phase, they then diffuse out axially into the bath.

Polymer-rich phase equations

Complete details of the model equation derivations and associated parameter evaluations are given elsewhere *(6)*.

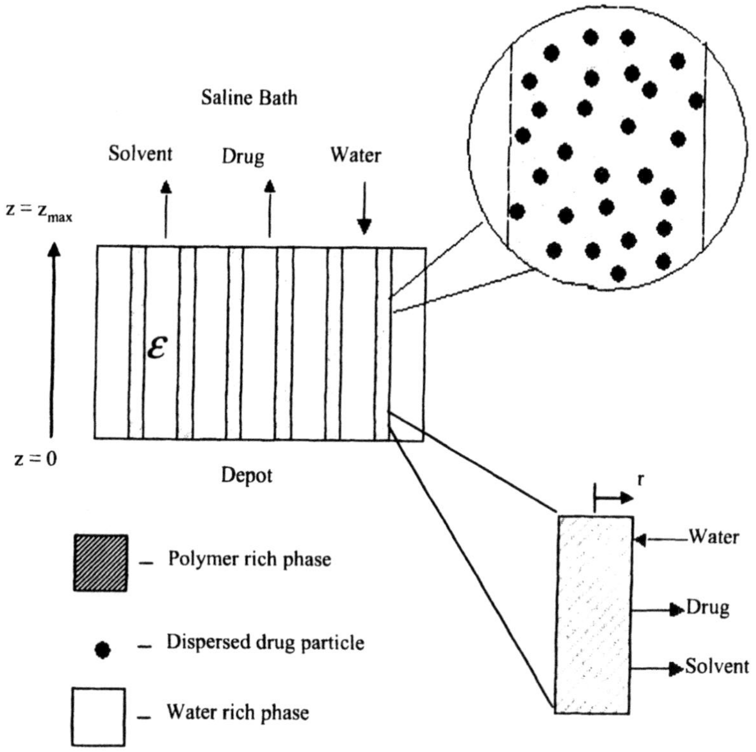

Figure 4. Schematic of depot model.

The subscript notation used is (1) water, (2) solvent, (3) polymer, and (4) dissolved drug. The drug dissolution reaction is assumed to be of the form:

$$\textit{1 mole dispersed drug} + \textit{b moles water} \rightarrow \textit{1 mole dissolved drug} \qquad (1)$$

Assuming that this reaction is first order with respect to water concentration, the balance equations for the independent components in polar coordinates become the following:

$$\frac{\partial C_{p1}}{\partial t}=\frac{1}{r}\frac{\partial}{\partial r}\left(rD_{p1}\frac{\partial C_{p1}}{\partial r}\right)-\left(bk''C_{p1}\right)\rho_p\left(4\pi r_d^2\right) \tag{2}$$

$$\frac{\partial C_{p2}}{\partial t}=\frac{1}{r}\frac{\partial}{\partial r}\left(rD_{p2}\frac{\partial C_{p2}}{\partial r}\right) \tag{3}$$

$$\frac{\partial C_{p4}}{\partial t}=\frac{1}{r}\frac{\partial}{\partial r}\left(rD_{p4}\frac{\partial C_{p4}}{\partial r}\right)+k''C_{p1}\rho_p\left(4\pi r_d^2\right) \tag{4}$$

$$\frac{dr_d}{dt}=-\frac{k''C_{p1}}{C_d} \tag{5}$$

C_{pi} is the molar concentration of dissolved component i, C_d is the molar concentration of pure drug, r is radial position, t is time and D_{pi} is the composition dependent diffusivity of component i; b and k'' are, respectively, the stoichiometric coefficient for water and the reaction rate constant per unit area (m/s) for the dissolution reaction of the dispersed drug particles; ρ_p is the number density of dispersed drug particles in $1/m^3$, and r_d is the radius of dispersed drug particles. The second term on the right side of equation (2) represents the consumption of water and is written as a loss term since the water remains associated with the drug it dissolves and is unavailable for further dissolution. Equation (5) implicitly depends on r through the radial dependence of the local water concentration (C_{p1}). In the derivation of these equations, bulk flow, axial diffusion, and cross diffusivities have been neglected.

The initial conditions for the above equations are given by:

$$C_{pi}=\left(C_{pi}\right)_o \quad i=1,2 \tag{6}$$

$$C_{p4}=0 \tag{7}$$

$$r_d=\left(r_d\right)_0 \tag{8}$$

Boundary conditions at $r = 0$ are fixed by the symmetry condition,

$$\left(\frac{\partial C_{pi}}{\partial r}\right)_{r=0}=0,\ i=1,2,4 \tag{9}$$

and at $r = r_c$, they are given in terms of a partitioning equilibrium with the water rich phase.

$$C_{pi}(r = r_c) = \frac{K_i \phi_i}{\tilde{V}_i}, \text{ for } i = 1, 2, 4. \tag{10}$$

In equation (10), ϕ_i is the volume fraction of component i in the water rich phase at the given axial position, z (Figure 4), $\tilde{V}_i$ is the molar volume of component i, r_c is the radius of the polymer rich cylinders, and K_i is the partition coefficient for the given component between the water-rich and polymer-rich phases. Since a similar partition coefficient for the drug is not readily available, K_1 is used as an approximation. Since the boundary conditions depend on z, equations (2-5) will also have a z-dependence.

Water-rich phase equations

Transport in the water-rich phase is also modeled as a diffusion-reaction system. The final form involves two independent diffusion equations for the solvent and drug, respectively, and an overall balance for the ternary system of nonsolvent, solvent, and drug.

$$\frac{\partial \phi_2}{\partial t} = D_{w2} \frac{\partial^2 \phi_2}{\partial z^2} - \mathrm{v}^{\bullet} \frac{\partial \phi_2}{\partial z} + (1 - \phi_2)\tilde{V}_2 R_2 - \phi_2 \left(\tilde{V}_1 R_1 + \tilde{V}_4 R_4 \right) \tag{11}$$

$$\frac{\partial \phi_4}{\partial t} = D_{w4} \frac{\partial^2 \phi_4}{\partial z^2} - \mathrm{v}^{\bullet} \frac{\partial \phi_4}{\partial z} + (1 - \phi_4)\tilde{V}_4 R_4 - \phi_4 \left(\tilde{V}_1 R_1 + \tilde{V}_2 R_2 \right) \tag{12}$$

$$\frac{d\mathrm{v}^{\bullet}}{dz} = \tilde{V}_1 R_1 + \tilde{V}_2 R_2 + \tilde{V}_4 R_4 \tag{13}$$

D_{wi} is the diffusivity of i in the water-rich phase, $\mathrm{v}^{\bullet}\,(= \phi_1 \mathrm{v}_1 + \phi_2 \mathrm{v}_2 + \phi_4 \mathrm{v}_4)$ is the volume-averaged bulk velocity in the axial direction. R_i is a "reaction" term that accounts for the fluxes between the polymer-rich and water-rich phases and is related to the interfacial fluxes, j_i, to or from the polymer-rich phase by:

$$R_i = j_i \frac{2\pi r_c \rho_c}{\varepsilon} \tag{14}$$

where, ε is the volume fraction of depot occupied by the water-rich phase, and ρ_c is the number of polymer-rich cylinders per unit cross-section of depot. The initial conditions for these equations are:

$$\phi_i = (\phi_i)_0 \, , \; i = 2, 4 \tag{15}$$

Since the polymer depot rests on a substrate in the dissolution device, it is assumed that no mass transfer occurs across the lower boundary, thus the boundary conditions there are:

$$\left(\frac{\partial \phi_i}{\partial z}\right)_{z=0} = 0 \;\; i = 2, 4; \;\; v^{\bullet}(z=0) = 0 \tag{16}$$

Transfer of solvent and drug at the bath interface at the top of the depot, i.e., $z = z_{max}$ (Figure 4), can be expressed in terms of a mass transfer coefficient, thus:

$$-D_{wi}\left(\frac{\partial \phi_i}{\partial z}\right)_{z=z_{\max}} = k\left(\phi_i - (\phi_i)_\infty\right) \; i = 2, 4. \tag{17}$$

k is the bath-side mass transfer coefficient and is assumed to be the same for the solvent and drug. Since as pointed out below, one expects a polymer-rich skin to form in the region near the bath interface, one might expect a difference between the mass transfer coefficient for the smaller, apolar solvent molecules and the large, water-soluble protein. However, we have found that using a mass transfer coefficient for the drug, which is larger than that of the protein by a factor of 3 (i.e. scaling with the diffusivity to the ½ power), produces a negligible change in results. Moreover, the difference in fluxes is to some extent accounted for through the explicit diffusivity dependence of the two species in equation (17). Similar to the polymer-rich phase, we assume that ternary cross diffusivities in the water-rich phase can be neglected. Further assumptions made in deriving these equations are that radial diffusion is negligible, i.e., concentrations are the same at all points at a given cross-section, and the diffusivities, D_{w2} and D_{w4}, are independent of composition.

Model parameters

The parameters in the model equations are in four categories: (i) transport properties (diffusivities, mass transfer coefficients, dissolution rate constants, and stoichiometry of the drug dissolution), (ii) thermodynamic properties of the ternary nonsolvent-solvent-polymer system (phase equilibrium concentrations, binary interaction parameters, and associated partition coefficients), (iii) geometric parameters (water-rich phase void volume, polymer-rich and water-rich phase cylinder radii) and (iv) component physical properties (specific volumes, molecular weights, and densities). In the case of the polymer-rich phase, diffusion coefficients are related to chemical potentials and associated thermodynamic parameters using friction coefficient formalism in combination with free volume theory and Flory-Huggins theory for the polymer-rich phase *(6)*. These dependencies can be expressed as follows

$$D_{p1}, D_{p2} = f\left(\zeta_{ij}, g_{ij}\right) \tag{18a}$$

$$\zeta_{ij} = f\left(K_{1i}, K_{2i} - T_{gi}, K_{1j}, K_{2j} - T_{gj}\right) \tag{18b}$$

ζ_{ij} are binary friction coefficients, g_{ij} are the Flory-Huggins interaction parameters, K_{ij} are free volume parameters, and T_{gi} are component glass transition temperatures. The parameters in equation (18b) are evaluated from viscosity data for the pure components and the composition dependent interaction parameters are determined using standard methods *(6)*. The protein diffusivity D_{p4} is estimated from the following expression:

$$D_{p4} = f_{24} D_{p2} \tag{19}$$

f_{24} is a constant based on a single-point measurement of D_{p4} and an estimate of D_{p2} *(6)*.

Numerical Results and Discussion

Values of the model parameters used in the calculations and discussions of their origins are given in a thesis reference *(6)*. The model ends up having just two free parameters, the volume fraction of the water-rich phase, ε, and the depot-bath mass transfer coefficient, *k*. For numerical purposes, the mass transfer coefficient is more conveniently evaluated in terms of the Nusselt number for mass transfer ($Nu = \frac{kl}{D_o}$) that arises from non-dimensionalizing the associated boundary condition (equation 17) in terms of a characteristic length dimension, *l*, and a reference diffusivity, D_o.

The bath-side mass transfer coefficient would normally be estimated from correlations available in the literature, however, it has been shown that fast phase inverting systems develop a skin at the interface, which is likely to decrease the mass transfer rate. This effect can be lumped into the external mass transfer coefficient. Thus the Nusselt number is chosen to be a fitting parameter. For systems with added Pluronic, the diffusivity of the drug in the water-rich phase D_{w4} is used as a fitting parameter. Determination of the water-rich void volume, ε , was done by fitting the release data using an iterative method *(6)*.

Parameter sensitivity

Calculations of lysozyme release rate show that for fixed Nu, the general shape of the drug release profile is insensitive to the depot porosity *(6)*. Profiles are characterized by a burst-type release followed by a virtual shut down, leaving drug trapped in the depot. Thus the water-rich phase releases all the drug contained in it and only a fraction of the drug from the polymer-rich phase is released. The polymer-rich phase initially contains a significant amount of solvent (NMP), which permits diffusion of the drug through the phase. Over time, NMP diffuses out of the polymer-rich phase, transforming it from a viscous liquid to a dense solid, causing the diffusivities in the polymer-rich phase to decrease by orders of magnitude, leading to a shut down of the release. Drug trapped in the polymer-rich phase would likely be released upon erosion of the depot on longer time scales.

Figure 5 illustrates model predictions of the effect of changing the Nusselt number on the drug release profile. Varying Nu leads to profiles ranging from burst-like to essentially zero order kinetics. Lowering the Nusselt number means that NMP leaves the depot slower and hence the polymer-rich phase solidifies slower, allowing more drug to diffuse out before the polymer-rich phase hardens. However, lowering the Nusselt number also lowers the rate at which the drug leaves the depot.

The curves in Figure 5 for Nu = 1.2 and Nu = 1.2×10^{-2} present an interesting asymptotic behavior in which the initial rate of drug release is the same in both cases, though a lower total release is predicted for the larger Nu. Similar patterns to those in Figure 5 are also found with decreasing D_{w4} at fixed Nu and ε *(6)*. This point is further illustrated in Figure 8 below.

Comparison to Experimental Data

PLGA-NMP System

Figure 6 shows the excellent fit of the lysozyme release profile predicted by the model to the experimental data of Brodbeck *et al.* *(4)*. The increased rate of drug release at long time (~20 days) most likely reflects the onset of erosive release *(2)*.

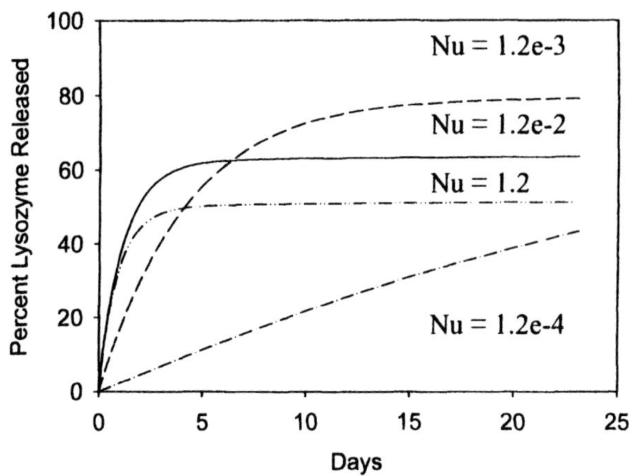

Figure 5. Model predictions for PLGA-NMP-Water systems for different Nusselt number for $\varepsilon = 0.52$ and $D_{w4} = 5 \times 10^{-11}\ m^2/s$.

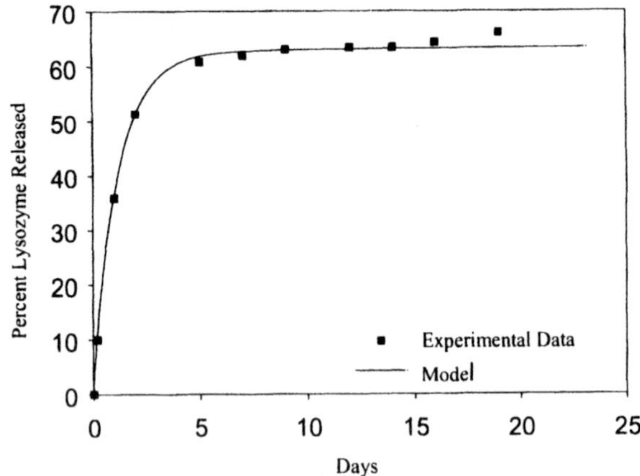

Figure 6. Comparison of model predictions and experimental release data for lysozyme from PLGA-NMP system. Model parameters used to fit the data are $\varepsilon = 0.52$, $Nu = 1.2 \times 10^{-2}$.

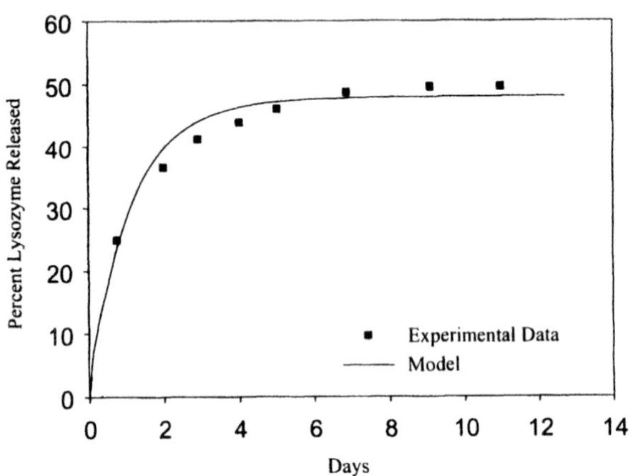

Figure 7. Comparison of model predictions and experimental release data for lysozyme from PLA-NMP system. Single model parameter used to fit the data is $\varepsilon = 0.39$. All other parameters same as Figure 6.

PLA-NMP System

Figure 7 shows a comparison between the model predictions and the lysozyme release data of DesNoyer and McHugh *(5)* from PLA-NMP solutions. In this case, since the physical characteristics of the polymers are similar, the value used for the Nusselt number was that obtained from the PLGA-NMP fit. Hence the only free parameter is ε, which in this case was found to be 0.39 for the fit shown. The agreement between the model and experimental data is good, considering that only one free parameter has been used.

PLA-NMP Systems with Pluronic

An important assumption in applying our model to the Pluronic systems is that the Pluronic does not affect the thermodynamics. Thus the rate of phase inversion and depot porosity remain the same as that for the pure PLA-NMP system. Likewise, the Nusselt number and water-rich phase volume fraction are expected to remain the same, leaving D_{w4} as the single fitting parameter. Figure 8 shows the comparison between the model and release data for the system with the highest concentration of additive in the formulation (7.2 wt% Pluronic). As seen, at this high concentration, the reduction in the diffusivity in the water-rich phase is greatest and leads to almost zero order release, consistent with experimental observations. Similar fits were found for the data at the lower Pluronic concentrations, with the fit value of the diffusivity showing a monotonic decrease with increasing Pluronic *(6)*.

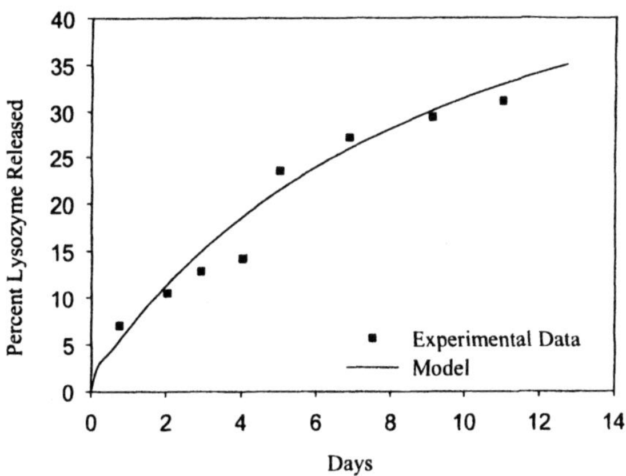

Figure 8. Comparison of model predictions and experimental release data for lysozyme from PLA-NMP system with 7.2 wt % Pluronic added. Single model parameter used to fit the data is $D_{w4} = 0.55 \times 10^{-11} m^2/s$, all other parameters are the same as Figure 7.

Conclusions

The drug release characteristics of injectable polymer solutions are profoundly influenced by the phase inversion that takes place on contact with the physiologic fluid. Release profiles vary from burst-type to zero-order, depending on the solvent and/or additives that preferentially migrate to the water-rich interface. Diffusion-based models are able to accurately capture the release behavior.

References

1. Brodbeck, K.J.; Gaynor-Duarte, A.T.; Shen, T.T.I. Gel composition and methods. US Patent 6,130,200, Oct 10, 2000.
2. Brodbeck, K.J.; Pushpala, S.; McHugh, A.J. Sustained release of human growth hormone from PLGA solution depots. *Pharmaceutical Res.* **1999,** *16*, 1825-1829.
3. Graham, P.D.; Brodbeck, K.J.; McHugh, A.J. Phase inversion dynamics of PLGA solutions related to drug delivery. *J. Control. Rel.* **1999,** *58*, 233-245.
4. Brodbeck, K.J.; DesNoyer, J.R.; McHugh, A.J. Phase inversion dynamics of PLGA solutions related to drug delivery. Part II. The role of solution thermodynamics and bath side mass transfer. *J. Control. Rel.* **1999,** *62*, 333-344.
5. DesNoyer, J.R.; McHugh, A.J. The effect of Pluronic on the protein release kinetics of an injectable drug delivery system. *J. Control. Rel.* **2003,** *86*, 55-24.
6. Raman, C. Modeling Drug Release from Fast Phase Inverting Injectable Drug Delivery Systems. M.S. Thesis, University of Illinois, Urbana, 2002.

Chapter 3

Poly(Ortho esters): Some Recent Developments

Jorge Heller[*] and John Barr

AP Pharma, 123 Saginaw Drive, Redwood City, CA 94063
[*]Corresponding author: jorgeheller2@aol.com

Poly(ortho esters) have been under development since the early 1970's and four distinct families have been investigated. The latest family, POE IV, is an autocatalyzed polymer that contains a latent acid in the polymer backbone, where erosion rates can be adjusted by varying the concentration of the latent acid, and mechanical properties of the polymer can be adjusted by choice of monomers. Thus, POE IV is a highly versatile material that is currently being commercialized for the treatment of post-operative pain and nausea.

Introduction

Poly(ortho esters) have been under development since the early 1970's and during that time, four distinct families have been described (*1*). The four families, shown in Scheme 1, have been prepared by two general reaction processes. In one process, the polymers are prepared by a transesterification reaction, which was used for the synthesis of POE I and POE III. However, this type of reaction involves long reaction times in order to drive the equilibrium towards polymer, is very difficult to scale up, and accurate control of molecular weight is virtually impossible. For these reasons these materials have not been commercialized, despite interesting properties, particularly for POE III, which has shown excellent ocular biocompatibility (*2*).

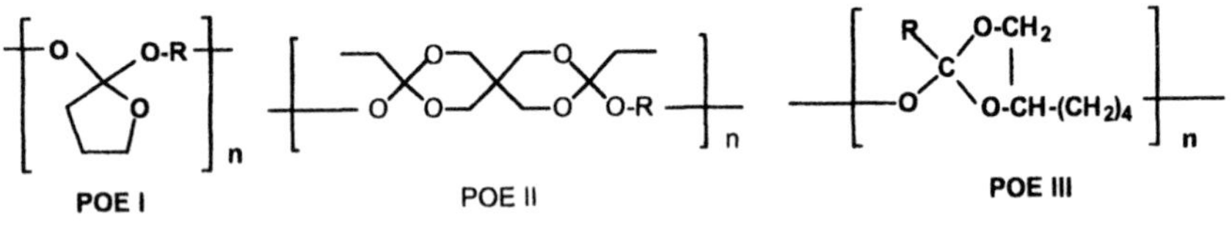

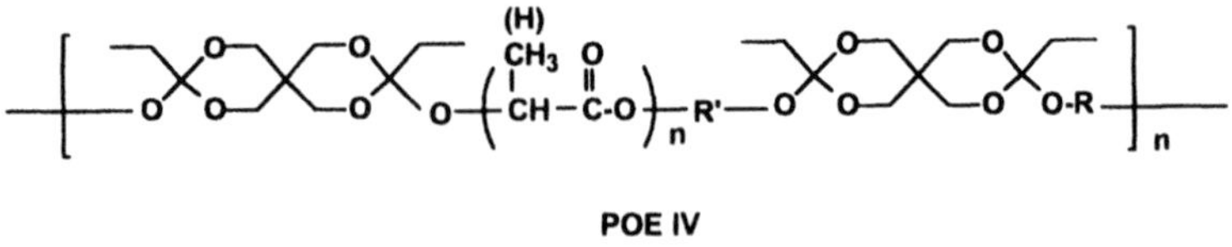

POE I POE II POE III

POE IV

Scheme 1. Chemical structures of the four families of poly(ortho esters).

An alternate means of preparing poly(ortho esters) is by the addition of diols to diketene acetals. The rationale for selecting the diketene acetal 3,9-bis(ethylidene-2,4,8,10-tetraoxaspiro[5.5]undecane) has been described elsewhere (*3*).

Results and Discussion

Polymer Synthesis

The synthesis shown in Scheme 2 has a number of significant advantages. Dominant among these are (i) ease of synthesis that allows easy scale-up and (ii) ability to vary polymer properties within very wide limits by choice of appropriate diol. This advantage will be elaborated further under polymer properties.

HO-R-OH +

Scheme 2. General scheme for the synthesis of poly(ortho esters).

This synthesis proceeds without the evolution of volatile by-products even though it is a condensation polymerization. It is thus possible to prepare dense crosslinked materials by first preparing a ketene acetal-terminated prepolymer and then crosslinking by using diols having at least one additional functionality (hydroxy group). This principle is shown in Scheme 3. The synthesis has been scaled-up to kilogram quantities, and polymers prepared under GMP conditions are now available from a contract manufacturer.

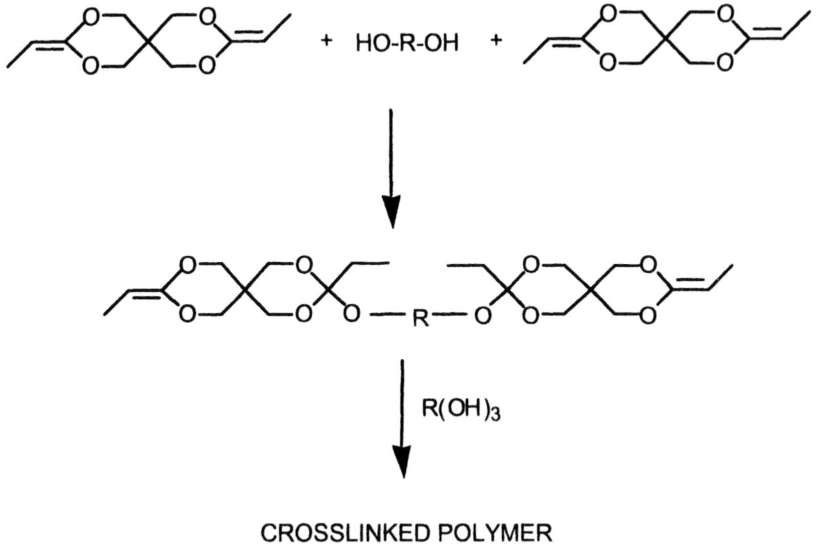

+ HO-R-OH +

R(OH)3

CROSSLINKED POLYMER

Scheme 3. General scheme for the synthesis of crosslinked poly(ortho esters) using alcohols with at least three functionalities.

In case of prepolymers with molecular weights low enough to exist as a viscous liquid at room temperature, drugs can be mixed into the prepolymer, followed by addition of the alcohol and crosslinking of the mixture by mild heating. Thus, sensitive materials can be incorporated without compromising their integrity. However, if the drug has reactive hydroxy groups, it will be chemically attached to the polymer. This procedure has been used for the incorporation of ivermectin as well as the preparation of a product with ivermectin covalently attached to the polymer (*4*).

Polymer Properties

Glass Transition Temperature

As already mentioned, polymer properties can be varied within very wide limits by an appropriate choice of the diol, or mixture of diols used in the synthesis. This process is illustrated in Figure 1, where the glass transition temperature of polymers prepared using a mixture of a rigid diol, *trans*-cyclohexanedimethanol, and a flexible diol, 1,6-hexane diol, was used (*5*). The data show a smooth change from a low of about 20°C for a polymer containing all 1,6-hexane diol to about 115°C for a polymer containing all *trans*-cyclohexanedimethanol.

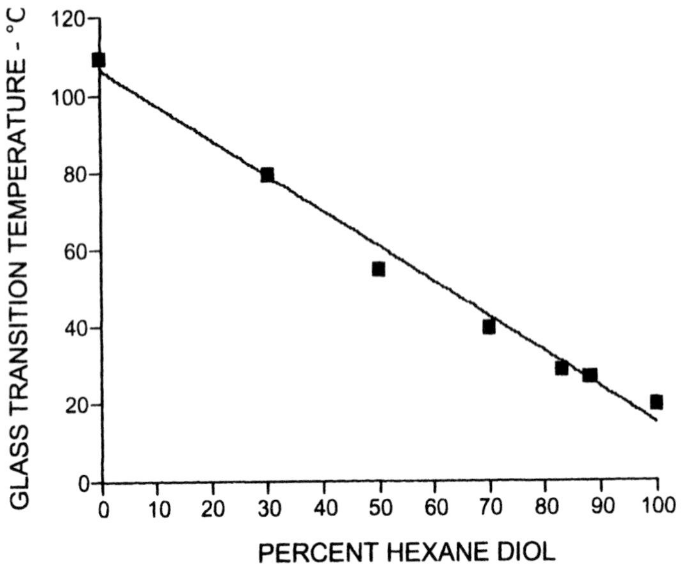

Figure 1. Glass transition temperature of 3,9-diethylidene-2,4,8,10-tetraoxaspiro [5.5] undecane, trans-cyclohexanedimethanol, 1,6-hexanediol polymer as a function of mole % 1,6-hexanediol. (Reproduced from Reference (5). Copyright 1983 MTP Press, Ltd.)

The glass transition can be lowered to values well below room temperature by increasing the flexibility of the diol, i.e., increasing the number of methylene

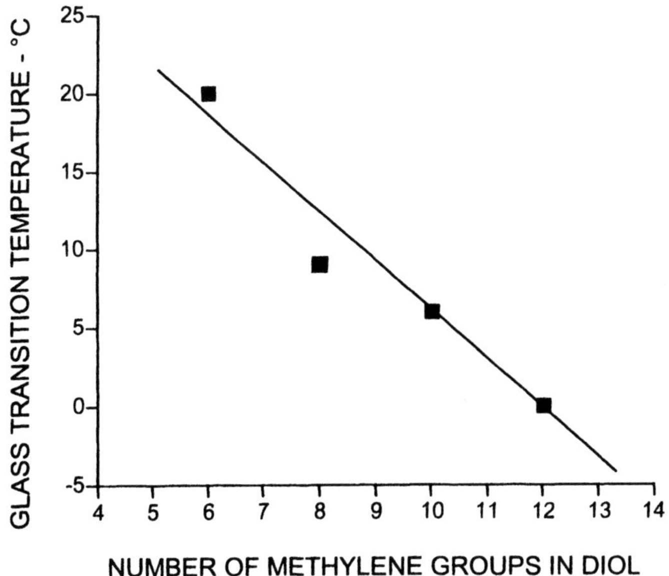

Figure 2. Effect of diol chain length on the glass transition temperature of polymers prepared from 3,9-diethylidene-2,4,8,10-tetraoxaspiro [5.5] undecane and α,ω-diols. (Reproduced with permission from Reference (6). Copyright 1995 Harwood Academic Publishers GmbH.)

groups separating the hydroxy groups. This effect of the number of methylene groups on the glass transition temperature is shown in Figure 2 (*6*).

Polymer Hydrolysis

Polymer hydrolysis as shown in Scheme 4 proceeds in three consecutive steps (*7*). In the first step, the lactic acid or glycolic acid segment in the polymer backbone hydrolyzes to generate a polymer fragment containing a carboxylic acid end-group that will lower the pH and catalyze ortho ester hydrolysis. A second cleavage produces free α-hydroxy acid, and this species provides the major catalytic effect for the hydrolysis of the ortho ester linkages. Further hydrolysis then proceeds in two steps to first generate the diol, or mixture of diols, used in the synthesis and pentaerythritol dipropionate, followed by ester hydrolysis to produce pentaerythritol and propionic acid.

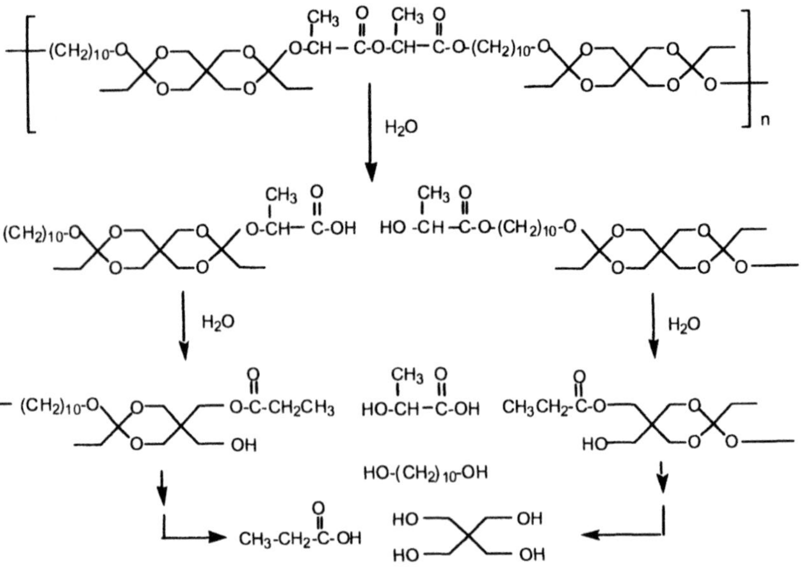

Scheme 4. *Stepwise hydrolysis of poly(ortho esters).*

Figure 3 shows polymer weight loss and release of lactic acid (*7*). The concomitant weight loss and release of lactic acid provides convincing evidence for an erosion-controlled process that is confined predominantly to the surface layers of the polymer matrix. However, because there is a significant drop in molecular weight of the remaining polymer, the process is clearly not pure surface erosion and some bulk erosion does take place. The water concentration in the bulk is very low because the polymer is very hydrophobic, and hence the rate of hydrolysis is limited by the amount of available water. In comparison, the water concentration in the outer eroding layers is high, and therefore, the rate of hydrolysis is high.

Stability

Even though ortho ester linkages are hydrolytically labile, the polymer is stable at room temperature, provided that moisture is rigorously excluded. Figure 4 shows room temperature stability of a somewhat hydrophilic polymer, containing 40 mole% latent acid that has been ground to microparticles, thus greatly increasing its surface area (*1*). This polymer completely erodes in only a few days when placed in aqueous buffer at pH 7.4 and 37°C. Clearly, despite the rapid erosion, it is completely stable when stored under anhydrous conditions.

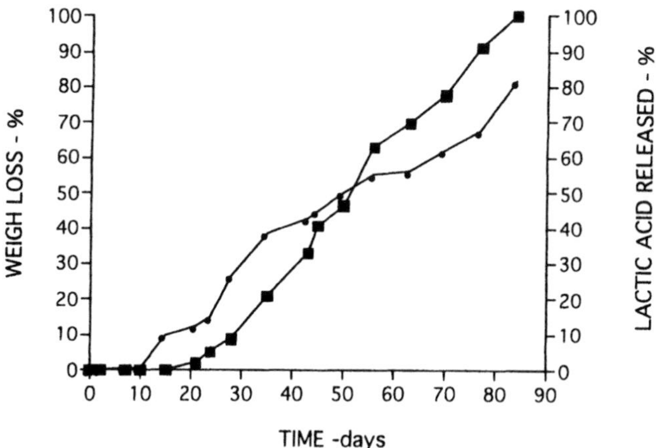

Figure 3. The relationship between lactic acid release (circles) and weight loss (squares) for a poly(ortho ester) prepared from 3,9-diethylidene-2,4,8,10-tetraoxaspiro [5.5] undecane, 1,10-decanediol and 1,10-decanediol lactide (ratio 100/70/30) (0.13 M sodium phosphate buffer at pH 7.4 and 37°C). (Reproduced from Reference (7). Copyright 1999 American Chemical Society.)

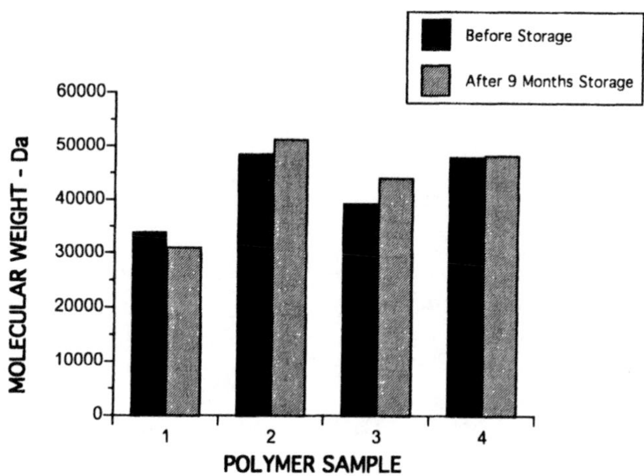

Figure 4. Stability of a polymer prepared from 3,9-diethylidene-2,4,8,10-tetraoxaspiro [5.5] undecane, cis/trans-cyclohexanedimethanol, triethylene glycol and triethylene glycol glycolide (ratio 100/35/25/40), stored at room temperature under anhydrous conditions. (Reproduced with permission from Reference (1). Copyright 2002 Elsevier Science B.V.)

Fabrication

Poly(ortho esters) are excellent thermoplastic materials that can be easily fabricated by conventional thermoplastic methods such as compression molding, extrusion or injection molding. Because glass transition temperatures are adjustable within wide limits, fabrication temperatures can be customized for specific drugs, taking into account their thermal stability. A detailed study of polymer extrusion and drug incorporation has been carried out (*8*). Furthermore, the polymer is soluble in solvents such as methylene chloride, tetrahydrofuran and ethyl acetate, and therefore, it is easy to prepare films by solution casting or coat devices such as cardiovascular stents. In addition, the vast array of microencapsulation methods is also available for polymer fabrication.

Polymer Forms

The ability to vary the glass transition temperature within very wide limits allows the preparation of two major types of materials, solid materials and materials that are gel-like at room temperature. The latter materials can directly be used for injection applications if properly formulated.

Solid Polymers

The versatility of solid polymers will be illustrated with two examples, the delivery of a small water-soluble molecule, 5-fluorouracil (5FU), and the delivery of a complex sensitive molecule, DNA.

5-Fluorouracil

The release of a low molecular weight, water-soluble drug can be illustrated with 5-fluorouracil (Figure 5) (*9*). The concomitant drug release and weight-loss of the device is consistent with an erosion-controlled process. In the data shown, the 5-FU material balance is a little low, but there is little doubt that the predominant mechanism controlling the release of 5FU is erosion and not diffusion. Furthermore, as shown in Figure 6, the rate of drug release is proportional to the drug loading (*10*). This observation is also consistent with a surface erosion process and provides additional evidence for the hydrolysis and erosion study already described.

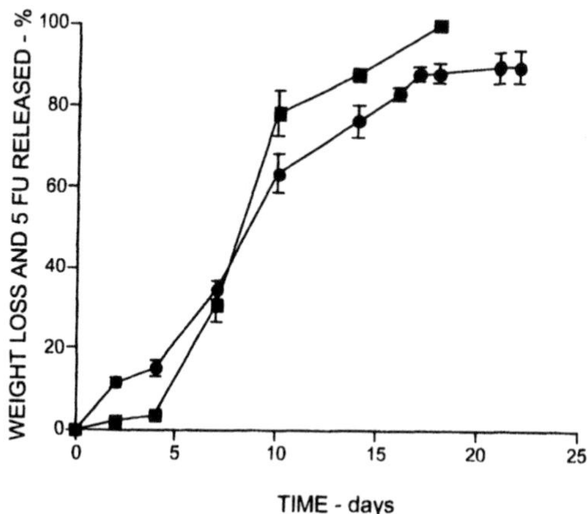

Figure 5. Polymer weight loss (circles) and 5-fluorouracil (5-FU) release (squares) from a polymer prepared from 3,9-diethylidene-2,4,8,10-tetraoxaspiro [5.5] undecane, 1,3-propanediol and triethylene glycol diglycolide (ratio 90/10). Drug loading 20 wt% (0.05 M phosphate buffer at pH 7.4 and 37°C). (Reproduced with permission from Reference (9). Copyright 2000 Elsevier Science B.V.)

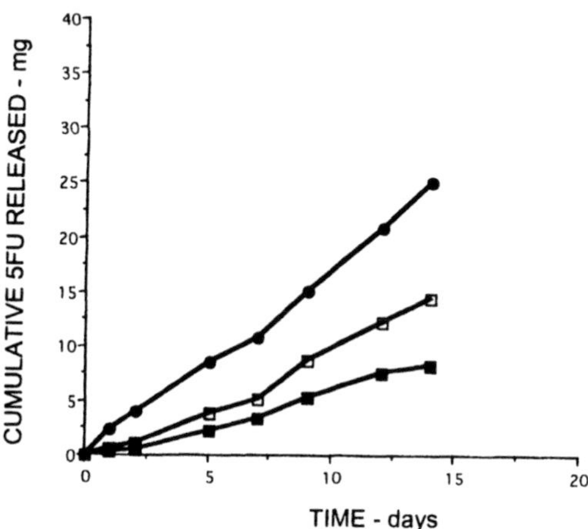

Figure 6. Effect of loading of 5-FU, released from a polyester prepared from trans-cyclohexanedimethanol, 1,6-hexanediol, triethylene glycol and triethylene glycol monoglycolide (ratio 15/40/40/5). 5.5 wt% (closed squares), 11.6 wt% (open squares), and 23.6 wt% (circles) of 5-FU. (Reproduced with permission from Reference (10). Copyright 2000 Elsevier Science B.V.)

DNA

Poly(ortho esters) are particularly well suited for the delivery of large, pH-sensitive materials since the internal pH during erosion is somewhat above 5, and drug release does not rely on diffusion but is controlled by matrix erosion. We have investigated the suitability of using microencapsulated plasmid DNA for vaccine delivery. In this application, an additional poly(ortho ester) was used, containing tertiary amine functionalities in the backbone (Scheme 5). The positively charged amino groups provide sites to complex with the negatively charged DNA (*11*).

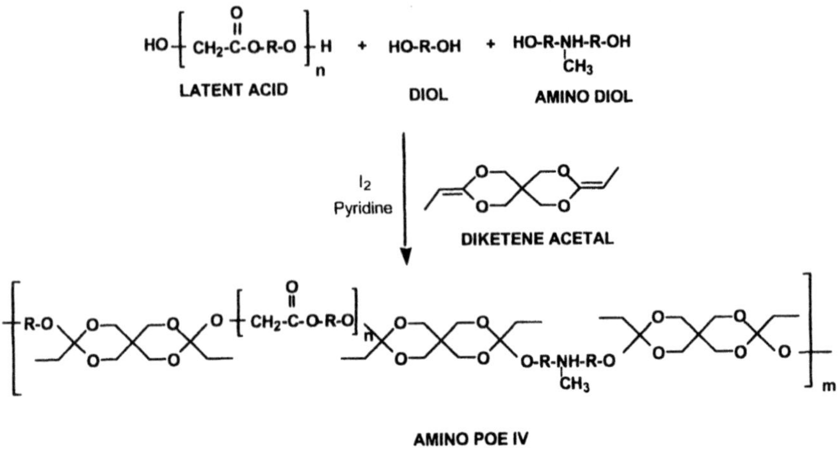

HO-[CH2-C(=O)-O-R-O]n-H + HO-R-OH + HO-R-N(CH3)-R-OH

LATENT ACID DIOL AMINO DIOL

I2 Pyridine

DIKETENE ACETAL

AMINO POE IV

Scheme 5. Synthesis of poly(ortho esters) capable of DNA complexation.

Figure 7 shows the release of plasmid DNA at pH 7.4, the cytoplasmic pH, and pH 5.5, the pH within phagosomes, to simulate microsphere uptake for both amino and non-amino polymers. There is clearly a significant difference between the two polymers in that the amino-containing polymer shows a 24-hour delay before DNA is released. This delay is very likely due to complex formation between the negatively charged DNA and the positively charged polymer, which prevents DNA release until the polymer has undergone significant erosion. This delay has been found to be important when 5 μm microspheres are taken up by antigen presenting cells (APC). We have shown that in a cancer model, the amino-polymer is significantly superior to the non-amino polymer (*11*).

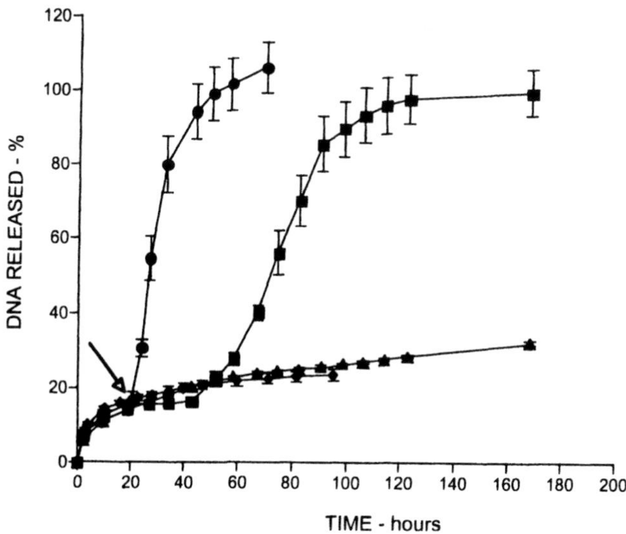

Figure 7. Release kinetics of DNA from a non-amino polymer at pH 7.4 (small filled circles), pH 5.0 (large filled circles) and an amino polymer at pH 7.4 (filled triangles) and pH 5.0 (filled squares). The arrow points to the time of pH change from 7.4 to 5.0. (Reproduced from Reference (11). Copyright 2004 Nature Publishing Group.)

Gel-like Materials

Suitably constructed polymers are gel-like, injectable materials at room temperature. One significant advantage in developing such systems is that drugs can be mixed into the polymer at room temperature, using a simple mixing protocol. In this way, even the most sensitive therapeutic agent can be incorporated without loss of activity. However, in order to allow injection, it is necessary to limit the viscosity by limiting the molecular weight to about 5 kDa. In addition, it is also necessary to use an excipient that further reduces the viscosity and renders the material less sticky.

Molecular Weight Control

Polymer molecular weight can be controlled by using an excess of diol, or by using a monofunctional alcohol that acts as a chain-stopper. The effect of adding a monofunctional alcohol to the reaction mixture on molecular weight is shown in Figure 8 (*12*).

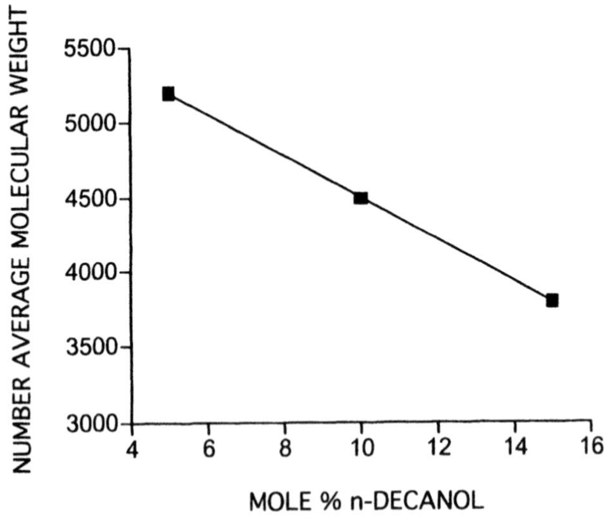

Figure 8. Effect of n-decanol on the molecular weight of a poly(ortho ester) prepared from 3,9-diethylidene-2,4,8,10-tetraoxaspiro[5.5]undecane, 1,10-decanediol and 1,10-decanediol lactide (ratio 100/70/30). (Reproduced with permission from Reference (12). Copyright 2002 Taylor & Francis.)

Development of a Formulation to Treat Post-Operative Pain

Gel-like poly(ortho esters), containing the drug mepivacaine, were used in an attempt to instill the material into a surgical incision prior to wound closure. The rationale for this treatment was to provide a high local concentration of mepivacaine within the incision, while at the same time maintaining a low systemic concentration. If the analgesic effect at the surgical site could be maintained, the patient dependence on orally administered opiates with their well-known side effects could be greatly reduced. Such materials are currently in the process of being commercialized, and a Phase II clinical trial using inguinal hernia repair is currently ongoing.

The structure of the polymer is shown in Scheme 6. The composition of the clinical formulation, designated as APF 112, is 77.6% polymer, 19.4% methoxy poly(ethylene glycol) 550 and 3% mepivacaine.

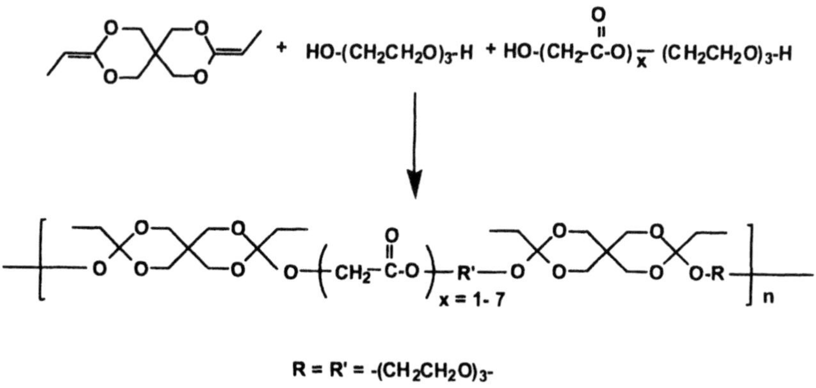

+ HO-$(CH_2CH_2O)_3$-H + HO-$(CH_2$-C(=O)-O$)_x$-$(CH_2CH_2O)_3$-H

x = 1- 7

R = R' = -$(CH_2CH_2O)_3$-

Scheme 6. Structure of the poly(ortho ester) used in clinical formulation APF 112.

Toxicology Studies in Preparation of an IND Filing

Two types of studies were carried out, (i) using a polymer hydrolysate and (ii) using the APF 112 formulation.

Polymer Hydrolysate

Hydrolyzation of the polymer into its hydrolysis products simulates the instantaneous erosion of an implant and thus represents a worse case scenario. The hydrolysate was prepared by hydrolyzing the polymer in phosphate buffered saline (PBS) at 80°C for 24 hours, adjusting the pH to 7.4 with NaOH, adding methoxy poly(ethylene glycol), mixing thoroughly, adding deionized water to adjust osmolarity, and finally filtering through a 0.45 µm filter. The solution was then injected subcutaneously into male and female Sprague-Dowley rats and into male and female beagle dogs. In the rat study, the doses used were 0, 1, 3 and 10 mL/kg and in beagle dogs, the doses were 0, 0.05, 0.1 and 0.2 mL/kg. Both animal species were observed for 14 days and no adverse effects by clinical observation and gross necropsies were found.

APF 112 Formulation

The following incisional wound instillation study was carried out. A 1-cm full thickness incision was made, creating a subcutaneous pocket by blunt dissection. The APF 112 formulation was administered into the subcutaneous pocket and the skin closed with 4-0 nylon sutures, which were removed after 7 days. The study was carried out using Sprague-Dowley male and female rats, using a 500 and 1000 μL single dose. The rats were sacrificed at day 8. Both doses were well tolerated, but the 1000 μL dose resulted in leakage and wound distension.

Conclusions

Synthesis and characterization of various poly(ortho esters) has been described. The physical behavior of this material can easily be changed by its composition and the choice of its constituent components. The suitability of poly(ortho esters) in medical applications was demonstrated in several examples. One poly(ortho ester) formulation, designated as APF 112, is currently in clinical trials.

References

1. Heller, J.; Barr, J.; Ng, S. Y.; Schwach-Abdellaoui, K.; Gurny, R. Poly (ortho esters): Synthesis, characterization, properties and uses. *Adv. Drug. Deliv. Rev.* **2002**, *54,* 1015-1039.
2. Einmahl, S.; Behar-Cohen, F.; Tabatabay, C.; Savoldelli, M.; D'Hermies, F.; Chauvaud, D.; Heller, J.; Gurny, R. A viscous bioerodible poly(ortho ester) as a new biomaterial for intraocular application. *J. Biomed. Mater. Sci.* **2000**, *50,* 566-573.
3. Heller, J.; Barr, J. Poly(ortho esters) – From concept to reality. *Biomacromol.* **2004**, *5,* 1625-1632.
4. Shih, C.; Seward, R. L.; *In vivo* and *in vitro* release of ivermectin from poly (ortho ester) matrices. I. Crosslinked matrix prepared from ketene acetal end-capped prepolymer. *J. Control. Rel.* **1993**, *25,* 155-162.
5. Heller, J.; Penhale, D. W. H.; Fritzinger, B. K.; Rose, J. E.; Helwing, R. F. Controlled release of contraceptive steroids from biodegradable poly(ortho esters). *Contracept. Deliv. Syst.* **1983**, *4,* 43-53.
6. *Trends and Future Perspectives in Peptide and Protein Drug Delivery;* Lee, V. H. L.; Hashida, M.; Mizushima, Y., Eds.; Harwood Academic Publishers: Switzerland, 1995; pp. 39-56.
7. Schwach-Abdellaoui, K.; Heller, J.; Gurny, R. Hydrolysis and erosion studies of autocatalyzed poly(ortho esters) containing lactoyl-lactyl acid dimers. *Macromol.* **1999**, *32,* 301-307.
8. *Scaffolding in Tissue Engineering;* Ma, P. X.; Elisseeff, J., Eds.; Marcel Dekker, New York, in press.
9. Heller, J.; Barr, J.; Ng, S. Y.; Shen, H-R.; Schwach-Abdellaoui, K.; Einmahl, S.; Rothen-Weinhold, A.; Gurny, R. Poly(ortho esters) – Their

development and some recent applications. *Eur. J. Pharm. Biopharm.* **2000,** *50,* 1221-128.

10. Ng, S. Y.; Shen, H-R.; Lopez, E.; Zherebin, Y.; Barr, J.; Schacht, E.; Heller, J. Development of a poly(ortho ester) prototype with a latent acid in the polymer backbone for 5-fluorouracil delivery. *J. Control. Rel.* **2000**, *65,* 367-374.
11. Wang, C.; Ge, Q.; Ting, D.; Shen, H-R.; Chen, J.; Eisen, H. N.; Heller, J.; Langer, R.; Putnam, D. Molecularly engineered poly(ortho esters) microspheres for enhanced delivery of DNA vaccines, *Nature Mater.* **2004**, *3,* 190-196.
12. Schwach-Abdellaoui, K.; Heller, J.; Gurny, R. Control of molecular weight for autocatalyzed poly(ortho esters) obtained by polycondensation reaction, *Int. J. Pol. Anal. and Charact.* **2002**, *7,* 145-161.

Chapter 4

Paclitaxel Delivery from Novel Polyphosphoesters: From Concept to Clinical Trials

Stephen K. Dordunoo[*], William C. Vincek, Mahesh Chaubal, Zhong Zhao, Rena Lapidus, Randall Hoover, and Wenbin Dang

Guilford Pharmaceuticals, Inc., 6611 Tributary Street, Baltimore, MD 21224
[*]Corresponding author: dordunoos@guilfordpharm.com

Novel and versatile biodegradable copolymers containing both phosphonate and lactide ester linkages in the polymer backbone (Polilactofate), which degrade in vivo into nontoxic residues, were investigated. Synthesis of Polilactofate (PLF) is achieved by bulk or solvent polymerization techniques. PLF is more hydrophilic than corresponding polylactides, takes up more than its own weight of water, swells and degrades with continuous mass loss over periods from weeks to in excess of 3 months. Paclitaxel, incorporated into PLF microspheres by standard methods was continuously released *in vitro* and *in vivo* over at least 60 days. Preclinical pharmacology studies have demonstrated that paclitaxel released from PACLIMER® microspheres has significant anti-tumor activity against various human cancer cells in murine models. The PACLIMER® formulation was tested for safety and toxicity in mice, rats and dogs. A no-adverse-effect level for PACLIMER® microsphere administrations could not be determined in these studies but tolerability was demonstrated and a maximum tolerated dose was identified. Pilot studies in human ovarian patients showed good tolerability of the drug, with continuous release of paclitaxel from PACLIMER® microspheres over at least 8 weeks. Paclitaxel plasma concentrations were far below those causing systemic toxicity.

Introduction

The present study relates to novel biodegradable copolymers, in particular those containing both phosphonate and lactide ester linkages in the polymer backbone (Polilactofate), which degrade *in vivo* into nontoxic residues. The copolymers are particularly useful as implantable medical devices and drug delivery systems.

Biocompatibility of terephthalate polymers (Dacron) is well known *(1)*. However, this polymer is non-biodegradable, which limits its clinical applicability. Incorporation of phosphoester linkages into the polymer converts this polymer into a biodegradable form. By careful selection of monomers, several novel polymers with different physical properties and various degradation profiles could be synthesized. Poly(lactide-*co*-ethylphosphate) (Polilactofate) is one class of linear phosphorus-containing copolymers made by extending the chains of low molecular weight polylactide prepolymers (made from lactide and propylene glycol monomers) with ethyl dichlorophosphate or ethyl phosphate monomers. A typical structure of Polilactofate is depicted in Figure 1.

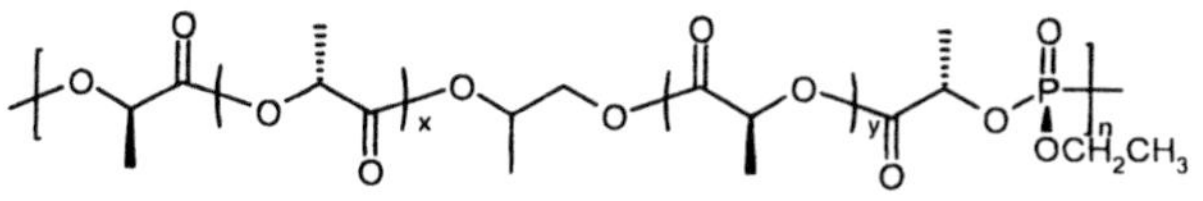

Figure 1. Structure of Polilactofate polymers, with (x+y)/2 =10 and n = number of repeating units.

The versatility of these polymers comes from the phosphorus atom, which is known for a multiplicity of reactions. Its bonding can involve the 3p orbitals or various 3s-3p hybrids; spd hybrids are also possible because of the accessible d orbitals. Thus, the physicochemical properties of poly(phosphoesters) can be readily changed by varying the pendent groups. The biodegradability of the polymer is due primarily to the physiologically labile phosphoester bond in the polymer backbone. By manipulating the backbone or the side chain, a wide range of biodegradation rates are attainable *(1,2)*.

An additional feature of poly(phosphoesters) is the availability of functional side groups. Because phosphorus can be pentavalent, drug molecules or other biologically active substances can be chemically linked to the polymer. For example, drugs containing carboxy groups may be coupled to phosphorus via a hydrolyzable ester bond. The P-O-C group in the backbone lowers the glass transition temperature of the polymer and, importantly, confers solubility in common organic solvents and a degree of hydrophilicity.

Drug delivery to solid tumors

Locally confined tumors are best treated by surgical excision, unless adjacent vital structures prevent complete removal *(3)*. Tumors in surgically difficult sites may be amenable to radiation therapy; however, tumors of certain histologic types are resistant to the effects of radiation. Current chemotherapeutic agents damage normally dividing cells as well as malignant cells. The dose of a chemotherapeutic agent that can be administered systemically is therefore limited by its toxic side effects to other organs, primarily hematopoietic, gastrointestinal, renal pulmonary and cardiac tissues. Although chemotherapy has achieved significant improvement of the treatment of hematological neoplasms, there has been, at best, marginal improvement in the treatment of solid tumors with chemotherapy alone.

The concept of intratumoral administration of antineoplastic agents was designed to circumvent some of the barriers to drug accumulation in the tumor *(3)*. Intratumor delivery of anticancer agents in a suitably designed controlled release system is postulated to optimize drug delivery to solid tumors by maintaining adequate concentration of the drug in the tumor for a considerable period of time while minimizing systemic exposure with concomitant reduction in toxic side effects. Several studies have demonstrated a moderate to high degree of success following the administration of drug directly into or near the tumor. The pharmacological response elicited following intratumor delivery of anticancer agents depends, among other factors, on the nature of the drug, the polymer type, formulation, site of implantation, the tumor type and the degree and rate of necrosis-necrobiosis.

Some arguments against the intratumor delivery include tumor inaccessibility, seeding of needle tracts, escape of fluids from the tumor due to high intratumor pressure and increased probability of inducing metastasis. Some or all of these arguments can also be made regarding tumor biopsy. However, fine-needle biopsy is an established diagnostic procedure that has been shown to be safe.

Paclitaxel – active ingredient

Paclitaxel is a novel anti-microtubule agent and the active ingredient in TAXOL®. Paclitaxel is a white to off-white crystalline powder with empirical formula $C_{47}H_{51}NO_{14}$ and a molecular weight of 853.9. The structure of paclitaxel is provided in Figure 2.

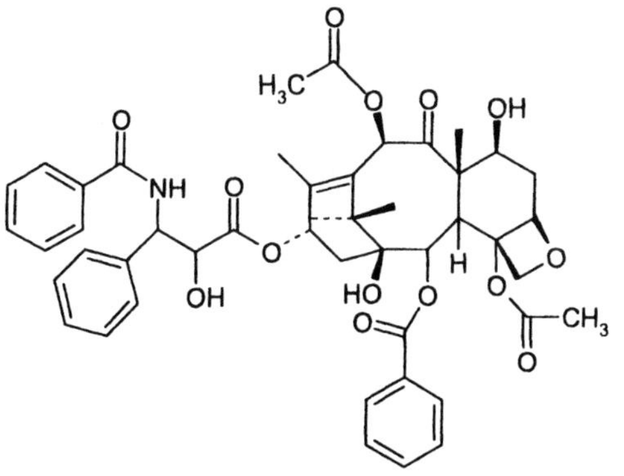

Figure 2. Structure of paclitaxel.

Paclitaxel is highly lipophilic, insoluble in water, and melts at 213-217°C. TAXOL® is approved by the U.S. Food and Drug Administration (FDA) for the treatment of (i) metastatic carcinoma of the ovary after failure of first-line or subsequent therapy, (ii) treatment of breast cancer after failure of combination chemotherapy for metastatic disease or relapse within six months of adjuvant therapy, and (iii) second-line treatment of AIDS-related Kaposi's sarcoma. TAXOL® is also approved for first line treatment for non-small cell lung cancer (NSCLC) in combination with cisplatin in patients who are not candidates for potentially curative surgery and/or radiation therapy.

We evaluated the use of peri- or intratumoral chemotherapy with controlled release paclitaxel microspheres (PACLIMER®) in various solid tumor models and in ovarian cancer patients.

Materials and Methods

General Synthesis of Polyester-polyphosphonate Copolymers

The most common general reaction in preparing a poly(phosphonate) is a dehydrochlorination between a phosphonic dichloride and a diol. A Friedel-Crafts reaction can also be used to synthesize poly(phosphonates) *(4)*. Polymerization typically is effected by reacting either bis(chloromethyl) compounds with aromatic hydrocarbons or chloromethylated diphenyl ether with triaryl phosphonates. Poly(phosphonates) can also be obtained by bulk condensation between phosphorus diimidazolides and aromatic diols, such as resorcinol and quinoline, usually under nitrogen or some other inert gas.

An advantage of bulk polycondensation is that it avoids the use of solvents and large amounts of other additives, thus making purification more straightforward. Somewhat rigorous conditions, however, are often required and can lead to chain acidolysis or hydrolysis in the presence of water. Thermally induced side reactions, such as crosslinking reactions, can also occur if the polymer backbone is susceptible to hydrogen atom abstraction or oxidation with subsequent macroradical recombination. To minimize these side reactions, the polymerization is preferably carried out in solution. Solution polycondensation requires that both the diol and the phosphorus component be soluble in a common solvent, typically a chlorinated organic solvent such as chloroform, dichloromethane, or dichloroethane. The solution polymerization is preferably run in the presence of equimolar amounts of the reactants and a stoichiometric amount of an acid acceptor, usually a tertiary amine such as pyridine or triethylamine. The product is typically isolated from solution by precipitation with a non-solvent and purified to remove the hydrochloride salt by conventional techniques such as washing with dilute hydrochloride. Reaction times tend to be longer with solution polymerization than with bulk polymerization. However, because overall milder reaction conditions can be used, side reactions are minimized, and more sensitive functional groups can be incorporated into the polymer. The disadvantage of solution polymerization is that the attainment of high molecular weights, i.e., MW higher than 20,000 Da, is less likely.

Interfacial polycondensation can be used when high molecular weight polymers are desired at high reaction rates. Mild conditions minimize side reactions. In addition, the dependence of high molecular weight on stoichiometric equivalence between diol and dichloride, inherent in solution methods, is avoided. However, hydrolysis of the acid chloride may occur in the alkaline aqueous phase. Sensitive dichlorides that have some solubility in water are generally subject to hydrolysis rather than polymerization. Phase transfer catalysts, such as crown ethers or tertiary ammonium chloride, can be used to bring the ionized diol to the interface to facilitate the polycondensation reaction.

Synthesis of Polilactofate

The detailed synthesis of the polymer has been discussed by Chaubal *et al.* *(5)*. Briefly, the lactide-phosphate copolymer was synthesized in two steps as shown in Figure 3. In the first step, a lactide prepolymer was prepared via a ring opening polymerization of D,L-lactide, using propylene glycol (PG) as the initiator and $Sn(Oct)_2$ as the catalyst. Unreacted lactide was removed by application of a vacuum at 125 °C for 2 hours. In the second step, the lactide prepolymer was reacted with ethyl dichlorophosphate at low temperature (-10 to -15 °C) in a solvent such as chloroform with triethylamine (TEA) and dimethylaminopyridine (DMAP) as acid acceptor and catalyst, respectively. At the end of the reaction, the solvent was removed using a rotary evaporator, and the residue was dissolved in acetone. The insoluble salts were filtered out and the residual base and catalyst were subsequently removed using a combination of acidic and neutral ion exchange resins. The final polymer was dissolved in dichloromethane and precipitated in a mixed solvent system of petroleum ether: ethyl ether (3:1, v/v), followed by drying in a vacuum oven to constant weight. The resulting copolymers were white amorphous solids, soluble in common organic solvents such as acetone, chloroform and dichloromethane. The structure of the resulting purified polymer was confirmed using one-dimensional ^1H and ^{31}P NMR and two-dimensional ^1H-^1H COSY NMR *(6)*.

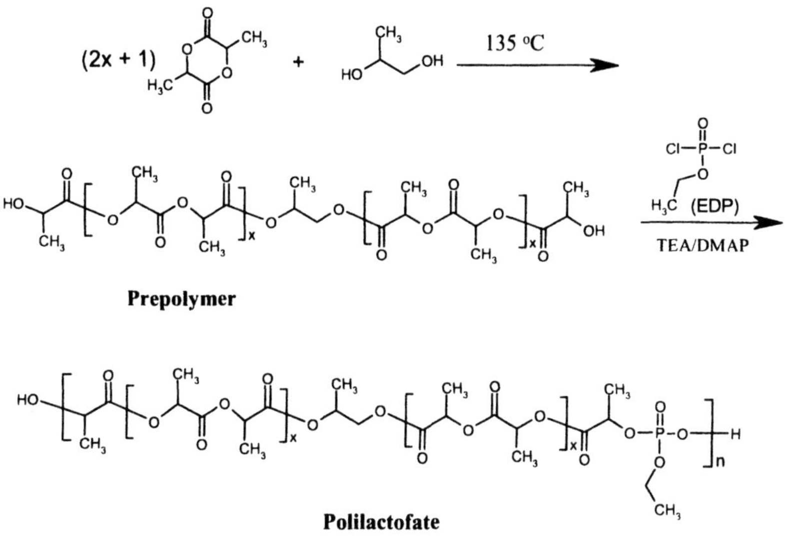

(2x + 1) + 135 °C

Prepolymer

$Cl-P(=O)-Cl$, H_3C (EDP), TEA/DMAP

Polilactofate

Figure 3. Two-step synthesis of poly(lactide-co-ethyl phosphate).

Polymer Hydrophilicity Measurements

The hydrophilicity of polymers was estimated using three parameters: water uptake, contact angle and surface polarity. For water uptake, polymer wafers produced by direct compression, films cast from polymer melt or polymer rods were immersed in 0.1 M phosphate buffer and incubated at 37 °C. Samples were retrieved at predetermined intervals from the incubation vials, blotted of excess water, and weighed to determine water uptake. As shown in Figure 4, Polilactofate rods absorb water up to 10-times their weight and swell before breaking up, demonstrating higher hydrophilicity compared to polylactic acid.

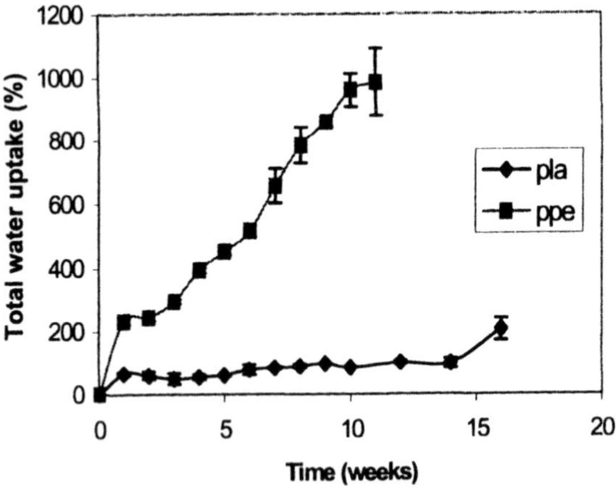

Figure 4. Water uptake and swelling behavior of Polilactofate, compared to polylactic acid (PLA).

Biodegradation Characteristics

Microspheres made from Polilactofate were studied under accelerated and normal *in vitro* degradation conditions *(7)*. Gel permeation chromatography (GPC), ^{1}H and ^{31}P-NMR, weight loss measurements, and differential scanning calorimetry (DSC) techniques were used to characterize the change of molecular weight (M_w), chemical composition, and glass transition temperature (T_g) of the degrading polymers. The results indicated that the copolymers degraded in a two-stage fashion, with cleavage of the phosphate-lactide linkages contributing mostly to the initial, more rapid degradation phase and cleavage of the lactide-lactide bonds being responsible for the slower, latter stage degradation (Figures 5 and 6). The decrease in the copolymer molecular weight was accompanied by a continuous mass loss (Figure 6) *(6)*.

A two-stage hydrolysis pathway was thus proposed to explain the degradation behavior of the copolymers. *In vivo* degradation studies performed in mice demonstrated a good *in vitro* and *in vivo* correlation for the degradation rates. *In vivo* clearance of the polymer was faster and without any lag phase. These copolymers are potentially advantageous for drug delivery and other biomedical applications, where rapid clearance of the polymer carrier and repeated dosing capability are essential to the success of the treatment.

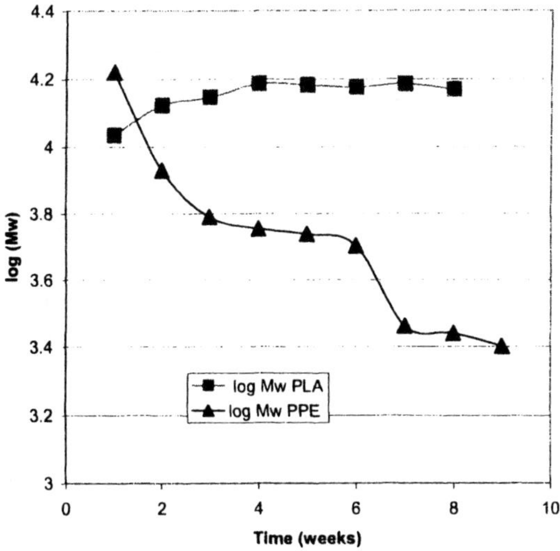

Figure 5. Reduction in molecular weight of Polilactofate (PPE) and poly(lactic acid) (PLA).

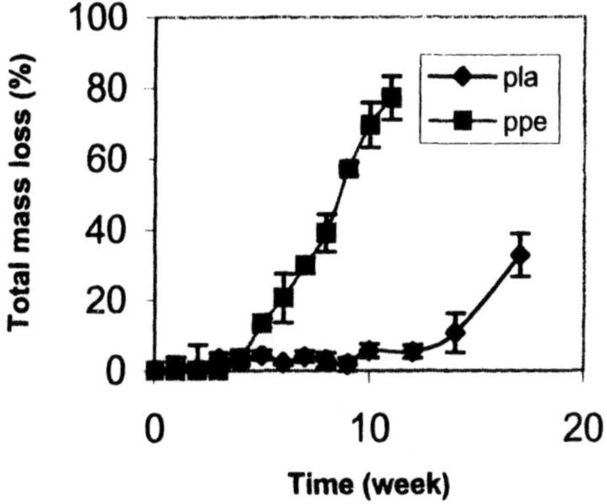

Figure 6. Degradation and mass loss behavior of Polilactofate (PPE) and poly(lactic acid) (PLA).

In vivo Biocompatibility of Polilactofate

About 100 mg polymer wafer was formed from Polilactofate and, as a reference, a copolymer of lactic and glycolic acid (PLGA RG755), which is known to be biocompatible. These wafers were inserted between muscle layers of the right limb of adult SPF Sprague-Dawley rats under anesthesia. The wafers were retrieved at specific times, and the surrounding tissues prepared for histopathological analysis by a certified pathologist. Similar tests were done after intramuscular or subcutaneous injection of various amounts of Polilactofate microspheres into S-D rats, tabulating implant site macrophage counts, as well as irritation scores. The phosphoester copolymer, Polilactofate, was shown to have an acceptable biocompatibility, similar to that exhibited by the PLGA reference wafer.

Preparation of PACLIMER® microspheres

Paclitaxel was incorporated into Polilactofate microspheres at 10-60% (w/w) loading by dissolving both substances in ethyl acetate and pumping the solution through an in-line homogenizer with 0.5% poly(vinyl acetate) solution to a container with an overhead stirrer. After the microspheres had hardened, they were filtered and lyophilized to give the PACLIMER® product. Particle size and size distribution of the resultant PACLIMER® microspheres were determined by a single particle optical sensing system (Model 770; Particle Sizing Systems, Langhome, PA).

The median diameter of the microspheres was 53 µm, with a range of 20–200 µm. Microspheres containing up to 40% paclitaxel are spherical in shape with smooth surfaces, while microspheres containing 50 or 60% drug were spherical but with rough surfaces (Figure 7).

In Vitro Release of paclitaxel from PACLIMER® microspheres

The *in vitro* release of paclitaxel was performed by incubating PACLIMER® microspheres in phosphate buffered saline (PBS) (pH 7.4) at 37°C. Paclitaxel has very low solubility and poor stability in PBS, and to maintain sink conditions, an octanol layer was placed on top of PBS to continuously extract the released paclitaxel. Paclitaxel concentration in octanol was analyzed at specific time points. As shown in Figure 10, the release of paclitaxel from PACLIMER® microspheres was slow, with <50% of the drug released in 2 months but was faster than the dissolution of paclitaxel as received.

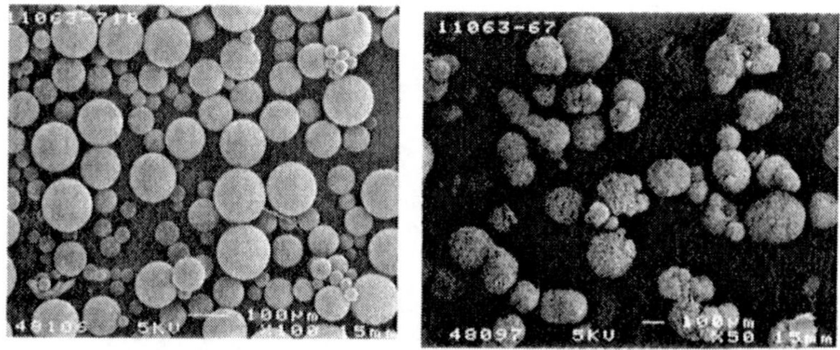

Figure 7. Scanning electron micrograph (SEM) of PACLIMER® microspheres containing 40% (left) and 60% (right) paclitaxel, showing spherical shapes with smooth and rough surfaces, respectively.

The X-ray diffraction pattern of paclitaxel as received showed peaks indicating crystalline material, while the pattern of PACLIMER® microspheres containing 10% of drug revealed the presence of amorphous material (Figure 8). PACLIMER® miscrospheres containing 50 or 60% of drug, however, showed the presence of crystalline material (Figure 9). Interestingly, PACLIMER® microspheres containing 40% initially did not show peaks in the X-ray but revealed peaks after the microspheres had been suspended in phosphate buffered saline at pH 7.4 and subsequently dried, suggesting physical state conversion of this composition in the presence of water.

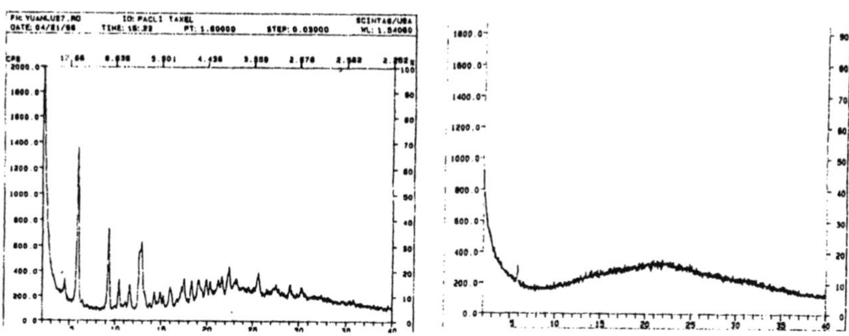

Figure 8. X-ray diffraction patterns of paclitaxel as received, showing crystalline material (left) and of PACLIMER® microspheres containing 10% drug, showing amorphous material (right).

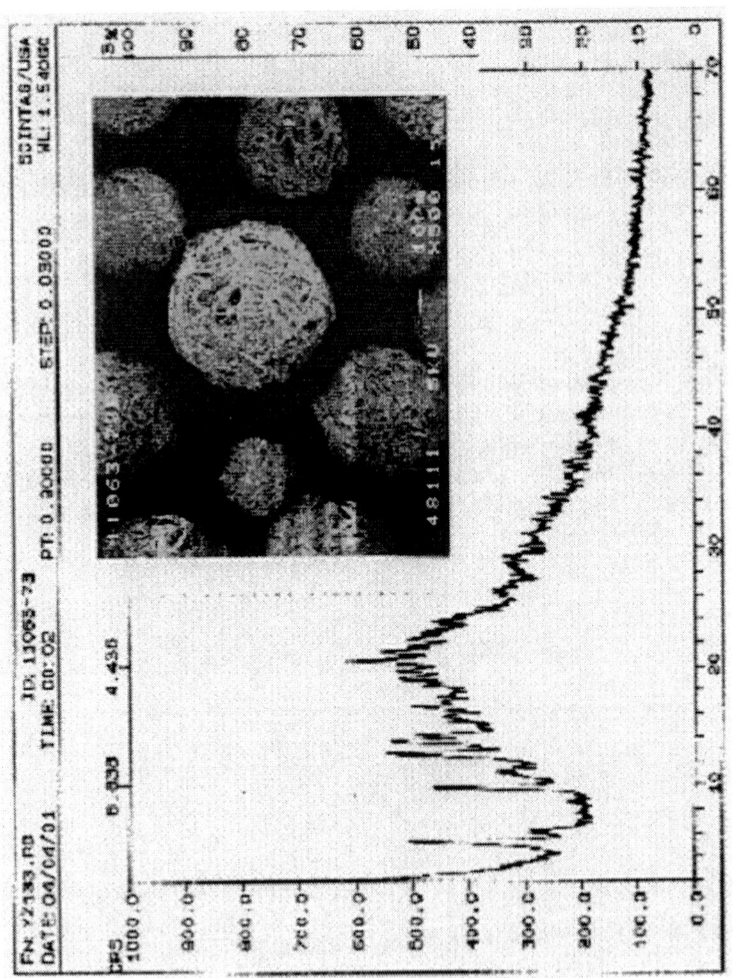

Figure 9. X-ray diffraction pattern and SEM of PACLIMER® microspheres containing 60% drug, showing crystalline material.

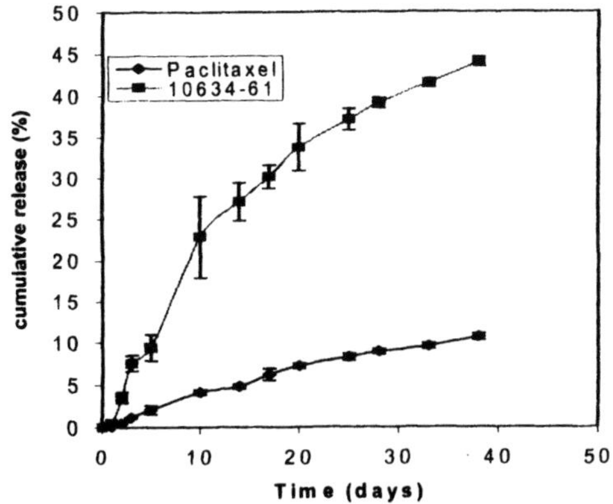

Figure 10. In vitro release of paclitaxel from PACLIMER® microspheres.

Results and Discussion

Nonclinical Pharmacology

Paclitaxel, the active pharmaceutical ingredient in PACLIMER® microspheres, has been shown to have multiple pharmacologic effects in human cancer cells *(6)*. These effects include increased polymerization and stabilization of microtubules, blockade of cells at the G_2/M phase of the cell cycle, inhibition of DNA synthesis, and induction of apoptosis *(7)*. Prolonged exposure (16-48 hours) of cells to paclitaxel *in vitro* was shown to result in increased cytotoxicity *(8)*. Paclitaxel is currently marketed in the USA under the trade name TAXOL®, a formulation providing immediate availability of paclitaxel. In contrast, PACLIMER® is a sustained release, biodegradable polymeric microsphere formulation containing 10% paclitaxel (w/w). The microsphere product is designed to release paclitaxel for a prolonged period of time and is intended for administration by injection to an area such as the intraperitoneal space. It is hypothesized that PACLIMER® microspheres will not only offer the pharmacological action of paclitaxel but may be more efficacious. The sustained release of paclitaxel from the PACLIMER® microspheres is expected to result in continuous exposure of tumor cells to the drug, thereby increasing its efficacy. To test this hypothesis, efficacy studies of PACLIMER® microspheres and paclitaxel alone have been performed in several mouse tumor models.

Preclinical Pharmacology

The pharmacological effects of PACLIMER® microspheres were tested in several murine models. The efficacy of various doses and administration schedules of PACLIMER® microspheres was compared to that of paclitaxel in severe combined immunodeficiency disease (SCID) mice bearing intraperitoneal (i.p.) xenografts of OVCAR-3 or OVCAR-5 human ovarian cancer cells. One additional experiment evaluated efficacy in nude mice bearing intraperitoneal xenografts of OVCAR-3 cells. A total of six experiments were performed.

The OVCAR-3 cell line was derived from the recurrent tumor of an ovarian cancer patient after failure of treatment with Adriamycin®, cisplatin and cyclophosphamide *(9)*. The OVCAR-5 cell line was derived from the untreated tumor of an ovarian cancer patient *(10)*. The NIH:OVCAR-3 possess the capacity to grow i.p. in female nude athymic mice. After i.p. injection of these cells, animals develop metastatic spread similar to that of clinical ovarian cancer. Disease progression is characterized by the development of massive ascites, extensive invasive i.p. tumors, and pulmonary metastases. The malignant ascites cells are transplantable, manifest cytoplasmic androgen and estrogen receptors, and express the ovarian cancer associated antigen CA125 (116,000 units/ml of ascites supernatant). The cells also have the same chromosome markers, which were present in the original cell line, NIH:OVCAR-3. Survival of nude athymic mice following i.p. passage of ascites is dependent on tumor cell inoculum, ranging from a median survival of 39 days with 40 million cells to 84 days for 11.5 million transplanted cells. The characteristics of this unique *in vivo* model make it well suited for the evaluation of new drugs and novel experimental therapies in ovarian cancer.

Study Design

OVCAR3 ascites were harvested from tumor bearing animals and prepared for i.p. injection at a concentration of ~0.2 ml 1xg packed ascites cells in 1 ml of saline. Each female SCID mice was injected i.p. with 1 ml of ascites suspension. Animals were randomized into placebo control, i.p. taxol treated and i.p. PACLIMER®-treated groups (n=6 per group). In the first experiment, animals were treated at day 4 with paclitaxel or PACLIMER® at 10 mg/kg and 40 mg/kg doses. In the second experiment, efficacy of PACLIMER® at 4, 10 and 40 mg/kg were compared with paclitaxel at 40 mg/kg. Another study was conducted using OVCAR3 animal model. The study design was similar to the previous study with the exception that treatment started at day 1 post cell inoculation. The treatment groups were: placebo control (n=20), PACLIMER® at 160 mg/kg (n=9), 80 mg/kg (n=10), 40 mg/kg (n=10), 4 mg/kg (n=10), paclitaxel at 40 mg/kg (n=10), 4 mg/kg (n=10) and 4 weekly treatment at 40 mg/kg (n=9). The results are shown in Table I and Plates 1 and 2. PACLIMER® at 40mg/kg gave 100% survival at day 100, which was better than

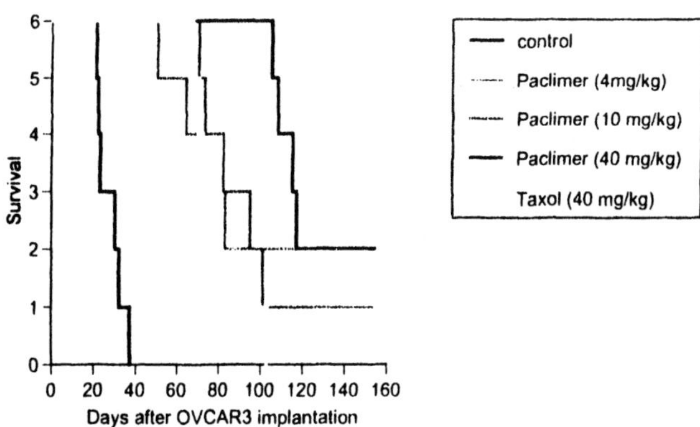

Plate 1. Representative survival plots of OVCAR-3 model in SCID mice. (See page 1 of color insert.)

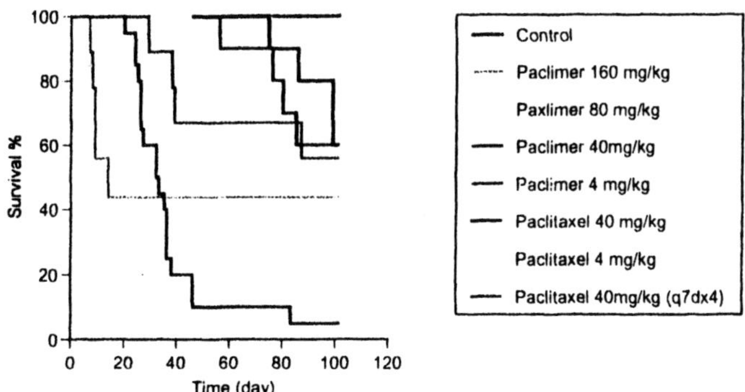

Plate 2. Representative survival plots of OVCAR-3 model in SCID mice. (See page 1 of color insert.)

survival (55%) for 4x 40mg/kg repeat doses of TAXOL at 7-day intervals (Plate 2). Early mortality was observed with the 160mg/kg indicating toxicity at this dose.

Table I. Summary of median survival in OVCAR3 experiment.

	Median survival (days)	
Paclitaxel dose (mg/kg)	Experiment 1	Experiment 2
0 (placebo control)	23	30
Taxol 10	64	N/A
Taxol 40	67	77
PACLIMER® 4	N/A	83
PACLIMER® 10	69	95
PACLIMER® 40	115	117

In summary, all preclinical pharmacology studies in OVCAR-3 models demonstrated that PACLIMER® microspheres had significant antitumor activity against human ovarian cancer cell lines in murine models. In these ovarian cancer model systems, survival was significantly increased after PACLIMER® microspheres administration compared to control microspheres. When compared to intraperitoneal paclitaxel alone, PACLIMER® microspheres appear to have greater efficacy in these experiments. In addition, higher total paclitaxel doses may be administered to rodents in the form of intraperitoneal PACLIMER® microspheres as compared to paclitaxel alone.

Efficacy in Intratumoral Models

Non-small Cell Lung Cancer: In vitro Study

The goal of this experiment was to test 6 non-small cell lung cancer (NSCLC) cell lines *in vitro* and to determine their sensitivity to paclitaxel. Briefly, the 6 NSCLC lines (A549, A427, H1299, H838, H358, H1650) were incubated with paclitaxel at concentrations ranging from 0.05 nM to 1000 nM for 48 hours. The cytoxicity was measured using Promega's Cell Titer 96 Aqueous MTS assay kit. As shown in Figure 11, all the cell lines tested are sensitive to paclitaxel.

Non-small Cell Lung Cancer In vivo Efficacy Study

Since all the cell lines were sensitive to paclitaxel, it was decided that the NSCLC line A549 was used for the first set of efficacy studies. NSCLC line H1299 was chosen for the 2nd efficacy experiment. Two experiments were

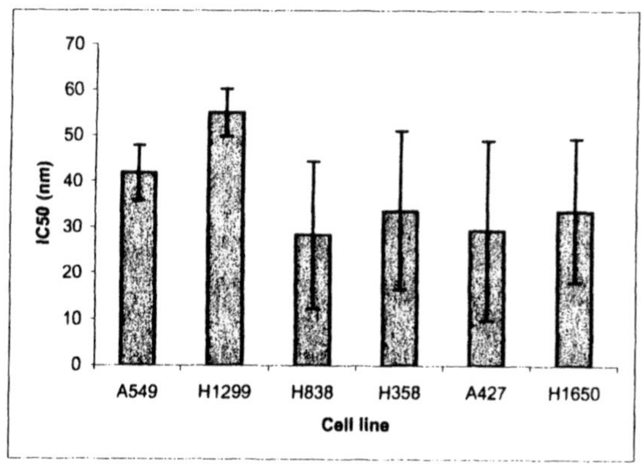

Figure 11. In vitro sensitivity of NSCLC lines to paclitaxel.

performed with A549 cell line. Briefly, A549 or H1299 cells were grown in culture to confluence and injected s.c. into nude mice. When the tumors reached ~200 mm^3, the following treatments were given:

Doses
Paclitaxel i.t: vehicle, 4, 12.5 and 24 mg/kg
Paclitaxel i.p.: vehicle, 24 mg/kg
PACLIMER® i.t.: polymer placebo, 4, 12.5 and 24 mg/kg

Figure 12 shows representative data from the two studies with A549, indicating that tumor size reduction was obtained with PACLIMER® at 24 mg/kg.

The tumor volume decreased to about a third of the initial volume after 8 days and remained at this level up to 30 days, when the experiment was terminated. At lower PACLIMER® dose (12.5 mg/kg), tumor size reduction was only noticed in the first 8 days following treatment. No efficacy was seen with any of the other treatment arms, with the exception of i.t. 4mg/kg Taxol arm in which case there seemed to be a delay in tumor growth. Other studies were performed, using PACLIMER® microspheres in brain, prostate and other cancers *(11-13)*.

Pre-clinical and Clinical Pharmacokinetics

The pharmacokinetics of paclitaxel following intraperitoneal, intramuscular or subcutaneous administration of PACLIMER® microspheres was investigated in mouse, rat, dog and humans. In one experiment, the pharmacokinetic profile of paclitaxel following administration of PACLIMER® was determined in female CD-1 mice. Mice received either 800 mg/kg PACLIMER® microspheres intraperitoneally or 1500 mg/kg subcutaneously. Plasma samples were collected at various times after dose administration. Five animals were sampled at each

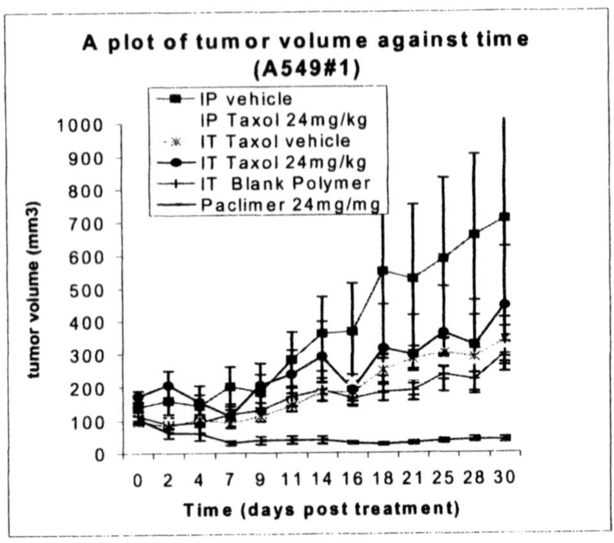

Figure 12. Effect of conventional TAXOL formulation and PACLIMER® microspheres on tumor volume.

time point for each route of administration; only one sample was collected from each animal. Samples were analyzed using a validated LC/MS/MS method. The limit of quantitation was approximately 0.293 nM. As shown in Figure 13 and Table II, plasma paclitaxel was measurable at all sampling times for both routes of administration. Higher plasma levels were obtained following i.p. adminitration compared with s.c. dosing. Following i.p. administration, plasma concentrations reached an average C_{max} of 321 nM at T_{max} of 0.17 hours. For subcutaneous administration, plasma C_{max} was about 25.3 nM at 0.17 hours. Plasma paclitaxel concentration was increased at Day 35 relative to earlier sampling times for both routes of administration, possibly due to degradation of the polymer. Therefore, a terminal elimination phase could not be calculated, and $AUC_{0-\infty}$, Cl/F and V_Z/F could also not be accurately determined.

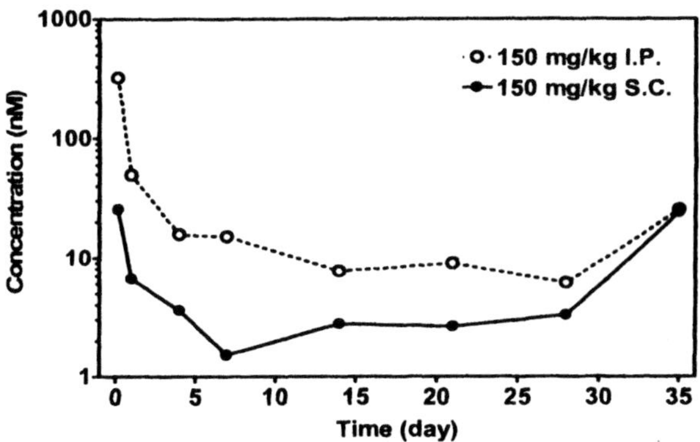

Figure 13. Plasma profiles of paclitaxel following IP and SC administration of PACLIMER® microspheres.

Table II. Effect of route of PACLIMER® microspheres administration on pharmacokinetic parameters of paclitaxel.

Route of Administration	AUC_{0-t} (nM*day)	T_{max} (day)	C_{max} (nM)
IP	564.6	0.17	321
SC	162.1	0.17	25.3

The pharmacokinetics of paclitaxel was also assessed following administration of PACLIMER® microspheres as part of a non-GLP toxicology study in rats. On Day 0, PACLIMER® microspheres were administered intraperitoneally to groups of rats at doses of 100, 200, and 300 mg/kg, and paclitaxel was administered at 30 mg/kg (N = 3/treatment/time point). A separate group of animals received a second PACLIMER® microspheres dose of 300 mg/kg on Day 28. Plasma samples were taken from three animals at each dose level at a number of time points. Only one sample was collected from each animal.

As observed in mice, following single doses of either paclitaxel or PACLIMER® microspheres, plasma paclitaxel was measurable at all sampling times post administration during the course of the study. Furthermore, neither exposure ($AUC_{0-\infty}$ or AUC_{0-t}) nor C_{max} was proportional to dose. Systemic exposure to paclitaxel following intraperitoneal paclitaxel dosing was greater than that observed following PACLIMER® microspheres administration. In the group of animals that received two doses of PACLIMER® microspheres (Days 0 and 28), plasma paclitaxel concentrations were also measurable at all sampling times post-dose (Tablet III). The average plasma concentration ranged from 4.74 to 24.8 nM across all time points, which was higher than observed following a single dose. The exposure to paclitaxel (AUC) was substantially higher than observed following a single dose.

Table III. Pharmacokinetic parameters in rat study

Dose[#] (mg/kg)	$AUC_{0-\infty}$ (nM*day)	AUC_{0-t} (nM*day)	C_{max} (nM)	T_{max} (day)	V_z/F (L/kg)
100	113.0	74.7	4.82	28	347
200	205.1	102.1	6.59	0.25	237
300	199.2	106.8	6.53	14	173
300**	489.8	443.6	24.8	10	34.1
30*	84.2	68.0	17.6	0.5	268

[#] *PACLIMER® microspheres doses are 10% paclitaxel (w/w).*
** *Group received a second dose of PACLIMER® microspheres on Day 28.* * TAXOL IV

The toxicokinetics of paclitaxel administered in PACLIMER® microspheres was assessed as part of a GLP toxicity study. Briefly, six male and six female beagle dogs received Polilactofate (90 mg/kg), paclitaxel (3 or 9 mg/kg), or PACLIMER® microspheres (15, 30, 90, or 120 mg/kg) (10% paclitaxel, w/w). Three males and three females per dose group were observed for 60 days and then subjected to an interim sacrifice; the other three males and three females were observed for 120 days and then sacrificed. Plasma samples were obtained at various times post dose and were assayed using a validated LC/MS/MS method, as in previous studies. The paclitaxel plasma concentration-time profiles for all animals are shown in Figure 14.

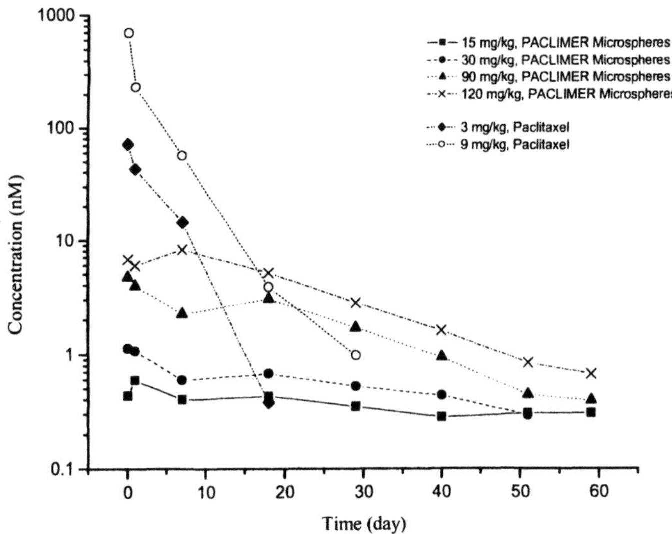

Figure 14. Plasma concentration of paclitaxel following intraperitoneal administration to female beagle dogs.

Plasma paclitaxel assay results indicated that all animals in the study were exposed systemically to paclitaxel following administration of PACLIMER® microspheres. Pharmacokinetic analysis of paclitaxel concentrations following administration of PACLIMER® microspheres indicated that exposure was relatively proportional to dose. Dose independence of pharmacokinetics was also inferred because clearance/bioavailability was consistent across all doses. However, dose proportionality was not observed between the 3 and 9 mg/kg paclitaxel dose groups. While the dose increased 3-times, exposure (AUC) increased approximately 7-fold and plasma C_{max} increased approximately 10-fold. In some animals, the terminal phase in plasma could not be determined due to either early sacrifice or increasing plasma concentrations over the last several time points. Average exposure (AUC_{0-t}) to paclitaxel following intraperitoneal administration of paclitaxel at a dose of 9 mg/kg was higher than that following a 90 mg/kg dose of PACLIMER® microspheres (272.1 v. 81.6 nM* day, respectively). Additionally, the C_{max} was approximately 59-times higher following the 3 mg/kg dose of paclitaxel and 140-times higher following the 9 mg/kg paclitaxel dose as compared to equivalent paclitaxel doses in PACLIMER® microspheres (Table IV).

Overall, the pharmacokinetic profile of paclitaxel administered as PACLIMER® microspheres indicates that plasma C_{max} is lower than that observed for equivalent doses of commercial paclitaxel formulation given i.p. This lower plasma C_{max} is important since acute systemic toxicity associated with paclitaxel (i.e., neutropenia) has been linked to exposure to plasma concentrations >500 nM. In fact, C_{max} was lower than 500 nM for all doses tested intraperitoneally in mouse, rat and dog. While C_{max} appears to correlate with dose for PACLIMER® microspheres and the commercial TAXOL®, there is no clear correlation between systemic paclitaxel exposure (as expressed by AUC) and dose. The plasma paclitaxel exposure appears to be reasonably dose proportional between 15 and 120 mg/kg in the dog; however, non-linearity was noted at higher doses, i.e., the 180-210 mg/kg dose range.

Table IV. Average plasma pharmacokinetic parameters of paclitaxel following intraperitoneal administration of PACLIMER® microspheres or paclitaxel to female beagle dogs.

	$AUC_{0-\infty}$ (nM*day)	AUC_{0-t} (nM*day)	Cl/F (L/day)	C_{max} (nM)	T_{max} (day)
Mean	28.7	9.6	71.5	0.624	3.45
SD	10.5	8.4	32.9	0.169	5.83
Mean	62.7	23.9	79.9	1.20	0.39
SD	44.7	9.6	45.0	0.370	0.45
Mean	210.5	81.6	106.2	4.96	1.81
SD	352.8	38.6	64.1	1.61	5.12
Mean	489	158.3	76.8	11.3	0.97
SD	660	145.2	56.6	15.7	1.95
Mean	44.4	52.7	86.2	71.8	0.25
SD	15.5	11.4	25.8	13.2	0.00
Mean	288.6	272.1	40.8	695	0.25
SD	101.9	108.7	15.1	338	0.00

The same observation was made in the rat study, where dose proportionality was not found between the 100 and 300 mg/kg doses of PACLIMER® micro-

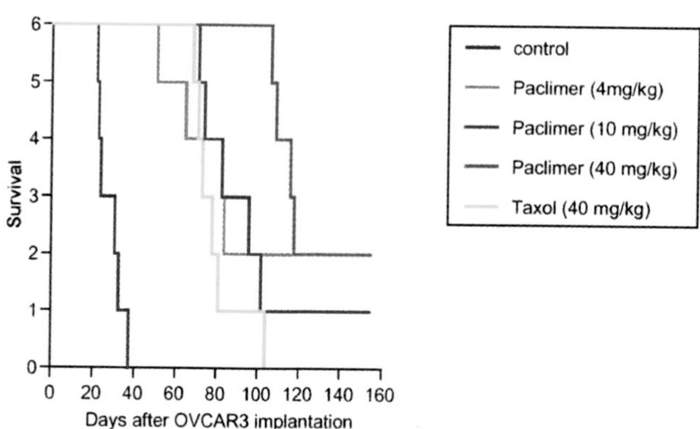

Plate 4.1. Representative survival plots of OVCAR-3 model in SCID mice.

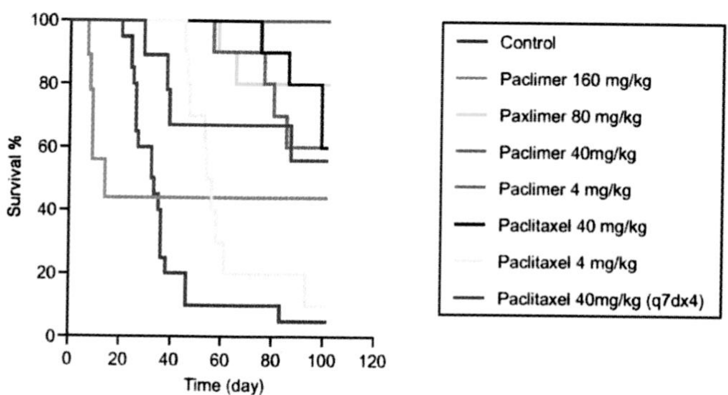

Plate 4.2. Representative survival plots of OVCAR-3 model in SCID mice.

spheres. However, a similar lack of dose proportionality for exposure was also noted for paclitaxel administered as TAXOL® formulation.

Phase 1 clinical trial with PACLIMER® Microspheres in Patients with Ovarian Cancer

Based upon the animal data, the maximal plasma paclitaxel concentrations achieved with proposed clinical doses of PACLIMER® microspheres (60 – 1800 mg/m^2) should be below the toxicity threshold. A Phase 1 clinical trial with PACLIMER® microspheres administered via intraperitoneal administration was conducted in patients with ovarian cancer. The objectives of this study were (1) to determine the safety and tolerability of intraperitoneal administration of PACLIMER® microspheres, (2) to identify and confirm the maximum tolerated dose (MTD) of PACLIMER® microspheres, and (3) to determine plasma paclitaxel concentrations at selected times after administration of PACLIMER® microspheres. This trial was conducted as a dose escalation study. An escalation occurs when the safety and tolerability of the previous dose level had been established for at least four weeks after administration. Each patient was to receive a maximum of two doses separated by eight weeks. Once the MTD had been identified, patients in a confirmatory cohort would receive two doses of PACLIMER® microspheres at the MTD, separated by eight weeks (or at the optimal interval based on pharmacokinetic or clinical considerations). The study duration for each patient was approximately 16 weeks. Patients were evaluated on the day of administration and weekly thereafter. The following parameters were assessed periodically throughout the study: physical examination, clinical assessment, GOG grade, vital signs, ECG, serum chemistry, hematology, urinalysis and CA 125 assessment. Plasma samples were obtained pre-administration, 2 and 24 hours after each PACLIMER® microspheres administration and at each weekly visit. To date, ovarian cancer patients have been treated with PACLIMER® microspheres at dose levels of 60 mg/m^2 to 1200 mg/m^2.

Figure 15 depicts plasma paclitaxel levels at various time points in ovarian cancer patients enrolled in the i.p. PACLIMER® microspheres study. Plasma samples were obtained weekly for up to 8 weeks following each dose. Plasma paclitaxel levels for the patient who received 60 mg/m^2 were below the limits of quantitation at each sampling interval. Plasma paclitaxel levels for patients in all other dose groups were measurable for eight weeks after treatment, indicating that paclitaxel was released from PACLIMER® microspheres in a sustained manner. Plasma paclitaxel levels for patients who received two doses of PACLIMER® microspheres revealed similar profiles following each dose. Paclitaxel levels for all patients at all time points studied were substantially below those associated with systemic toxicity.

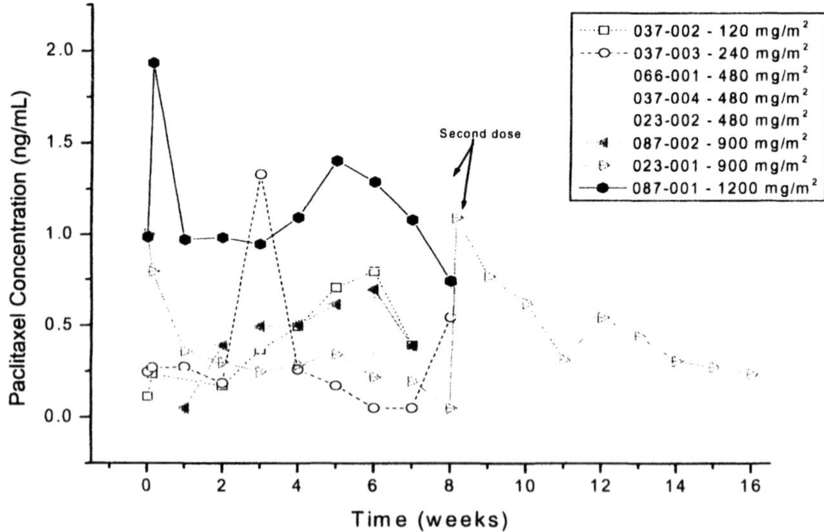

Figure 15. Plasma concentrations of paclitaxel following intraperitoneal administration to female ovarian cancer patients.

Conclusions

PACLIMER® microspheres is a novel polymer-based formulation of paclitaxel, a potent antineoplastic agent with significant activity in patients with ovarian cancer as well as other tumors. PACLIMER® microspheres (10% paclitaxel) has antitumor activity in preclinical murine *in vivo* models of ovarian cancer (OVCAR-3 and OVCAR-5 cell lines grown intraperitoneally in immunocompromised mice). PACLIMER® microspheres is being developed for the intraperitoneal treatment of patients with Stage III ovarian cancer. Paclitaxel is released over a prolonged period of time as the polymer microsphere is hydrolyzed. *In vitro* release extends over 80 days or longer. In toxicokinetic studies, paclitaxel could be measured in the plasma for periods up to 63 days post dose, indicating paclitaxel was being released from the polymer over the entire period. Additionally, the plasma concentrations produced in those studies were below those associated with acute systemic toxicities (>500nM), such as bone marrow depression. The average maximal concentration measured in dogs was 11.6 nM at a human-equivalent dose of 2400 mg/m^2. Therefore, PACLIMER® microspheres is hypothesized to deliver paclitaxel locally (e.g. within the peritoneal cavity) at therapeutic concentrations for a prolonged period of time without reaching plasma levels associated with systemic toxicity.

Acknowledgements

The contributions of the following researchers are hereby acknowledged: Thomas Hamilton, Eric Burak, Eric Spicer, Harris Holland, Kory Engelke, Robert Garver, and Jim English.

References

1. Kadiyala, S.; Lo, H.; Ponticello, M.S.; Leong, K.W. Poly(phosphoesters): Synthesis, Physicochemical Characterization and Biological Response. In Biomedical Application of Synthetic Biodegradable Polymers, Hollinger, J.O., Ed., **1995**, 33-57.
2. Chaubal, M.V.; Su, G.; Spicer, E.; Dang, W.; Branham, K.E.; English, J.P.; Zhao Z. In vitro and in vivo degradation studies of a novel linear copolymer of lactide and ethylphosphate. *J. Biomater. Sci. Polym. Ed.* **2003**, *14*, 45-61.
3. Jain, R.K. Delivery of molecular and cellular medicine to solid tumors. *Adv. Drug Deliv. Rev.* **1997**, *26*, 71-90.
4. Mao, H.; Kadiyala, I.; Leong, K.W.; Zhao, Z.; Dang, W. Biodegradable polymers: poly(phosphoester)s. In "Encyclopedia of Controlled Drug Delivery", Vol.1, Mathiowitz, E., Ed., Wiley-Interscience, **1999**, 45-60.
5. Chaubal, M.V.; Gupta, A.S.; Lopina, S.T.; Bruley, D.F. Polyphosphates and other phosphorus-containing polymers for drug delivery applications. *Crit. Rev. Ther. Drug Carrier Syst.* **2003**, *20*, 295-315.
6. Millenbaugh, N.; Gan, Y.; Au, J. Cytostatic and apoptotic effects of paclitaxel in human ovarian tumors. *Pharm. Res.* **1998**, *15*, 122-127.
7. Innocenti, F.; Danesi, R.; Di Paolo, A.; Agen, C.; Nardini, D.; Bocci, G.; Del Tacca, M. Plasma and tissue disposition of paclitaxel (TAXOL) after intraperitoneal administration in mice. *Drug Metab. Dispos.* **1995**, *23*, 713-717.
8. Markman, M.; Francis, P.; Rowinsky, E. et al. Intraperitoneal TAXOL (paclitaxel) in the Management of Ovarian Cancer. *Ann. Oncol.* **1994**, *5*, S55-58.
9. Hamilton, T.C.; Young, R.C.; Louie, K.G.; Behrens, B.C.; McKoy, W.M.; Grotzinger, K.R.; Ozols, R.F. Characterization of a xenograft model of human ovarian cancer which produces ascites and intraabdominal carcinomatosis. *Cancer Res.* **1984**, *44*, 5286-5290.
10. Johnson, S.W.; Laub, P.B.; Beesley, J.S.; Ozols, R.F.; Hamilton, T.C. Increased platinum-DNA damage tolerance is associated with cisplatin

resistance and cross-resistance to various chemotherapeutic agents in unrelated human ovarian cancer cell lines. *Cancer Res.* **1997**, *57*, 850-856.

11. Harper, E.; Dang, W.; Lapidus, R.G.; Garver, R.I., Jr. Enhanced efficacy of a novel controlled release paclitaxel formulation (PACLIMER delivery system) for local-regional therapy of lung cancer tumor nodules in mice. *Clinical Cancer Research* **1999**, *5*, 4242-4248.
12. Lapidus, R.G.; Dang, W.; Rosen, D.M.; Gady, A.M.; Zabelinka, Y.; O'Meally, R.; DeWeese, T.L.; Denmeade, S.R. Anti-tumor effect of combination therapy with intratumoral controlled-release paclitaxel (PACLIMER microspheres) and radiation. Prostate (New York) **2003**, *58*, 291-298.
13. Li, K.W.; Dang, W.; Tyler, B.M.; Troiano, G.; Tihan, T.; Brem, H.; Walter, K.A. Polilactofate microspheres for paclitaxel delivery to central nervous system malignancies. *Clinical Cancer Research* **2003**, *9*, 3441-3447.

Chapter 5

Novel Nanogels for Drug Binding and Delivery

P. Somasundaran*, Fang Liu, Soma Chakraborty, Carl C. Gryte, Namita Deo, and T. Somasundaran

NSF IUCR Center for Advanced Studies in Novel Surfactants, Langmuir Center for Colloid and Interfaces, Columbia University, New York, NY 10027
***Corresponding author: ps24@columbia.edu**

There is a vital need for efficient ultrasmall, biocompatible vehicles for extraction and release of drugs, toxics as well as sensory attributes etc. Appropriately modified crosslinked polymeric nanogels and liposomes meet this need as they are porous or vesicular in nature. Poly(acrylamide) (PAM) and poly(acrylic acid)(PAA) and liposomes nanogels with narrow size distribution have been tested in this work for drug binding and release. For enhanced efficiency, systematic modifications of the nanogels by a two-step postgrafting strategy were carried out successfully. Drug encapsulation experiments using amitriptyline and bupivacaine as target drug molecules were conducted with these nanogels as well as liposomes in order to determine their use for scavenging the overdosed drugs. The modified nanogels and liposomes showed marked capability to extract overdosed drugs. Results show nanogels and liposomes to be powerful vehicles for drug binding and delivery types of applications if modified appropriately.

Introduction

Drug toxicity in humans is one of the major health care problems, which can be induced by therapeutic miscalculation, illicit drug usage, or suicide attempt. For example, amitriptyline is an antidepressant drug, and excessive use of amitriptyline is a suicide method in the United States *(1)*. Similarly bupivacaine, which is used to provide anesthesia during surgical procedures, causes cardiotoxicity if injected in excess. There is a need for scavenging systems for detoxifying overdosed patients by removing as much of the drug as possible within hours or less. Such systems should be either small enough or biodegradable in order to be excreted from the human body after the removal of the drug. Since biodegradable polymeric nanogels and liposomes fulfill these requirements, they can be considered as potential candidates for overdosed drugs scavenging.

Nanogels discussed here are crosslinked macromolecular networks made from water-dispersible, biocompatible polymers. They are promising candidates for biomedical applications that include controlled drug delivery systems and toxic scavengers *(2,3)*. Appropriately modified nanogels can be designed to rapidly absorb large quantity of desired molecules from the body fluid. Because of their small size, they can also respond faster (seconds *versus* days for hydrogels) than bulk gels *(4-8)*. Since they can go through even the tiniest blood capillaries, nanogels have been explored for drug binding and delivery applications *(2,3)*.

Poly(acrylamide) (PAM) and poly(acrylic acid) (PAA) were tested for building nanogels for the study of amitriptyline and bupivacaine extraction because they are water soluble, biocompatible and nontoxic *(7,8)*. These nanogels were synthesized by inverse microemulsion polymerization in the form of very small particles (40-100 nm) *(9)*. A series of chemical modifications were carried out by copolymerizing N-acryloxysuccinimide material into PAM and PAA and then using the activity of succinimide towards nucleophilic molecules to substitute various chemical functions into the nanogels structure. To generate ionic polymers, the amide groups on PAM were hydrolyzed to pendant carboxylic acid groups. This modification permitted the introduction of hexyl and decyl hydrophobic chains, glycine and acrylic (anionic) groups as well as combinations of these groups. In each case, the degree of substitution was limited to about 10% because of the instability of the inverse microemulsion systems towards the fourth component, i.e., N-acryloxysuccinimide.

The synthesized nanoparticles were characterized in terms of size, crosslink density and micropolarity. The potential of the nanogels to extract overdosed drugs was studied by evaluating amitriptyline and bupivacaine under different physical conditions. Interaction of these nanogels with the drugs was studied also in presence of the surfactant sodium dodecylsulphate. Furthermore, the interaction of liposomes, made from phosphatidyl choline and phosphatidic acid, was also investigated.

Materials and Methods

Materials

Acrylamide and N,N'-bismethyleneacrylamide (99%) were obtained from Aldrich Chemical Co. and used as received. N-acryloxysuccinimide was obtained from Sigma Company and recrystallized from a dry solvent mixture of n-hexane and ethyl acetate. Aerosol OT (AOT, sodium bis 2-ethylhexyl sulfosuccinate) was obtained from Fluka Chemical. A ^{137}Cs source (600 rad/min) was used for irradiation of polymerization. Span 80(sorbitan monooelate), tween 80 poly(oxyethylene sorbitan monooelate), glycine, n-hexylamine, dimethylformamide (DMF), triethylamine (TEA), sodium dodecyl sulfate (SDS), hexadecyl trimethylammonium bromide (HTAB), and pyrene (99%) were obtained from Aldrich and used as received. Toluene (HPLC grade) and other organic solvents were also used as received. Water was distilled twice before use. Amitriptyline and bupivacaine were obtained from Aldrich and used as received.

Syntheses of Modified Poly(acrylamide) Nanogels and Their Copolymers

Synthesis of Poly(acrylamide) Nanogels (PAM) (1-d)

Acrylamide and N,N'-methylenebisacrylamide were dissolved in distilled water, and the mixture was added to a solution of AOT in toluene. A region with the water phase in the form of spherical droplets (i.e., water-to-AOT = 14.1 mole ratio) was used to create the inverse microemulsion. AOT was selected as the surfactant because of its robust character and high degree of colloidal stability. Nitrogen was bubbled through the microemulsion for 15 minutes in order to remove any dissolved air. The microemulsion was then immediately irradiated using γ-ray from a ^{137}Cs source (600 rad/min). γ-ray irradiation was used to initiate polymerization because thermal initiators have the tendency to destabilize the system. After irradiation for a period of 20 minutes, the flask was removed from the radiation chamber and the content was precipitated into a large excess of methanol, washed, and freeze-dried.

Synthesis of Acrylamide—N,N'-methylenebisacrylamide—N-Acryloxysuccinimide Copolymers

The procedure was essentially the same as above, except that the recrystallized N-acryloxysuccinimide, along with acrylamide and N,N'-methylenebisacrylamide, were dissolved in cold distilled water, and the cold monomer solution was quickly added to a cold solution of AOT in toluene. The contents were precipitated, filtered, rinsed, and freeze-dried. The fine white powder was stored over $MgSO_4$ in a desiccator under vacuum.

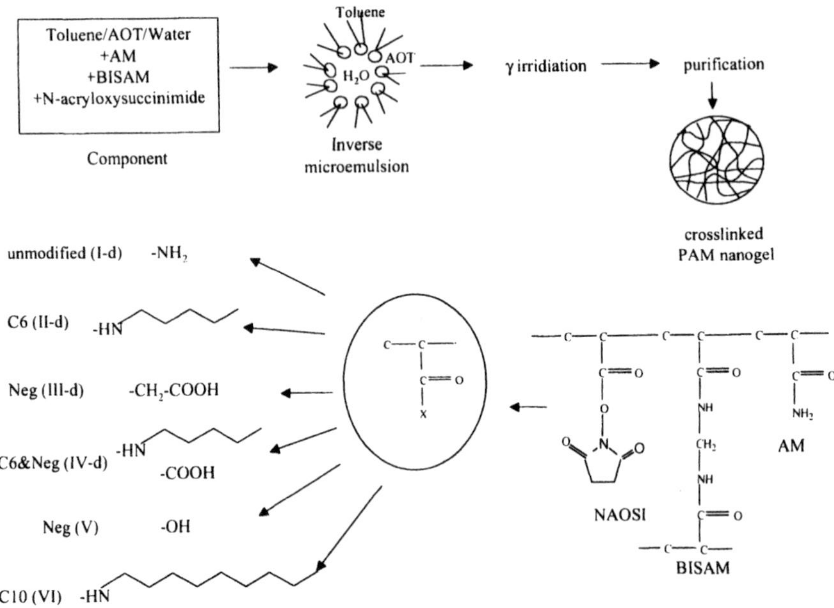

Figure 1. Synthesis of poly(acrylamide) and modified poly(acrylamide) nanogels by inverse microemulsion technique.

Synthesis of Nanogels with Hexyl Groups (II-d)

Modification of the poly(acrylamide) structure was accomplished by copolymerizing N-acryloxysuccinimide monomer and further by replacing it with aliphatic amine or glycine. N-acryloxysuccinimide, an active monomer, was incorporated into the nanogel backbone through inverse microemulsion polymerization. Such a species offers the easiest way for the modification of pendant groups by nucleophilic substitution, using appropriate groups. Succinimide is replaced by aliphatic amine or glycine in the postgrafting step. Since N-acryloxysuccinimide is sensitive to premature hydrolysis in water, all the processes were performed under ice-cold conditions. In this work, bisacrylamide was 5% and N-acryloxysuccinimide was 5%. Figure 1 illustrates the reaction scheme. In this case, n-hexylamine in DMF and trace amounts of TEA was added to a water suspension of the nanogels, prepared in the previous step. The system was stirred at 0-5°C for 12 hours, then precipitated into THF, filtered, rinsed with THF a couple of times, and freeze-dried.

Synthesis of Nanogels with Decyl Groups (VI)

N-decylamine was used for the synthesis of nanogels with pending decyl groups, following the same procedure as sample II-d.

Synthesis of Ionic Nanogels (III-d)

Glycine and trace amounts of TEA were added to a water suspension of nanogels prepared in step (2), and the system was stirred for 12 hours, precipitated into THF, washed with THF for several times, and freeze-dried.

Synthesis of Ionic Nanogels with High Charge Densities (V)

Ionic nanogels with high charge densities (V) were prepared by hydrolyzing PAM nanogels at 60°C for 30 minutes. The polymers were precipitated into THF, purified, and dried under vacuum. The resultant nanogels were used for comparison with the corresponding nanogels, containing both hexyl and high ratio of carboxylic acid groups (IV-d).

Synthesis of Nanogels with Hydrophobic and Ionic Groups (IV-d)

Nanogels with hexyl groups (II-d) were dissolved in 20 ml of water, then hydrolyzed at 60°C for 30 minutes. The resultant products were precipitated into THF, washed, and freeze-dried.

Synthesis of Poly(acrylic acid) Nanogels

Span 80 and Tween 80 were solubilized in hexane. Acrylic acid and N,N' methylenebisacrylamide dissolved in water were added to the surfactant system. Nitrogen was bubbled through the microemulsion for 15 minutes in order to remove any dissolved air. The microemulsion was then immediately irradiated, using γ-ray from a ^{137}Cs source (600 rad/min) for a period of 20 minutes. The resulting product was precipitated into a large excess of acetone and freeze-dried.

Drug Binding Experiments

The degree of drug binding to different nanogels at room temperature was determined over a wide range of drug concentrations (0–4000 μM). Drug solutions were prepared in distilled water or in PBS, followed by dispersion of dry nanogel powder (5 mg/ml) in the same solvent. The system was stirred at 25°C for four hours, and the nanogels were separated from the drug solution using a 20-nm ultrafiltration filter (Anotop 25 syringe filter, Whatman). The drug concentration in the filtrate was measured by UV absorbance, and the bound drug amount was estimated from the difference between the initial and final drug concentrations in the solution. The structures of the drugs under investigation are shown in Figure 2.

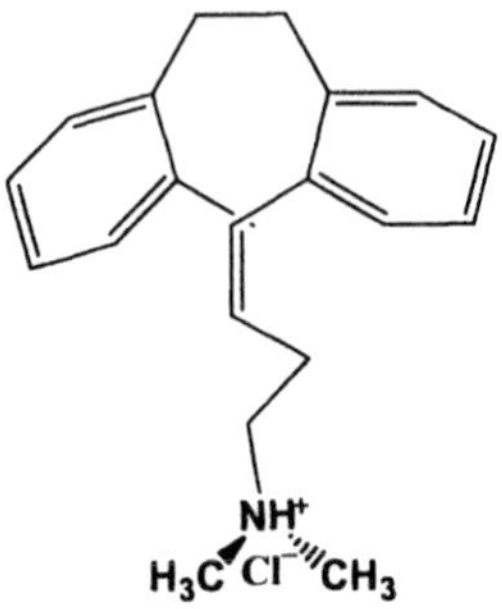

NH+
H_3C Cl− CH_3

Amitriptyline, pKa = 9.4

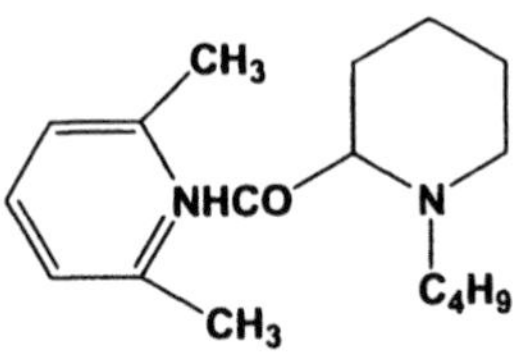

CH_3
NHCO
N
C_4H_9
CH_3

Bupivacaine, pKa = 8.24

Figure 2. Drugs used in the extraction process.

Characterization of Nanogels

Dynamic Light Scattering

A Brookhaven BI 200 goniometer and BI 900 digital correlator were used to perform dynamic light scattering measurements. An argon-ion laser (Lexel, Model 95), operating at 488 nm wavelength, was used as the light source. Time-averaged scattered light intensities were measured at scattering angles of 20 to 150° at different nanogel concentrations. The samples were cleaned of dust particles by filtering through 0.5 μm pore size membrane filters (Millipore) in a closed filtration circuit.

Zeta Potential

Zeta potential measurements were carried out with a ZetaPlus meter, manufactured by Brookhaven Instruments Corp. (USA). During the measurements, the ZetaPlus meter was set to maintain the measuring cell at 25°C. The nanogels were dispersed to make 1 mM suspensions in 10^{-3} M KNO_3.

Fluorescence

Nanogel powder was added to a saturated pyrene solution in water to make a series of nanogel suspensions at different concentrations. Pyrene fluorescence spectra were recorded using SPEX FluoroMax-2 spectrofluorometer with an excitation wavelength of 335 nm. The slit widths were 2.0 nm for excitation and 0.1 nm for emission. The spectra were corrected for Rayleigh and Raman bands in blank nanogel suspensions. Three scans were averaged, and the emission spectra were recorded. The ratio of the intensities of the third (383 nm) to the first (373 nm) vibronic peak (I_3/I_1) in the emission spectra of the monomer pyrene was used to estimate the polarity of the pyrene environment.

Results and Discussion

Stability of Nanogels in Different Media

The colloidal stability of unmodified and modified PAM nanogels in different solvents was first determined (Table I). Regardless of the modification, all nanogels used in this work formed stable and easily dispersible suspensions in water. Unmodified nanogels (I-d) formed clear dispersions, while dispersions of all the other nanogels were cloudy. In solvents such as methanol, only the negatively charged nanogels (III-d & V) formed stable dispersions. The least stable systems were sample (IV-d) and sample (VI), which precipitated in all the solvents except water. Sample II-d was stable because the hexyl (C_6) chains were buried in the core of the nanogel particle. When the attached hydrophobe was increased to decyl (C_{10}; sample IV), the nanogel tended to form aggregates between the particles.

Table I. Dispersion properties of poly(acrylamide) and modified poly(acrylamide) nanogels.

5% Crosslinked Nanogels Identification	Water	Phosphate buffered saline	Ethanol	Methanol	Acetone	THF
(III-d) Neg (low COOH)	+	+	+	+	+	
(V) Neg (high COOH)	+	+	+		+	
(I-d) Unmodified	+	+	+/-	-	-	-
(II-d) C6	+	+	+/-	-	-	-
(IV-d) C6/Neg	+		-	-	-	-
(VI) C10	+	-	-	-	-	-

+: stable colloidal dispersion (6 months +); -: precipitated; +/-: dispersed only at diluted concentration

Zeta Potential of Nanogel Particles

The zeta potential was measured in 10^{-3} M KNO_3 solution. All nanogels containing carboxylic acid groups (III-d, IV-d) showed negative zeta potential due to the hydrolysis of the carboxylic acid group. The unmodified nanogels (I-d), hexyl nanogels (II-d) and decyl nanogels (VI) were almost neutral since there was no significant hydrolysis of the poly(acrylamide). KNO_3 was added in order to keep the conductivity of the medium constant in the experiments.

Fluorescence analysis for hydrophobicity

Pyrene was used as the probe to monitor the hydrophobicity of the nanogels. In the pyrene fluorescence spectrum, the intensity ratio of the third to the first vibronic peaks (I_3/I_1) is considered to be a polarity parameter that is sensitive to the hydrophobicity of the environment. An I_3/I_1 value of 0.6 corresponds to aqueous environment, while a value of 0.85 indicates the micellar interior. At a constant pyrene dosage, the hydrophobicity of the unmodified nanogels (I-d) was found to be 0.75, which suggests that the nanogel interior is somewhat more hydrophobic than the exterior environment. Hexyl (II-d) and decyl (VI) nanogels were only slightly more hydrophobic than the corresponding unmodified nanogels (I-d). This observation suggests that the chemically bound hydrophobic groups do not form micelle-type domains because the crosslinked structure prevents this reorganization. For carboxylic acid groups introduced into the nanogels, the I_3/I_1 was as low as 0.58 – 0.6, indicating the charged nanogels (III-d) to be polar.

Nanogel Particle Size and Size Distribution

Effect of Functional Groups

Unmodified PAM nanogels (I-d) have a hydrodynamic radius of 30 nm, while all the modified nanogels exhibited larger particle sizes. Thus the functional groups can be believed to play a key role in determining the swelling abilities of the nanogels. The modified nanogels showed a significantly higher swelling ratio than the unmodified neutral poly(acrylamide) nanogels. In the case of nanogels containing hexyl (II-d) or decyl (VI) chains, the increase in the swelling ratio obtained suggests that the hydrophobes do not form significant hydrophobic domains. Instead, there is the possibility of repulsion between the hydrophilic backbone and the hydrophobic side chains, causing the network to swell. The swelling ratios of the charged nanogels (III-d and V), compared to the unmodified PAM nanogels (I-d), can mainly be attributed to the presence of the carboxylic acid groups.

Effect of Crosslink Density

Nanogels with crosslinking densities below 1% swelled significantly in deionized water. In the case of materials with high crosslinking densities, the particle size increase was negligible, suggesting the formation of a compact structure.

Temperature Effects

The hydrodynamic radius of PAM nanogels was found to increase gradually with temperature with a larger increase around 55-65°C, possibly due to the temperature-induced hydrolysis of the amide group.

Interaction of Poly(acrylamide) (PAM) Nanogels with Amitriptyline

The potential of the poly(acrylamide) nanogels, synthesized by inverse microemulsion polymerization, was evaluated to scavenge amitriptyline from aqueous dispersion. The effect of crosslink density, functional groups, temperature and the dispersion medium on the extraction process was studied systematically.

Functional Groups

The drug binding behavior of PAM nanogels with various functional groups is plotted in Figure 3 as a function of crosslinking density. Introduction of the functional groups into the nanogels shows marked enhancement in drug binding ability.

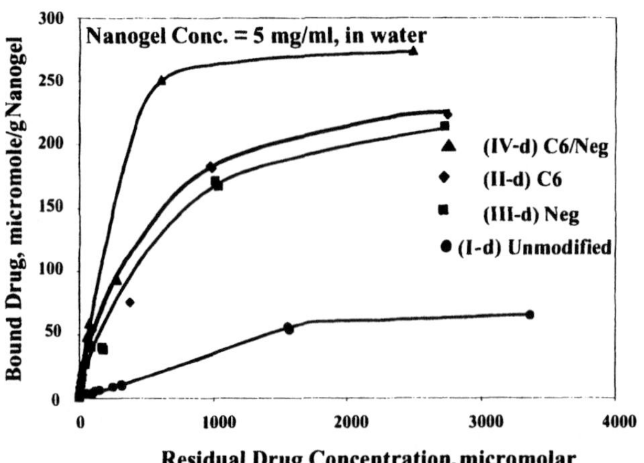

Figure 3. Drug binding ability of PAM and modified PAM as a function of residual drug concentration.

Binding interactions between nanogels and drug depend on their respective functionalities, which in turn depend on the pH of the solution. The pH of deionized water is between 6 and 7. In this pH range, amitriptyline is cationic and possesses hydrophobic characteristics. Since the unmodified nanogel (I-d) has a relatively neutral polymer backbone, the binding between the polymer backbone and amitriptyline is minimal. In the case of the negatively charged nanogels, extraction is higher due to the electrostatic interaction with the positively charged drug molecules. Hydrophobic interaction between the drug and the hydrophobic moiety of the nanogels also results in enhanced extraction of drug by the hexyl nanogels. The dually modified nanogels (IV-d), containing both hexyl and carboxylic acid groups and the maximum binding obtained with it, is attributed to the combined effects of ionic exchange and hydrophobic interactions.

Effect of Crosslink Density

At a fixed drug concentration in water, an increase in crosslinking density decreases the drug binding, with a nearly constant minimum value for all densities above 1% (Figure 4). The highly crosslinked nanogels with lower swelling ratios are less open for amitriptyline molecules to diffuse in, and in this case the drug binding process is proposed to take place mainly on the outer layer of the nanogels. When the crosslinking density is lower than 1%, the internal space is more open to the drug molecules, and the overall drug binding by the nanogel is higher.

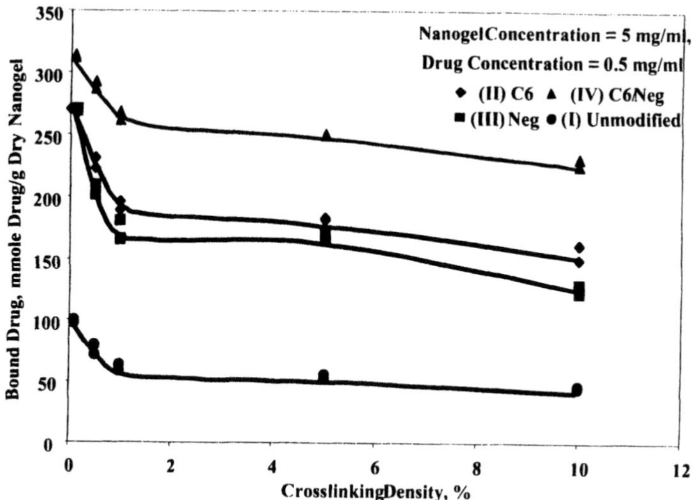

Figure 4. Effect of crosslink density on the drug binding ability of PAM and modified PAM.

Effect of the Chain Length

Nanogels with attached decyl groups (VI) were subjected to drug binding experiments in order to investigate the effect of the hydrophobic chain length. The decyl nanogels showed higher binding capacity in water than the hexyl analogues, indicating that the longer hydrophobic chain can attract more drug molecules. However, in saline, the decyl nanogels became unstable and precipitated out form the solution. Thus, although longer hydrophobic chains may enhance the drug loading properties, the systems are effective only if they can be kept dispersed in solution. Optimization of the chain length for nanogels is essential for maximum drug loading.

Effect of Media

The drug binding experiment was repeated with phosphate buffer saline (PBS; ion concentration = 0.15 M) and the results obtained are given in Figure 5. Significant differences in the drug binding ability in saline and water were observed for the unmodified nanogels, presumably due to the nonionic nature of the unmodified samples. The drug binding capacity of all nanogels containing functional groups is affected in PBS due to the high ionic strength of the medium, screening both the charged groups on the nanogels and the amitriptyline molecules. Ion exchange between the drug molecules and the charged nanogels is therefore limited, causing a decrease in the drug loading capacity for these ionic nanogels. In the case of the dually modified nanogels (IV-d), the particles can form aggregates and precipitate out in saline, resulting in a decrease of the drug binding capacity of these nanogels. The total binding efficiency of hydrophobic nanogels is still much higher than those of other nanogels. Hence hydrophobic modification is identified as a promising method for improving the drug binding in physiological systems.

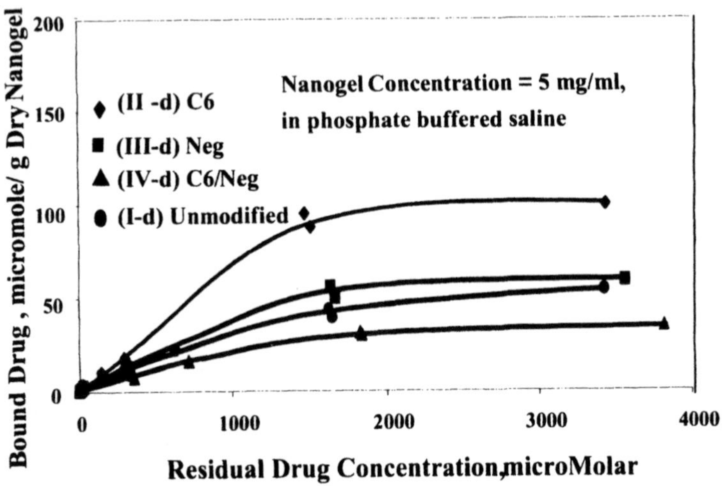

Figure 5. Drug binding ability of PAM and modified PAM as a function of residual drug concentration in presence of PBS.

pH Effect

Since amitriptyline and the carboxylic acid groups are completely ionized only within the pH 7 to 9 range, all the nanogels absorbed the maximum amount of drug within this range, whereas the capacity became lower below pH 7 and above pH 9.

Drug Binding by Poly(acrylic acid) (PAA) Nanogels

Nanogels with anionic groups, based on poly(acrylic acid), were also tested for their effectiveness as drug scavenger. The drug binding ability of PAA was investigated using bupivacaine and amitriptyline as the model drugs. It was observed that the extraction was markedly enhanced at pH 7 and pH 9, as compared to pH 4 (Figures 6,7). At pH 7 and 9, the drugs are positively charged and strongly interact with the pendant carboxylate ions on the polymer backbone. In acidic medium, the drugs are not ionized and the extent of ionization of the carboxylic acid groups is reduced, causing a decreases in the binding efficacy at pH 4.

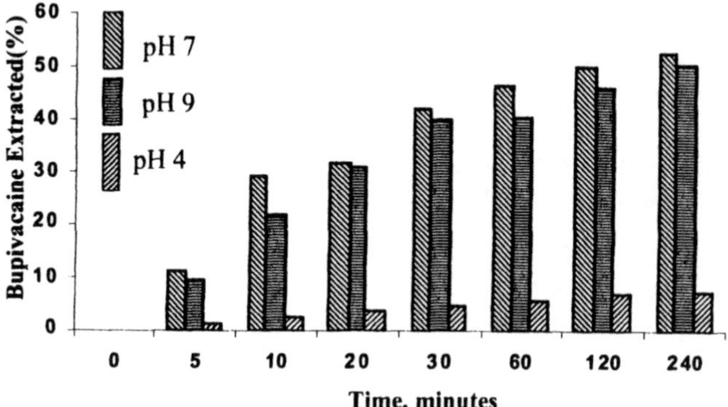

Figure 6. Dependence of extraction of bupivacaine by poly(acrylic acid) nanogels on pH.

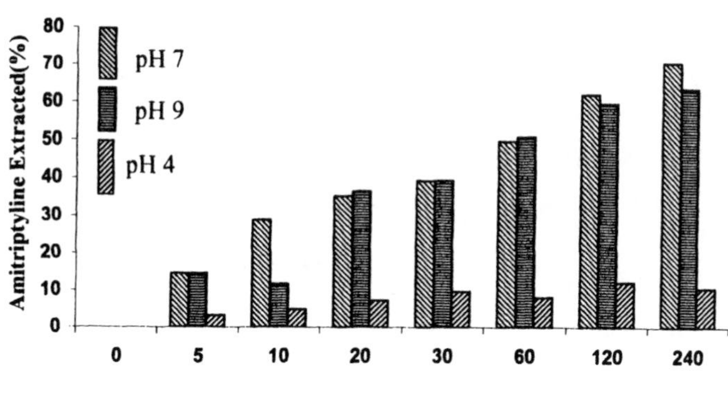

Figure 7. Dependence of extraction of amitriptyline by poly(acrylic acid) nanogels on pH.

Interaction of Amitriptyline with Lipid Bilayers

Liposomes are a vehicle similar to nanogels with hydrophobic domains but with different permeation properties. Their potential to extract amitriptyline was also evaluated. The liposomes used for the study were made from a 1:1 mixture of phosphatidylcholine and phosphatic acid. The drug concentration in the lipid bilayers after interaction of the liposomes with amitriptyline was determined.

The result is shown in Figure 8. In the absence of salt, the drug concentration in the lipid bilayers increased linearly with increase in the bulk drug concentration. In water, the surface charge of the lipsomes is highly negative due to the presence of anionic groups on both phosphatidic acid and phosphatidylcholine. This favors electrostatic interactions between the liposomes and the cationic drug molecule (Figure 9). In the presence of phosphate buffer (ionic strength 0.1 M), however, a significant decrease in the partitioning of the drug into the lipid bilayers was observed due to the decrease in electrostatic interaction. The maximum drug partitioning into the liposomes in buffer solution was observed at 6 mM drug concentration. Complete destruction of the liposome structure was detected on further increase in drug concentration.

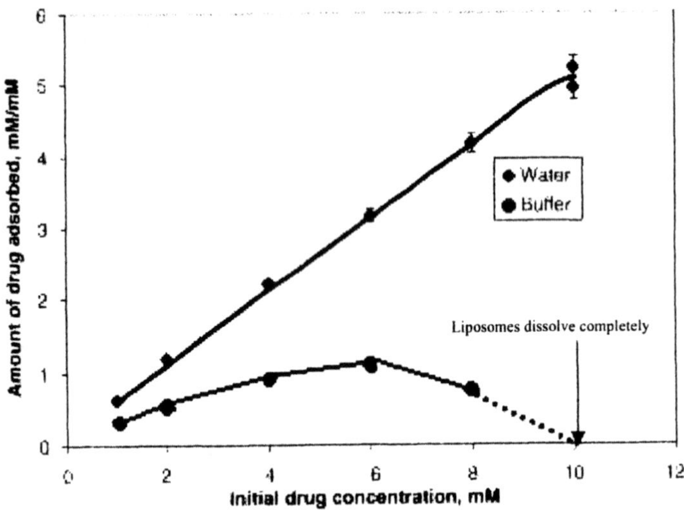

Figure 8. Drug extraction by liposome as a function of initial drug concentration in water and in buffer.

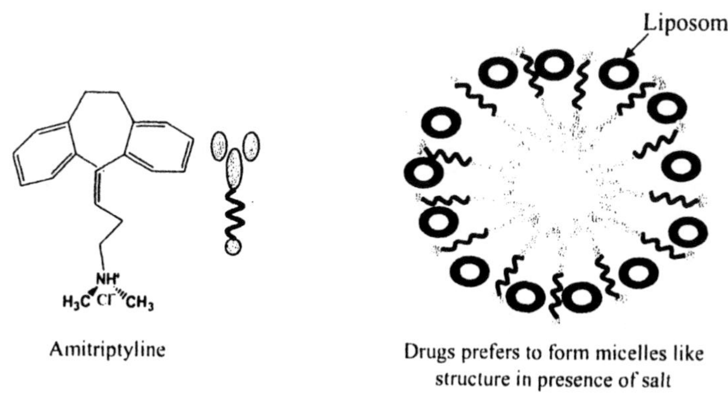

Figure 9. Interaction of amitriptyline with oppositely charged liposomes.

Interaction of Poly(acrylamide) Nanogels with Surfactant

In contrast to enhancing the hydrophobicity of the nanogels by chemical modification, poly(acrylamide) nanogels can also be modified by absorbing free surfactants on the polymer backbone. Compared to PAM nanogels alone, these complexes are expected to have higher hydrophobicity, which can be used for loading drug and varying other properties of the formulation under physiological conditions. The hydrophobicity of PAM and modified PAM in the presence of sodium dodecyl sulphate (SDS) has been calculated. Such study can provide insight into the interaction of biocompatible surfactants with the nanogels.

Binding Isotherm

All the nanogels used in this experiment were 5% crosslinked. The degree of SDS binding to all types of polymeric nanogels was measured after equilibrating the nanogels in SDS solutions. The final residual SDS concentration was determined by titrating with HTAB (hexydecyl trimethylammonium bromide). The binding isotherm was recorded in both distilled water and in 0.1 M NaCl solution.

Effect of Functional Groups

As shown in Figure 10, all nanogels could bind SDS molecules through the pendant functional groups. The modified nanogels (II-d, III-d and IV-d) showed a higher binding ability than the unmodified nanogels. This difference is attributed to the hydrophobic interactions between SDS molecules and the functional groups. In all the binding isotherms a critical point exists, which corresponds to the critical micelle concentration (CMC) of SDS. The slope of the isotherm undergoes a sharp transition above the CMC. The difference in the slopes indicates that the attractive interaction forces between the nanogel and SDS micelles are much stronger than those between the nanogels and SDS molecules. An interesting phenomenon is the binding of the anionic SDS to the negatively charged nanogels (sample III-d). Since both sample III-d and SDS contain negative charge, the resultant binding process could not be attributed to electrostatic forces.

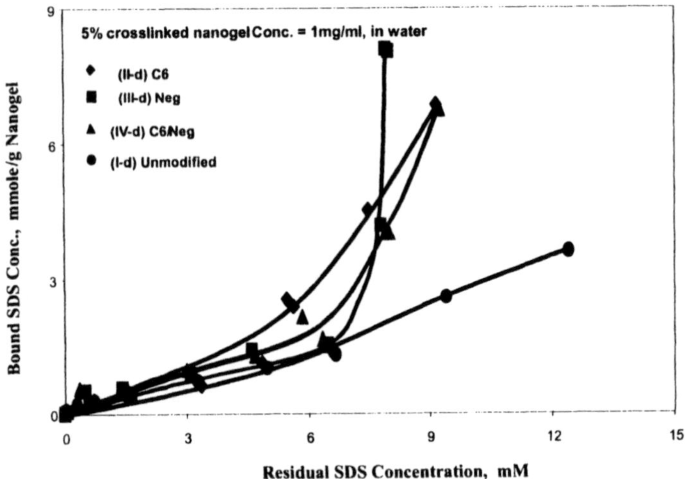

Figure 10. Interaction of PAM and modified PAM with SDS as a function of residual drug concentration.

Formation of Internal Hydrophobic Domains inside Nanogel Particle

Fluorescence experiments were carried out, using pyrene as a probe, to monitor the hydrophobic properties of the nanogel/SDS mixtures as a function of SDS concentration in the system. If pyrene molecules are in an aqueous environment, the I_3/I_1 is about 0.6, whereas this value can increase up to 1.2-1.3 in a nonpolar organic solvent such as hexane. For surfactant micelles, the ratio is usually between 0.8 and 1 *(9)*. The I_3/I_1 values are plotted in Figure 11 as a function of SDS concentration in water.

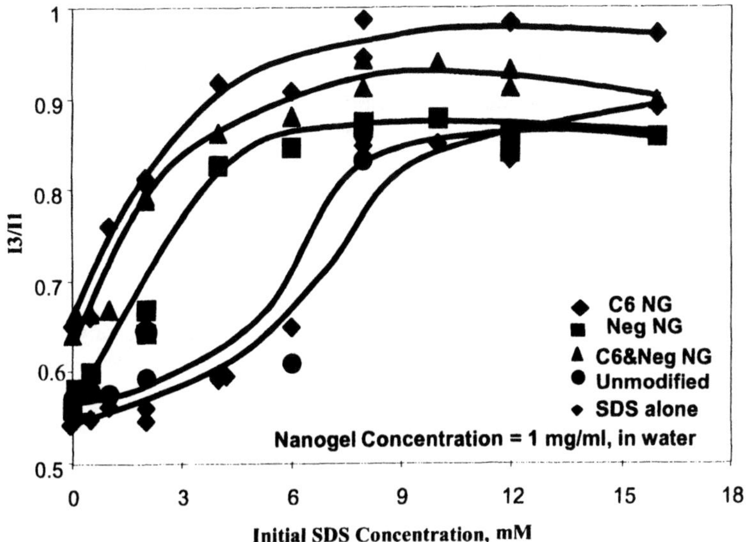

Figure 11. Hydrophobicity of PAM and modified PAM as a function of initial SDS concentration.

For SDS alone, the sharp transition of I_3/I_1 takes place at a SDS concentration of 8 mM, which is due to the formation of micelles (CMC). As unmodified nanogels (I-d) are added to the system, the I_3/I_1 values overlap with that of SDS alone, indicating no hydrophobic domain below the CMC of SDS. Although some SDS molecules are bound to the polymer backbone, the local SDS concentration is not high enough to form a micellar-type structure below the CMC. However, in the case of hexyl nanogels (II-d), the I_3/I_1 transition occurs at a SDS concentration of 2 mM, which is much lower than the CMC of SDS. The presence of the hexyl groups on the polymer backbone is proposed to induce the SDS aggregation around the hydrophobe and cause the formation of hydrophobic domain at a much lower CMC. This nanogel/SDS complex has a higher hydrophobicity than the nanogel alone, which is expected to be more effective in solubilizing water-incompatible substances.

In the case of the negatively charged nanogels (III-d), the CMC of the SDS/nanogel system is observed to shift to the low SDS concentration range. Since both the nanogel and SDS have similar charge, the major interaction in this case should be due to hydrophobic forces. It is likely that the bound charges on the nanogel backbone produce a high ionic strength environment with SDS molecules in the nanogel domain, thus forming micelle-like structure inside the nanogels at lower concentrations due to the salting out effect.

For nanogels containing both hexyl and carboxylic acid groups (IV-d), the hydrophobic domains could be generated at SDS concentrations below the CMC, which is due to the combined effect of hydrophobic and ionic interactions. Thus combining the chemical modifications and the surfactants, a series of polymer/surfactant supramolecular structures can be created, which can generate novel carriers for drug extraction and delivery.

Conclusions

Novel poly(acrylamide) and poly(acrylic acid) nanogels with a narrow size distribution were synthesized, using the inverse microemulsion polymerization technique and were modified by attaching hydrophobic and ionic groups to the nanogel backbone. By introducing functional groups into the nanogel particles, the nanoenvironment of the nanogels was modified with a significant impact on the interaction between the nanogels and the drug molecules.

Modified nanogels showed significant enhancement in drug binding compared to the unmodified nanogels in water. In phosphate buffered saline, although the binding efficiency of all types of nanogels decreased, the performance of the hydrophobic nanogels is satisfactory compared to other nanogels. Hydrophobic modification appears to be more promising for drug binding applications.

The crosslinked structure of nanogels has an influence on the drug distribution inside the nanogels. The low crosslinking density increased the overall drug binding of the nanogels due to increase in the internal space of the nanogels. Poly(acrylic acid) nanogels showed pH-sensitive drug extraction. The extraction was better at pH 7 and pH 9 than at pH 4. Liposomes made of phosphatidylcholine and phosphatic acid also extracted amitriptyline in water.

Modified and unmodified PAM interacted with SDS differently. The poly(acrylamide) backbone of nanogel particles alone showed strong interactions with surfactants. As the functional groups were introduced to the polymer backbone, the interactions between the nanogels and SDS further increased. Thus nanogels and liposomes with appropriate modification offer a promising route for both drug extraction and delivery.

Acknowledgements

The authors acknowledge the financial supports from NSF Industry/University Research Center for Advanced Surfactants at Columbia University and NSF Engineering Research Center for Particle Science and Technology at University of Florida.

References

1. Anthony A., Drug Overdose, http://www.emedicine.com/aaem/topic169.htm.
2. English, A.; and Mafe, S.; Manzanares, J.; Yu, X.; Grosberg, A.; Tanaka, T. "Equilibrium swelling properties of polyampholytic hydrogels", *J. Chem. Phys.* **1996**, *104*, 8713-8720.
3. Kiser, P.; Wilson, G.; Needham, D. "A synthetic mimic of the secretory granule for drug delivery", *Nature* **1998**, *394*, 459-462.
4. Philippova, O. E.; Hourdet, D.; Audebert, R.; Khokhlov, A. R., "pH-Responsive Gels of Hydrophobically Modified Poly(acrylic acid)", *Macromolecules* **1997**, *30*, 8278-8285.
5. Eichenbaum, G.; Kiser, P.; Dobrynin, A.; Simon, S.; Needham, D., "Investigation of the Swelling Response and Loading of Ionic Microgels with Drugs and Proteins: The Dependence on Cross-Link Density", *Macromolecules* **1999**, *32*, 4867-4878,
6. Eichenbaum, G.; Kiser, P.; Shah, D.; Simon, S.; Needham, D., "Investigation of the Swelling Response and Drug Loading of Ionic Microgels: The Dependence on Functional Group Composition", *Macromolecules* **1999**, *32*, 8996-9006,.
7. Kriwer, B.; Walter, E.; Kissel, T., " Synthesis of bioadhesive poly(acrylic acid) nano- and microparticles using an inverse emulsion polymerization method for the entrapment of hydrophilic drug candidates" *J. Controlled Rel.* **1998**, *56*, 149-158.
8. Peppas, N. *Hydrogels in Medicine and Pharmacy*: CRC Press: Boca Raton, FL, 1986.
9. Wilk, R., Synthesis and characterization of pyrene-labeled high molecular weight polyacrylamide polymers and nanogels. Doctoral dissertation, Columbia University, 1994.

Chapter 6

Nanogel Engineered Designs for Polymeric Drug Delivery

Nobuyuki Morimoto[1], Shin-ichiro M. Nomura[1,2], Naomi Miyazawa[1,2], and Kazunari Akiyoshi[1,2,*]

[1]Institute of Biomaterials and Bioengineering and [2]Center of Excellence Program for Frontier Research on Molecular Destruction and Reconstruction of Tooth and Bone, Tokyo Medical and Dental University, 2–3–10, Kanda-Surugadai, Chiyoda-ku, Tokyo 101–0062, Japan
[*]Corresponding author: akiyoshi.org@tmd.ac.jp

Functional nanogels and hydrogels, based on self-assembly of associating polymers, were designed for novel drug delivery applications. In particular, hydrophobized polysaccharides such as cholesteryl-bearing pullulans form physically cross-linked nanogels by self-assembly. These nanogels trap hydrophobic molecules (antitumor drugs), proteins (enzymes, insulin, antigen protein), and nucleic acids (DNA plasmids), and therefore, can be used as polymeric nanocarriers in cancer chemotherapy, protein delivery, and artificial vaccines. Stimuli-responsive nanogels such as pH-responsive, thermoresponsive, and photoresponsive nanogels were also designed using a similar self-assembly method. Macrogels with well-defined nanostructures were obtained by self-assembly, utilizing these nanogels as building blocks. The self-assembly method based on associating polymers is an efficient and versatile technique for the preparation of functional nanogels and hydrogels.

Introduction

Polymer hydrogels are being widely used for biotechnological and biomedical applications *(1,2)*. In particular, nanometer-sized (<100 nm) polymer hydrogels (nanogels) with characteristics of both nanoparticles and hydrogels have attracted interest in drug delivery applications such as protein and gene delivery *(3,4)*. Compared with other nanoparticles that are mostly used for their unique surface characteristics, some additional properties are expected from these newly designed nanogels, for example trapping biomolecules inside the gel and responding rapidly to an external stimulus. In general, chemically crosslinked nanogels have been synthesized by microemulsion polymerization or a crosslinking reaction of intra-associated polymer molecules *(4-7)*. We reported a novel method for physically crosslinked nanogels by controlled association (self-assembly) of hydrophobically modified polymers in water *(8)*.

Macromolecular systems, using self-assembly of polymeric amphiphiles, have attracted attention as unique vehicles for drug delivery *(9)*. Amphiphilic block copolymers, which consist of both hydrophobic and hydrophilic segments, form nanosized polymeric micelles in water and have been extensively studied as nanocarriers *(10)*. Amphiphilic graft copolymers or random copolymers that have both hydrophobic and hydrophilic groups also form micelle-like structures in water *(11)*. They have been extensively studied not only for their biological relevance such as enzyme models, but also in other applications. Hydrophobically modified, water-soluble polymers (hydrophobized polymers), modified with less than 5 mol% hydrophobic groups, are called "associating polymers" because they increase solution viscosity by intermacromolecular associations.

We have found that polysaccharides that are partially modified (less than 3 mol%) by hydrophobic groups such as cholesteryl show unique associative behavior. In 1993, we reported that cholesteryl-bearing pullulans (CHP) form stable, monodisperse hydrogel nanoparticles (20-30 nm) by intermacromolecular self-association in dilute aqueous solution *(8)*. These nanoparticles are considered to be microscopic hydrogels (nanogels), in which the association of hydrophobic groups provide crosslinking points within the polymer network *(12)*, and therefore, present a novel method for preparing monodisperse hydrogel nanoparticles. It has been proven that many other hydrophobically modified polymers form nanoparticles by self-assembly in water, for example alkyl group-modified poly(*N*-isopropylacrylamide) (PNIPAM) *(13)*, deoxycholic acid-modified chitosan *(14)*, bile acid-bearing dextran *(15)*, cholesteryl-bearing poly(amino acids) *(16)*, and spiropyrane-bearing pullulans *(17)*.

Results and Discussion

Characteristics of Nanogels

Formation of Nanogels of Hydrophobized Pullulans

Typical characteristics of two nanogels, cholesteryl-bearing pullulan (CHP) and alkyl group-bearing pullulan (C16P) (Figure 1), determined by size exclusion chromatography-multi angle laser light scattering (SEC-MALS) and dynamic light scattering (DLS), are shown in Table I. Transmission electron microscopy (TEM) and atomic force microscopy (AFM) show that the spherical particles have a narrow size distribution.

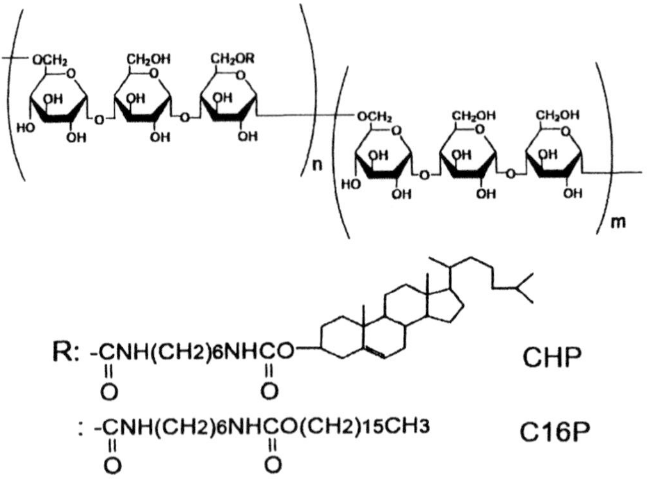

Figure 1. Chemical structure of cholesterol-bearing pullulan (CHP) and alkyl group-bearing pullulan (C16P).

The average polymer density (Ø) of the nanoparticles can be calculated from the hydrodynamic radius (R_H) and the molecular weight of a nanoparticle, M_W, using the following equation:

$$\Phi = \frac{M_W}{N_a} \bullet \left(\frac{4}{3}\pi R_H^3\right)^{-1}$$

where *Na* is Avogadro's number.

Table I. Characteristics of CHP and C16P nanogels determined by SEC-MALS and DLS.

Sample	*Mw* *($x10^5$)*	*Mw/ Mn*	*Aggregation Number*	R_H *(nm)*
CHP	**4.7**	**1.02**	**4.2**	**12.9**
C16P	**5.1**	**1.01**	**4.6**	**15.5**

The polysaccharide density in one particle of CHP was 0.08 g/mL, while it was 0.05 g/mL in C16P. These values correspond to 8 and 5 wt% hydrogels. It was found that one hydrophobic domain consists of approximately four cholesteryl groups per CHP nanoparticle and twelve alkyl groups per C16P nanoparticle, determined by fluorescence quenching methods. The association of cholesteryl as well as alkyl groups is limited by steric hindrance of the polysaccharide chain. One nanoparticle of CHP consists of approximately 44 hydrophobic groups, while one nanoparticle of C16P contains 35 hydrophobes. Thus, there are fewer crosslinking points in a C16P hydrogel nanoparticle (three) than in a nanoparticle of CHP (eleven). A CHP nanoparticle has a relatively compact structure compared to a C16P nanoparticle because of the high hydrophobicity of a cholesteryl group compared to a hexadecyl group. The size, density, and number of crosslinking domains of these nanoparticles are regulated by the number of hydrophobic groups as well as by their structures.

A unique property of self-assembled nanogels is that their formation and collapse can be dynamically controlled by host-guest interaction of their cholesteryl groups with cyclodextrin *(18)*. Since the cholesteryl group is a suitable guest for ß-cyclodextrin (ß-CD) *(19)*, the CHP nanogels dissociated upon complexation with ß-CD to yield a dissociated CHP-CD complex. The monodisperse nanogels of the CHP self-aggregate were regenerated after addition of 1-adamantancarboxylic acid (ADC), which is a better guest molecule in ß-CD than cholesterol.

The viscosity of semi-dilute solutions (above ~2% w/w) of hydrophobically modified pullulan is drastically increased. At higher concentrations, macroscopic gels are formed. CHP gave a structure in which the nanogels were linked, while hexadecyl group-bearing pullulan (C16P) macrogels mainly consisted of a fibrous network structure. Thus, we successfully controlled the association of hydrophobically modified pullulans from the molecular to the nanoscale and macroscopic levels (Figure 2).

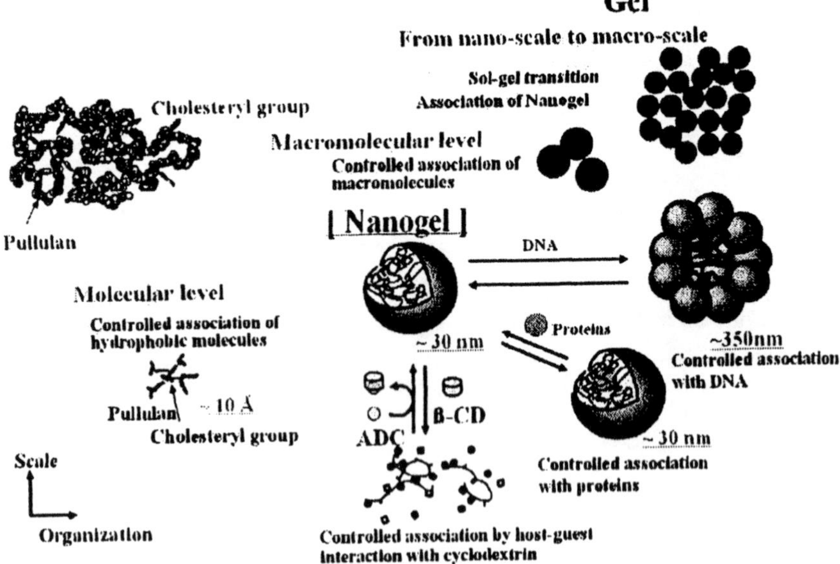

Figure 2. Hierarchy of self-assembly of hydrophobized polymers.

Interaction of Nanogels with Hydrophobic Molecules

The CHP nanogels can bind various hydrophobic substances such as fluorescent probes, bilirubin (BR) *(20)*, and antitumor adriamycin *(21)*. These hydrophobic substances are located in the polysaccharide domain as well as in the hydrophobic cholesteryl domain of the CHP nanogels. The driving force for the complexation of these substances primarily relies on their hydrophobicity. In addition, CHP nanogels have chiral binding sites due to the chiral centers on the polysaccharides.

Interaction of Nanogels with Proteins

CHP nanogels selectively interact with guest proteins as hosts and are regarded as useful drug carriers. Various water-soluble proteins have been complexed by CHP nanogels. The nanogels of the complexes showed excellent colloidal stability without precipitation. In addition, no dissociation of the proteins from the complexes has been observed, even after a week at pH 7.2 and 25°C. The complexation between a guest molecule (protein) and a host molecule (CHP nanogel) can be determined by the number of complexation sites

(n) and the complexation constant (K). One CHP nanogel complexes with about one BSA ($K = 8 \times 10^6$, MW 66,000 Da), two α-chymotrypsin ($K = 2 \times 10^5$, MW 25,000 Da), two myoglobin ($K = 2 \times 10^4$, MW 17,800 Da), four molecules of cytochrome c ($K = 8 \times 10^3$, MW 12,500 Da), and five molecules of insulin ($K = -2 \times 10^6$, MW 5,735 Da) *(22)*. The maximum amount of protein complexed by CHP nanogels depends on the molecular weight (or size) of the protein. The hydrogel structure of the self-aggregate also influences the complexation. This is an interesting example of host-guest interaction in a macromolecular system (Figure 3).

Most native enzymes are complexed at room or physiological temperatures and loose their enzyme activity in the complex *(23)*. However, lipase was complexed in its native form and retained its activity in the complex. The complexed proteins gained substantial thermal stability due to the fixation of the proteins or enzymes in the hydrogel matrix of the nanogels.

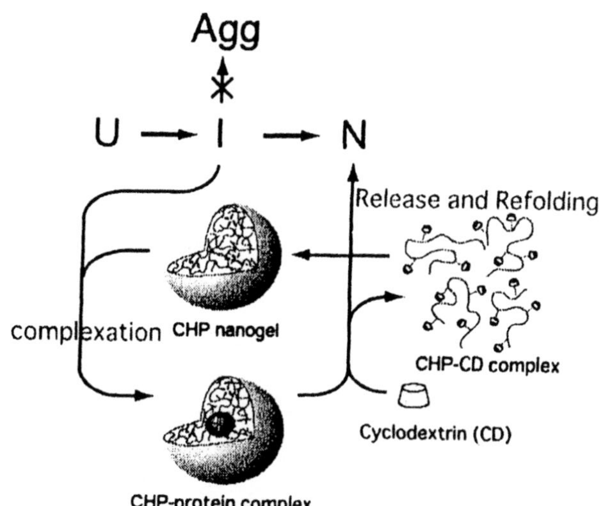

Figure 3. Schematic representation of controlled association between a CHP nanogel and a protein.

Artificial Chaperone

Nanogels effectively prevent protein aggregation (e.g. CAB and CS) by forming nanogel-protein complexes during protein refolding from GdmCl denaturation. In most cases, nanogels interacted more strongly with denatured proteins than native proteins. In living systems, molecular chaperones are known to trap denatured proteins selectively. It is noteworthy that simple amphiphilic nanogels can simulate this function.

Another interesting and important finding is that complexed proteins were released in their refolded, native form upon dissociation of the nanogels in the presence of ß-CD. This is similar to the two-step mechanism of a molecular chaperone; i.e., capture of a heat-denatured protein and release of the refolded protein. Utilizing this property, this system provided heat-shock protein-like activity such as thermal stabilization of enzymes. The present nanogel system was also effective in assisting the refolding of proteins and the renaturation of the inclusion body of a recombinant protein.

Interaction of Nanogels with DNA

Cationic nanogels were synthesized by introducing an amino group into the cholesteryl-bearing pullulan (CHP-NH_2). CHP-NH_2 formed CHP-like nanogels. The basic unit of chromatin is the nucleosome, which consists of a core of 146 base pairs (bp) of DNA wrapped around a histone octamer. The total charge of a CHP-NH_2 nanogel particle is close to that of a histone octamer. After the nanogel was mixed with plasmid DNA (5-7 kbp), globular particles (100-200 nm) were formed by complexation between nanogel and DNA. Hybrid gel particles, consisting of both nanogels and nucleic acids, were obtained (Figure 4). The self-assembly of the nanogels was controlled by using DNA as template. DNA exhibits a stepwise folding transition with the ordered ring structure *(26)*. This system may also be interesting for gene delivery applications.

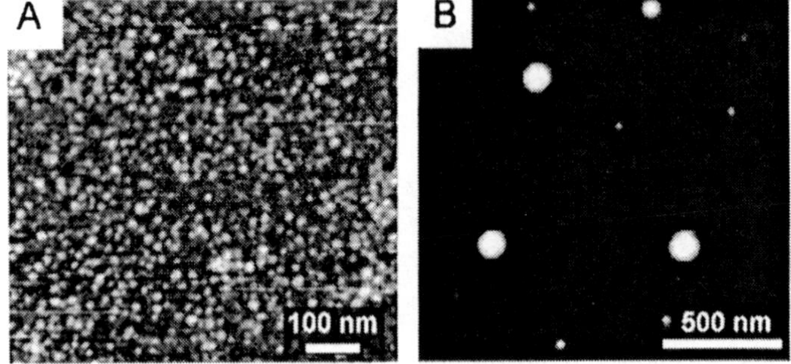

Figure 4. Atomic force microscopic images of CHP-NH_2 nanogels (A) and the nanogel / DNA complexes (B).

Application to Drug Delivery Systems

Cancer Chemotherapy

These CHP nanogels complex various antitumor drugs such as adriamycin *(21)*, cisplatin and its derivatives *(27)*, and neocarzinostatin chromophore *(28)*. For example, CHP nanogels complexed adriamycin (ADR) by simply mixing of the two compounds in phosphate-buffered saline (PBS) at pH 7.4 and 25°C. One nanogel complexed more than one hundred ADR molecules. The chemical stability of ADR was greatly improved after complexation. Spontaneous release of ADR from the complex was sufficiently slow, with more than 70% of ADR still remaining in the complex after 7 days at pH 7.4 and 25°C. The release of ADR was facilitated by decreasing the pH of the medium. Thus, CHP nanogels are useful as cancer drug carriers.

Insulin Delivery by Nanogels

Insulin (Ins) spontaneously forms complexes with CHP nanogels in water, resulting in very stable colloids with nanogel particles of 20-30 nm in diameter. Thermal denaturation and subsequent aggregation of Ins were effectively suppressed upon complexation *(29)*. The complexed Ins were significantly protected from enzymatic degradation. Notably, spontaneous dissociation of Ins from these complexes was barely observed in buffer solution. The CHP-Ins complex nanogels lowered the blood glucose level (BGL) to 50-60% of the initial value within 30 minutes after the i.v. injection. This decrease was similar to that observed for free Ins. This observation indicates that the original physiological activity of complexed Ins was preserved *in vivo* after i.v. injection. The complexed Ins was displaced from the nanoparticles by BSA by an exchange mechanism, without significant loss of physiological activity. Therefore, these nanogels behave as excellent protein drug carriers.

Artificial Vaccine: Nanogels Adjuvant for antigen Protein

The induction of a specific immune response against tumor cells is a highly achievable goal in immune therapy for cancer. Shiku and coworkers reported a novel hydrophobized polysaccharide nanogel/oncoprotein complex vaccine that can induce strong cellular and humoral immune responses against HER2 expressing tumors *(30-33)*. A truncated protein consisting of the 147 N-terminal amino acids of the proto-oncogene c-erbB2/neu/HER2 (HER2) was complexed with CHP nanogels to form complexed nanogels. In mice, immunized with the complexes, HER2-specific $CD8^+$ cytotoxic T lymphocytes (CTL) could be generated and could prevent the growth of subsequently inoculated HER2 expressing tumors. In addition, mice immunized with CHP-HER2 complexes could produce an extremely high titer of IgG antibodies against HER2 protein, indicating a possible activation of helper $CD4^+$ T cells. The complete rejection of tumors was observed when CHP-HER2 complex nanogels were applied on day 3 after inoculation.

Dendritic cells (DC), as bone marrow-derived APC, were able to elicit host immune responses by stimulating the proliferation of $CD4^+$ T cells and $CD8^+$ T cells after exposure to CHP-HER2 complex nanogels. The complete rejection of tumors also occurred when CHP-HER2-pretreated DC was administered as vaccine 10 days after tumor inoculation. Therefore, bone marrow-derived DC, pretreated with CHP nanogel/oncoprotein complex, is a powerful tool for enhancing the effectiveness of oncoprotein for antitumor vaccination, opening a wide range of options for immune cell therapy.

Design of Functional Self-Assembled Nanogels and Hydrogels

A variety of functional nanogels were designed, based on self-assembly of associating polymers. For example, functional CHP nanogels were prepared by chemical modifications with amino groups; carboxyl groups; cell-specific saccharides such as galactose *(34)*; poly(ethylene oxide) (PEO) *(35)*; and thermoresponsive polymers such as Pluronic® (poly(ethylene oxide)-poly(propylene oxide)-poly(ethylene oxide) block copolymer) *(36)*, as shown in Figure 5.

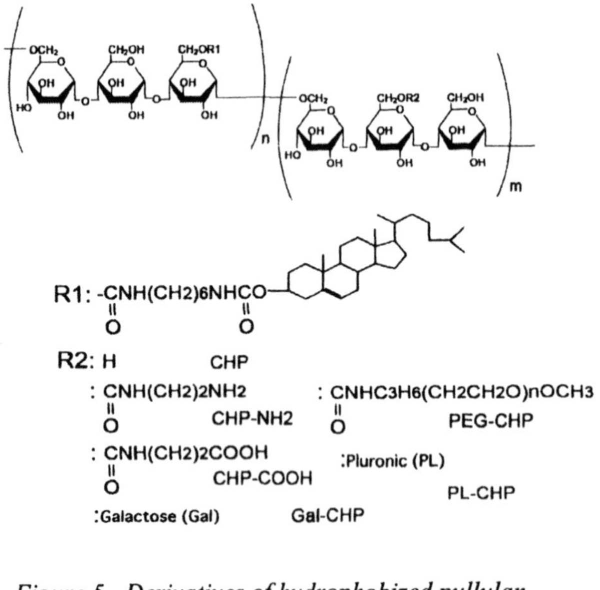

Figure 5. Derivatives of hydrophobized pullulan.

pH-responsive nanogels with a helical structure were obtained by partial modification of poly(L-lysine) with hydrophobic cholesteryl groups *(16)*. Novel photoresponsive nanogels with particle size of 200 nm were prepared by self-assembly of spiropyrane-bearing pullulan (SpP), in which spiropyrane groups were attached to pullulan as assembling units instead of cholesteryl groups *(17)* (Figure 6). The solution properties of these gels could be controlled by photo-stimulation via isomerization between hydrophobic spiropyrane and hydrophilic merocyanine. The nanogels showed molecular chaperone-like activity in protein refolding. Thermoresponsive nanogels were prepared by self-assembly of two different hydrophobically modified polymers, namely cholesterol-bearing pullulan (CHP) and the copolymer of *N*-isopropylacrylamide (NIPAM) and *N*-[4-(1-pyrenyl)butyl]-*N*-*n*-octadecylacrylamide] (PNIPAM-C18Py).

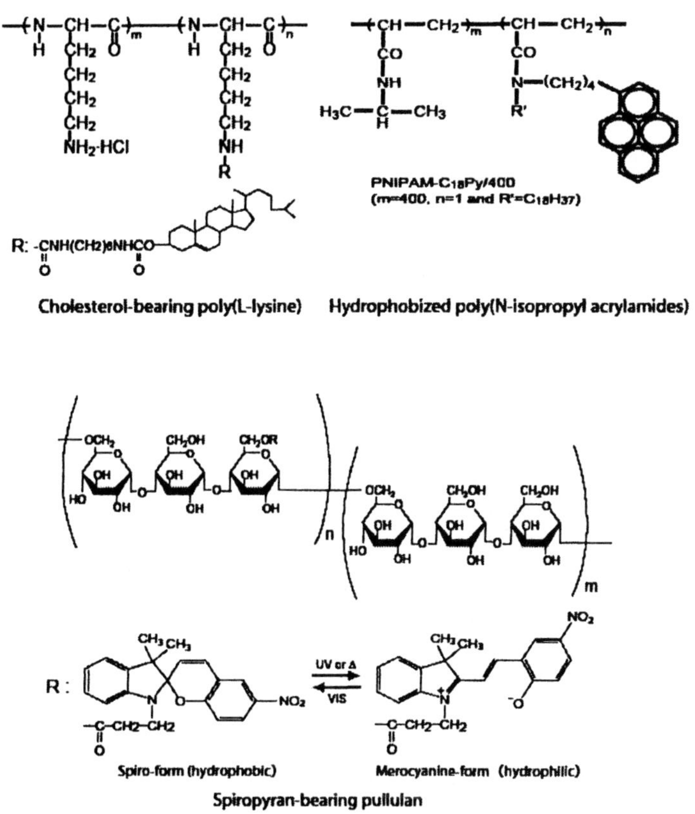

PNIPAM-C_{18}Py/400
(m=400, n=1 and R'=$C_{18}H_{37}$)

Cholesterol-bearing poly(L-lysine) Hydrophobized poly(N-isopropyl acrylamides)

Spiro-form (hydrophobic) Merocyanine-form (hydrophilic)

Spiropyran-bearing pullulan

Figure 6. Structure of various associated polymers.

The association of two polymers via their hydrophobic groups represents a new method of preparing stable functional nanogels. Recently, we developed nanogels based on polymerizable precursors. For example, a methacryloyl group-bearing CHP (CHPMA) formed nanogels (particle size 20 nm) in water, similar to CHP nanogels. Hybrid nanogels composed of polysaccharides and 2-methacryloyloxyethyl phosphorylcholine (MPC) polymer were prepared by radical copolymerization with MPC in dilute aqueous concentration of the nanogels, using CHPMA nanogels as a primary seed. In a semi-dilute aqueous solution, nanogels formed network structures to yield macrogels, which were connected by polymerization between nanogel macromers and MPC. CHPMA nanogels acted as crosslinkers for gelation. TEM observation showed that the nanogel structures were retained after gelation and that the nanogels were well dispersed in the macrogel. In other words, the nanogels were immobilized in MPC macrogels. Hybrid hydrogels with nanogel crosslinkers acted as well-defined matrices for controlled release of proteins.

Conclusions

Functional nanogels and hydrogels, based on self-assembly of associating polymers, were designed for novel drug delivery applications. These nanogels trap hydrophobic molecules such as antitumor drugs, proteins such as enzymes, insulin, and antigen protein, as well as nucleic acids (DNA plasmids). Therefore, these nanogels can be used as nanocarriers in cancer chemotherapy, protein delivery, and artificial vaccines. Stimuli-responsive nanogels such as pH-responsive, thermoresponsive, and photoresponsive nanogels were also designed using this self-assembly method. Macrogels with well-defined nanostructures were obtained by self-assembly, utilizing these nanogels as building blocks. The self-assembly method based on associating polymers is an efficient and versatile technique for the preparation of functional nanogels and hydrogels.

References

1. Kopecek, J. Polymer chemistry - Swell gels. *Nature* **2002**, *417*, 388.
2. Byrne, M.E.; Park, K.; Peppas, N.A. Molecular imprinting within hydrogels. *Adv. Drug Delivery Rev.* **2002**, *54*, 149-161.
3. Vinogradov, S.V.; Bronich, T.K.; Kabanov, A.V. Nanosized cationic hydrogels for drug delivery: preparation, properties and interactions with cells. *Adv. Drug Delivery Rev.* **2002,** *54*, 135-147.
4. McAllister, K.; Sazani, P.; Adam, M.; Cho, M.J.; Rubinstein, M.; Samulski, R.J.; DeSimone J.M. Polymeric nanogels produced via inverse

microemulsion polymerization as potential gene and antisense delivery agents. *J. Am. Chem. Soc.* **2002,** *124*, 1519815207.

5. Gan D.J.; Lyon L.A. Tunable swelling kinetics in core-shell hydrogel nanoparticles. *J. Am. Chem. Soc.* **2001,** *123*, 7511-7517.
6. Huang, H.; Kowalewski, T.; Remsen, E.E.; Gertzmann, R.; Wooley, K.L. Hydrogel-coated glassy nanospheres: A novel method for the synthesis of shell cross-linked knedels. *J. Am. Chem. Soc.* **1997,** *119*, 11653-11659.
7. Kuckling, D.; Vo, C.D.; Wohlrab, S.E. Preparation of nanogels with temperature-responsive core and pH-responsive arms by photo-cross-linking. *Langmuir* **2002,** *18*, 4263-4269.
8. Akiyoshi, K.; Deguchi, S.; Moriguchi, N.; Yamaguchi, S.; Sunamoto, J. Self-aggregates of hydrophobized polysaccharides in water-Formation and characteristics of nanoparticles. *Macromolecules* **1993,** *26*, 3062-3068.
9. Duncan, R. The dawning era of polymer therapeutics. *Nature Reviews Drug Discovery* **2003,** *2*, 347-360.
10. Kataoka, K.; Harada, A.; Nagasaki, Y. Block copolymer micelles for drug delivery: Design, characterization and biological significance. *Adv. Drug Deliv. Rev.* **2001,** *47*, 113-131.
11. *Associative Polymers in Aqueous Solutions*; Glass, J.E., Ed.; ACS Symposium Series, Vol. 765, American Chemical Society: Washington, DC, 2000.
12. Akiyoshi, K.; Deguchi, S.; Tajima, H.; Nishikawa, T.; Sunamoto, J. Microscopic structure and thermoresponsiveness of a hydrogel nano-particle by self-assembly of a hydrophobized polysaccharide. *Macro-molecules* **1997,** *30*, 857-861.
13. Yamazaki, A.; Song, J.M.; Winnik, F.M.; Brash, J.L. Synthesis and solution properties of fluorescently labeled amphiphilic (N-alkylacryl-amide) oligomers. *Macromolecules* **1998,** *31,* 109-115.
14. Lee, K.Y.; Jo, W.H.; Kwon, I.C.; Kim, Y-H.; Jeong,S.Y. Structural determination and interior polarity of self-aggregates prepared from deoxycholic acid-modified chitosan in water. *Macromolecules* **1998,** *31*, 378-383.
15. Nichifor, M.; Lopes, A.; Carpov, A.; Melo, E. Aggregation in water of dextran hydrophobically modified with bile acids. *Macromolecules* **1999,** *32*, 7078-7085.
16. Akiyoshi, K.; Ueminami, A.; Kurumada, S.; Nomura, Y. Self-association of cholesteryl-bearing poly(L-lysine) in water and control of its secondary structure by host-guest interaction with cyclodextrin. *Macromolecules* **2000,** *33*, 6752-6756.
17. Hirakura, T.; Nomura, Y.; Aoyama, Y.; Akiyoshi, K. Photoresponsive nanogels formed by the self-assembly of spiropyrane-bearing pullulan that

act as artificial molecular chaperones. *Biomacromolecules* **2004**, *5*, 1804-1809.

18. Akiyoshi, K.; Sasaki, S.; Kuroda, K.; Sumanoto, J. Controlled association of hydrophobized polysaccharide by cyclodextrin. *Chem. Lett.* **1998**, 93-94.
19. Breslow, R.; Zhang, B.J. Cholesterol recognition and binding by cyclodextrin dimmers. *J. Am. Chem. Soc.* **1996**, *118*, 8495-8496.
20. Akiyoshi, K.; Deguchi, S.; Tajima, H.; Nishikawa, T.; Sunamoto, J. Self-assembly of hydrophobized polysaccharide-Structure of hydrogel nanoparticle and complexation with organic compounds. *Proc. Japan Acad. Sci. B* **1995**, *71*, 15-19.
21. Akiyoshi, K.; Taniguchi, I. Fukui, H.; Sunamoto, J. Hydrogel nanoparticle formed by self-assembly of hydrophobized polysaccharide, stabilization of adriamycin by complexation. *Eur. J. Pharma. Biopharm.* **1996**, 42, 286-290.
22. Akiyoshi, K.; Sunamoto, J. *Supermolecular Sci.* **1996**, 3, 157.
23. Nishikawa, T.; Akiyoshi, K.; Sunamoto, J. Supramolecular assembly between nanoparticles of hydrophobized polysaccharide and soluble protein complexation between the self-aggregate of cholesterol-bearing pullulan and alpha-chymotrypsin. *Macromolecules* **1994**, *27*, 7654-7659.
24. Akiyoshi, K.; Sasaki, Y.; Sunamoto, J. Molecular chaperone-like activity of hydrogel nanoparticles of hydrophobized pullulan: Thermal stabilization with refolding of carbonic anhydrase B. *Bioconjugate Chem.* **1999**, 10, 321-324.
25. Nomura, Y.; Ikeda, M.; Yamaguchi, N.; Aoyama, Y.; Akiyoshi, K. Protein refolding assisted by self-assembled nanogels as novel artificial molecular chaperone. *FEBS Lett.* **2003**, *553*, 271-276.
26. Miyazawa, N.; Sakaue, T.; Yoshikawa, K.; Akiyoshi, K. Biomolecular Chemistry, Proceedings of the ISBC2003, Maruzen Co., Ltd. Japan, 2003, 68.
27. Satoh, E.; Kanda, K.; Akiyoshi, K.; Kidani, Y.; Shinohara, A.; Chiba, M.; Yanagie, H.; Takamoto, S.; Eriguchi, M. Proceeding of 15th International Congress on Anti-Cancer Treatment, France 2004; p. 356.
28. Taniguchi, I.; Fujiwara, M.; Akiyoshi, K.; Sunamoto, J. Substitution for apoprotein of neocarzinostatin by self-aggregate of cholesterol-bearing pullulan. *Bull. Chem. Soc. Jpn.* **1998**, *71*, 2681-2685.
29. Akiyoshi, K.; Kobayashi, S.; Shichibe, S.; Mix, D.; Baudys, M.; Kim, S. W.; Sunamoto, J. Self-assembled hydrogel nanoparticle of cholesterol-bearing pullulan as a carrier of protein drugs: Complexation and stabilization of insulin. *J. Control Rel.* **1998**, *54*, 313-320.
30. Gu, X-G.; Schmitt, M.; Hiasa, A.; Nagata, Y.; Ikeda, H.; Sasaki, Y.; Akiyoshi, K.; Sunamoto, J.; Nakamura, H.; Kuribayashi, K.; Shiku, H. A

novel hydrophobized polysaccharide/oncoprotein complex vaccine induces in vitro and in vivo cellular and humoral immune responses against HER2-expressing murine sarcomas. *Cancer Res.* **1998**, *58*, 3385-3390.

31. Wang, L.; Ikeda, H.; Ikuta, Y.; Schmitt, M.; Miyahara, Y.; Gu, X.; Nagata, Y.; Sasaki, Y.; Akiyoshi, K.; Sunamoto, J.; Nakamura, H.; Kuribayashi, K.; Shiku, H. Bone marrow-derived dendritic cells incorporate and process hydrophobized polysaccharide/oncoprotein complex as antigen presenting cells. *Int. J. Oncl.* **1999**, *14*, 695-701.
32. Shiku, H.; Wang, L.J.; Ikuta, Y.; Okugawa, T.; Schmitt, M.; Gu, X.G.; Akiyoshi, K.; Sunamoto, J.; Nakamura, H. Development of a cancer vaccine: peptides, proteins, and DNA. *Cancer Chemotherapy Pharm.* **2000**, *46*, S77-S82.
33. Ikuta, Y.; Katayama, N.; Wang, L.; Okugawa, T.; Takahashi, Y.; Schmitt, M.; Gu, X.; Watanabe, M.; Akiyoshi, K.; Nakamura, H.; Kuribayashi, K.; Sunamoto, J.; Shiku, H. Presentation of a major histocompatibility complex class 1-binding peptide by monocyte-derived dendritic cells incorporating hydrophobized polysaccharide-truncated HER2 protein complex: implications for a polyvalent immuno-cell therapy. *Blood* **2002**, *99*, 3717-3724.
34. Taniguchi, I.; Akiyoshi, K.; Suda, Y.; Sunamoto, J.; Yamamoto, M.; Ichinose, K. *J. Bioactive and Compatible Polymers* **1999**, *14*, 195.
35. Taniguchi, I.; Akiyoshi, K.; Sunamoto, J. Effect of macromolecular assembly of galactoside-conjugated polysaccharides on galactose oxidase activity. *Macromol. Chem. Phys.* **1999**, *200*, 1386-1392.
36. Deguchi, S.; Akiyoshi, K.; Sunamoto, J. Solution property of hydrophobized pullulan conjugated with poly(ethylene oxide) poly(propylene oxide) poly(ethylene oxide) block-copolymer: Formation of nanoparticles and their thermosensitivity. *Macromol. Rapid Commun.* **1994**, *15*, 705-711.
37. Akiyoshi, K.; Kang, E.-C.; Kurumada, S.; Sunamoto, J.; Principi, T.; Winnik, F.M. Controlled association of amphiphilic polymers in water: Thermosensitive nanoparticles formed by self-assembly of hydrophobically modified pullulans and poly(N-isopropylacrylamides). *Macromolecules* **2000**, *33*, 3244-3249.

Chapter 7

Triblock PLLA–PEO–PLLA Hydrogels: Structure and Mechanical Properties

Sarvesh K. Agrawal[1], Heidi A. Sardhina[1], Khaled A. Aamer[2], Naomi Sanabria-DeLong[2], Surita R. Bhatia[1], and Gregory N. Tew[2,*]

[1]Department of Chemical Engineering, University of Massachusetts at Amherst, 686 North Pleasant Street, Amherst, MA 01003
[2]Department of Polymer Science and Engineering, University of Massachusetts at Amherst, 120 Governors Drive, Amherst, MA 01003
[*]Corresponding author: tew@mail.pse.umass.edu

Triblock copolymers made from poly(L-lactide)-poly(ethylene glycol)-poly(L-lactide) have attracted attention recently because of their ability to form elastic gels, which have potential applications in drug delivery and tissue engineering. We have perfomed rheology studies on several of these gels, formed with varying lengths of the hydrophobic (PLLA) blocks. The elastic moduli of these gels were found to be greater than 10,000 Pa, matching well with the moduli measured for several native human tissues. The strength of the gels is seen to be strongly dependent on the PLLA block length, thus offering a mechanism to control the mechanical properties as desired for particular applications. The gel strength is dependent upon the network structure, which in turn governs the degradation behavior of the gels and hence the release rate of bioactive molecules. Hence we establish the usefulness of these materials for producing tailor-made hydrogels, suitable for specific biomedical applications.

Introduction

Biodegradable and biocompatible polymeric materials have attracted widespread interest during the past decade for potential applications as drug carriers for delivery systems (*1,2*) and cell scaffolds for tissue engineering (*3,4,5*). Microspheres, nanoparticles and hydrogels made from amphiphilic polymers can be used to encapsulate hydrophobic drugs and other bioactive molecules, which can be released in a controlled manner over a long period of time. Encapsulation in polymeric biomaterials reduces the interaction of the drugs with the biological environment and prevents the drug from being quickly removed by the body's reticuloendothileal system (*6*). The polymer eventually degrades and is removed from the human body over time, thereby releasing the drug without leaving any traces of the polymer at the application site.

In particular, materials capable of forming hydrogels are of specific interest for biomedical applications, because hydrogels can be implanted in the body along with biological molecules or cells in a minimally invasive manner. In addition, hydrogels closely mimic natural tissues in water content, interfacial tension, and mechanical properties (e.g., soft and rubbery) (*3,5*). Bioactive molecules and drugs can be physically entrapped in the gel matrix or chemically attached to the polymer. Gel formation can occur through several mechanisms, including crosslinking of the polymer matrix, and physically linking through hydrogen bonding, crystallization, or hydrophobic interactions.

Poly(lactic acid) (PLA) has been extensively investigated for application in drug delivery systems (*7,8*). It is biodegradable, adapts well to biological environments, and does not have adverse effects on blood and tissues. Poly(ethylene glycol) (PEO) has the advantage of being hydrophilic, non-ionic, biocompatible and has also been used in various biomedical applications for increasing the hydrostability of blood contacting materials (*9,10*). Due to such unique properties, copolymers of poly(ethylene glycol)-*co*-poly(lactide) with an AB and ABA architecture have been extensively studied for biomedical applications (*11-25*). In addition, these polymers are known to form hydrogels at varying concentration of the polymer, depending on the exact architecture, with transitions near body temperature. It has been observed in gel-forming PLA-PEG diblock systems that the ratio of the blocks can be effectively modulated to tune a number of important physical properties, including the permeability of the gel and its degradation rate (*26,27*), which are crucial in controlling the release characteristics from these gels. The structural properties of the gel matrix, such as its micromorphology and pore size, are directly related to the mass transport of water into the gel and transport of drug out of the polymer. Hence the chemical composition and microstructure design can be used to adapt the structure-property relationship and produce tailor-made polymer matrices. For biodegradable systems, drug release takes place through a combination of processes, involving diffusion of drug from inside of the matrix to outside and dissolution and surface erosion of the hydrogel. The degradation behavior of a gel can be well understood with knowledge of the gel network microstructure. It is also known that physical properties and mechanical strength of a gel are strongly related to the density of network junctions in the gel (*28*). Through our study, we expect to obtain some insight into the structure-property relationships of the PLLA-PEO-PLLA gels, including the dependence of gel strength on the chemical composition and concentration of polymer, which in turn relates to the gel structure. This insight will give us a handle on controlling the degradation behavior, release profiles and mechanical properties of the gels. In this study we have synthesized polymers made with constant PEO block length (MW=8,900) and varying length of the hydrophobic (PLLA) block and have characterized them in gel phase using mechanical rheology.

Materials and Methods

Materials

L-lactide (Aldrich) was purified by recrystallization in dry ethyl acetate and by sublimation prior to polymerization. The α,ω dihydroxy poly(ethylene glycol) macro initiator with molecular weight 8,900 (PEG 8K, Aldrich) was dried at room temperature under vacuum prior to polymerization. Stannous (II) 2-ethyl hexanoate (Alfa Aesar) was used without further purification.

Synthesis of PLLA-PEO-PLLA Triblock Copolymer

PLLA-PEO-PLLA triblock copolymers were synthesized by bulk polymerization (Figure 1). PEG was introduced into a dried polymerization tube. The tube was purged with nitrogen, and placed in an oil bath at 150 °C. Stannous (II) 2-ethylhexanoate was introduced under nitrogen to the molten PEG and stirred for 10 minutes, followed by addition of L-lactide to the macroinitiator/catalyst melt. The polymerization was carried out at 150 °C for 24 hours with stirring, after which the reaction was quenched by methanol. The product was dissolved in tetrahydrofuran and precipitated into n-hexane. The process of dissolution/reprecipitation was carried out three more times. The copolymer was dried under vacuum at room temperature for two days.

Sample Preparation and Instrumentation

The copolymer polydispersities were measured versus poly(ethylene oxide) standards, using GPC (HP 1050 series, a HP 1047A differential refractometer, and three PLgel columns (5 µm 50A°, two 5 µm MIXED-D) in dimethylformamide as eluting solvent at a rate of 0.5 ml/min at room temperature. The copolymer compositions were determined by ^1H NMR (Bruker, DPX300, 300MHz spectrometer, d-chloroform).

Gels were prepared by slow addition of dried polymer sample to a fixed volume of DI water (15 mL), followed by stirring and heating. Gels were then transferred to a Bohlin CVO rheometer for oscillatory measurements. A cone-and-plate geometry with a 4° cone, 40 mm diameter plate, and 150 mm gap was used for all experiments on hydrogels. For liquid samples with a low viscosity, a couette geometry was used. Stress amplitude sweeps were performed to ensure that subsequent data was collected in the linear viscoelastic regime. Frequency sweeps were performed at a constant stress (0.1 - 2.0 Pa, depending on the sample) in the frequency range 0.01 - 100 Hz. At high frequencies, a resonant frequency of the rheometer motor was observed; thus, data are reported up to a frequency of approximately 10 Hz, depending on the particular sample.

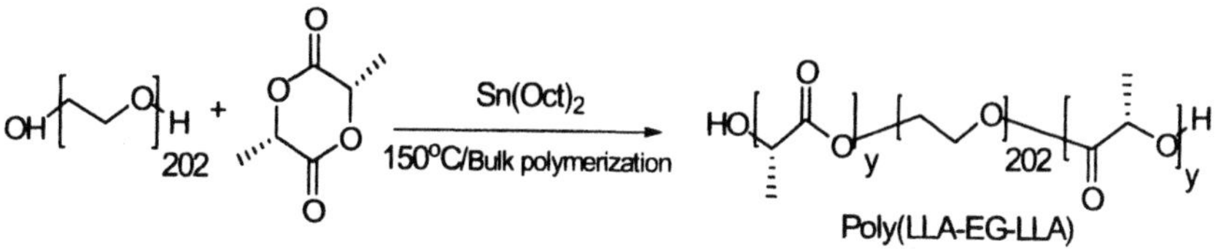

Figure 1. Bulk polymerization of PEG with L-lactide at 150 °C.

Results and Discussion

Block Copolymer Synthesis

The copolymers shown in Table I were prepared by ring-opening polymerization of L-lactide at 150 °C in the bulk, using stannous (II) 2-ethylhexanoate as catalyst. This method is known to limit racemization of the stereocenter, and produce polymers of significant molecular weight and narrow polydispersity (*11*). The macroinitiator, PEO, had a molecular weight (MW) of 8,900 Daltons. Four different polymers were prepared with increasing PLLA block lengths, varying from a total DP of 26 to 74 so that the total lactide composition is always smaller than PEO. ^1H-NMR integration was used to establish the M_w for PLLA blocks as opposed to GPC standards (*14,15*). In all cases, the polymerization was not run to completion since this is known to broadens the molecular weight distribution.

Table I. Molecular weight characteristics of PLLA-PEO-PLLA triblock copolymers.

Sample	M_{nPEG} [a]	M_{nPLLA} [b]	M_{Total}	*MWD* [c]	*Total* DP_{PLLA} [b]	$Wt\%_{PEO}$	$Wt\%_{PLLA}$
1	8,900	1872	10,772	1.31	26	82.6	17.4
2	8,900	3168	12,068	1.15	44	73.7	26.3
3	8,900	3744	12,644	1.24	52	70.4	29.4
4	8,900	4320	13,220	1.18	60	67.3	32.7
5	8,900	5328	14,228	1.20	74	62.6	37.4

[a] Determined by MALDI_TOF and GPC; [b] Determined by ^{1}H NMR; [c] Determined by GPC. (Adapted with permission from reference *(29)*. Copyright 2004).

Hydrogel Properties

We were able to make hard, physically associated gels in aqueous solutions. These associated network gels are analogous to reversible network gels formed from hydrophobically-modified telechelic polymers. Triblock copolymers with an ABA architecture and with hydrophobic endblocks have been commonly observed to form flowerlike micelles in dilute solutions. As the concentration of the polymer increases, the micelles become more packed and come closer to each other. The sticky hydrophobic end groups of the polymer attach themselves to adjacent micelles, leading to bridging between them (Figure 2), which gives rise to an entropic attraction. As the concentration is increased further, the bridging density increases, leading to a randomly crosslinked phantom network structure. The strength of the network is proportional to the number of random "crosslinks" or hydrophobic junctions in the network.

Characteristics of gel formation were seen to change as the length of hydrophobic groups for polymers varied. Solutions of the triblock with the smallest hydrophobes (**1**) did not gel up to concentrations of 20 wt% polymer, while the remaining triblocks formed gels at moderate concentrations (16-20 wt%). Figure 3a shows a series of aqueous solutions of $PLLA_{22}PEO_{202}PLLA_{22}$ (**2**) at concentrations ranging from 10-20 wt%. The polymer is seen to form viscous solutions at 10, 12 and 14 wt%, while gelation occurs above 14 wt%. The gels formed were strong enough to support their own weight and the stir bar weight when kept upside down for a long period of time (Figure 3b).

Rheology

All samples were characterized by dynamic mechanical rheology *(29)*. Figure 4 shows the elastic modulus *G'* and the viscous modulus *G"* versus frequency for 16 wt% of **5** at 25 °C. The gel displays a high degree of elasticity, with *G'* only weakly dependent on frequency and greater than *G"* over the entire frequency range. The elastic modulus of the sample was found to be greater than 10,000 Pa, which is almost an order of magnitude higher than that reported by other groups for polymers based on combinations of similar chemical blocks, for example stereocomplexed systems formed by mixing PLLA-PEG and PDLA-PEG (*30,31*). Biomaterials with moduli in the kPa range are of widespread interest since many native tissues have moduli in this range, although most have nonlinear response to strain. For example, human nasal cartilage (234 ± 27 kPa) (*32,33*), bovine articular cartilage (990 ± 50 kPa) (*32,33*), pig thoracic aorta (43.2 ± 15 kPa) (*34*), pig adventitial layer (4.72 ± 1.7 kPa) (*34*), right lobe of human liver (270 ± 10 kPa) (*35*), canine kidney cortex and medulla (~10 kPa) (*36*), and nucleus pulposus and eye lens (~10^3 Pa) (*36*) have moduli in this range. Moreover, for scaffolding applications, it is often desirable to "match" mechanical properties of the polymer matrix to those of the surrounding tissue (*37*). Similar properties are useful for building '*in-situ*' hydrogel implants for drug delivery or for making drug-releasing wound dressings.

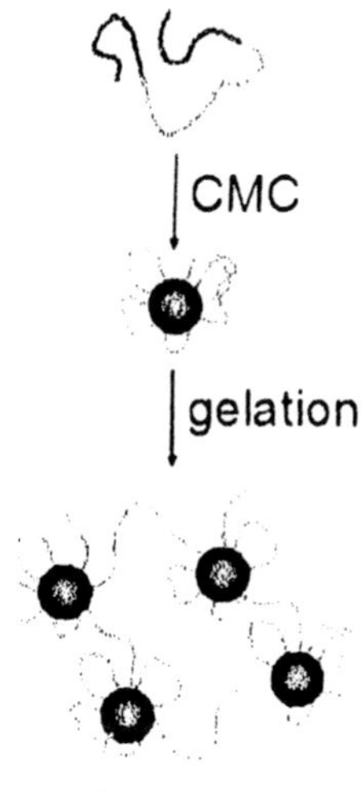

Figure 2. Bridging between micelles leading to gel network formation.
Continued on next page.

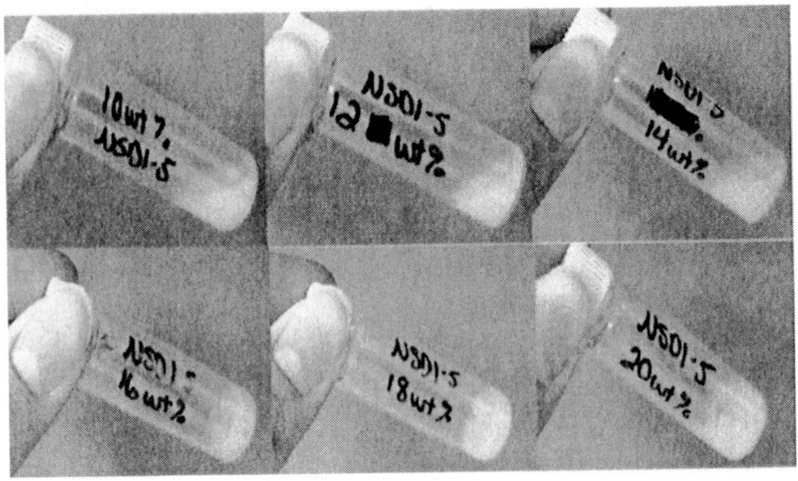

Figure 2. Continued.

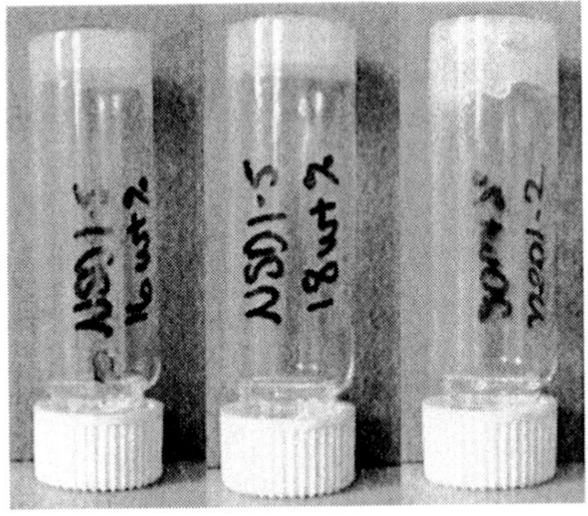

Figure 3. Upper panel: Hydrogel formation occurs for $PLLA_{22}PEO_{202}PLLA_{22}$ for concentrations as low as 16 wt%. Lower panel: At 16, 18 and 20 wt% (left to right), these gels are strong enough to support the weight of the stir bar along with their own weight.

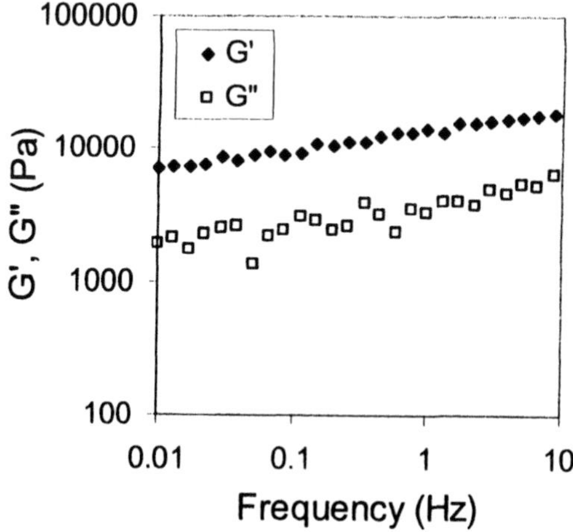

*Figure 4. Elastic modulus G' (filled diamonds) and viscous modulus G' (open squares) as a function of frequency for a gel of sample **5** at 16 wt% polymer and T = 25 °C. (Reproducd with permission from reference (29). Copyright 2004.)*

The elastic modulus G' of gels formed from the different polymer systems **3**, **4** and **5** at concentrations of 20, 16 and 16 wt%, respectively, is shown in Figure 5. The measurements were taken at 25 °C (filled symbols) and 37 °C (open symbols). The polymers represent a group of molecules differing in the length of the hydrophobic moieties, with **3** having the shortest PLLA blocks. Although the gel of **3** was prepared at higher concentration (20 wt%) compared to the 16 wt% for the two other samples, it still forms a weaker gel. It is clear from these results that the hydrophobic length or DP has a pronounced effect on the elastic modulus. The strength of the gels was seen to improve by more than two orders of magnitude, from ~100 Pa to 10,000 Pa, as the length of the hydrophobic block is increased from 26 (or 52 since there are two PLLA blocks per molecule) to >30. For comparison, the complex modulus G^* of a 20 wt% solution of **1** at 25 °C is also shown in Figure 5. This sample, with hydrophobic block lengths of 13, forms a viscoelastic liquid with $G^* < 1$ Pa. Qualitatively similar trends have been observed for the high-frequency limit of G' for PEO, containing alkyl hydrophobe end-caps (*38*). However, the dependence of G' on hydrophobe length is much weaker for these alkyl-capped systems than for our PLLA-PEO-PLLA gels, while the high-frequency elastic moduli of gels of fluoroalkyl-capped PEO were found to be insensitive to the hydrophobe length (*39,40*). Thus, the strong dependence of G' on the PLLA block length should offer a straightforward way to tune the rheological response of our gels for specific bioapplications.

The elastic modulus at 37 °C for these gels is also shown on Figure 5. All three gels display a decrease in *G'* with temperature; however, with a much larger decrease for **3** than either **4** or **5**. This decrease in *G'* with increasing temperature is common for gels composed of PLA-PEO-PLA molecules (*41*). It is important to note that, despite this decrease, the value of *G'* for **5** remains high at physiological temperatures (roughly 10,000 Pa), and that both **4** and **5** are still elastic gels with *G'* nearly independent of frequency.

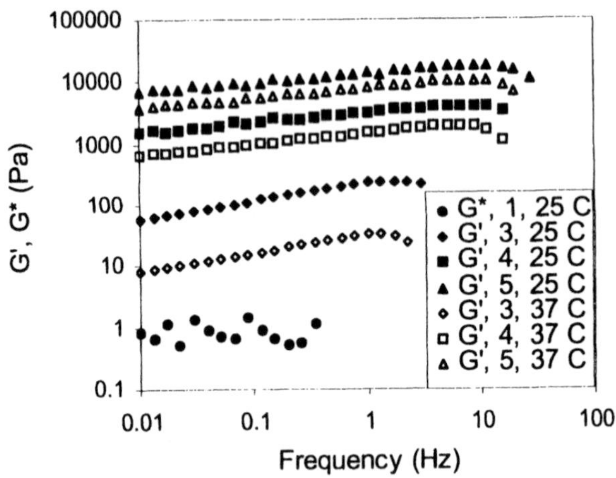

Figure 5. Elastic moduli versus frequency for gels formed from samples 3, 4, and 5 at concentrations of 20, 16, and 16 wt%, respectively; at 25 °C (filled symbols) and 37 °C (open symbols). For comparison, the complex modulus G of a 20 wt% solution of sample 1 is also shown. (Adapted with permission from reference (29). Copyright 2004.)*

Figures 6 and 7 show the influence of temperature on *G'* for samples **3** and **5**. Interestingly, *G'* for sample **3** decreases with temperature as expected; however, near 40 °C the sample undergoes a transition toward increasing *G'*. At 50 °C, the elastic modulus is even more linear with temperature and significantly higher at low frequency than found at 37 °C. The physical and structural properties of the gel responsible for these changes are not understood at this time.

However, it is possible that at 25 °C, the initial measurement, the hydrogel is composed of slightly hydrated micelles, but that these micelles begin to melt and destabilize as the temperature is increased near 40 °C (T_g of PLLA is approximately 52 °C and should be lowered by the presence of water as a weak plasticizer). Then, as the temperature is increased further, the PLLA domains may dehydrate in a manner similar to that observed for the PPO segments of Pluronics *(42)*. The observed behavior is consistent with this desolvation transition but more work must be performed to prove this assumption.

Similar changes in phase behavior with temperature have been reported for similar systems, i.e., PEG-PLLA, PEG-PDLA *(43,44)* and PLGA-PEO-PLGA *(45)*. Alternatively, **5** behaves in a more expected manner (Figure 7), with a steady decrease in *G'* as the temperature is increased. This behavior would be expected because water is less likely to hydrate larger PLLA blocks. However, at 70 °C, the data shows a slight increase in modulus consistent with PEO dehydration.

Sol-gel transitions can be defined as $G' > G''$, and Figure 8 shows that for **3** at 37 °C $G' \approx G''$ for the entire frequency range. Interestingly, the observed increase in *G'* at 50 °C for sample **3** is confirmed in the inset of Figure 8, where *G'* is much larger than *G"*. The frequency-dependence of *G'*, observed for sample **3** at 25 °C and 37 °C, as opposed to the frequency-independent behavior observed at 50 °C, may indicate the presence of a slow relaxation mechanism at lower temperatures.

The data on **3** at the gel point (37 °C) can be fit to a power law scaling of the form $G' \sim G'' \sim \omega^{\Delta}$ *(46)*. Here, ω is frequency and Δ is an exponent, whose

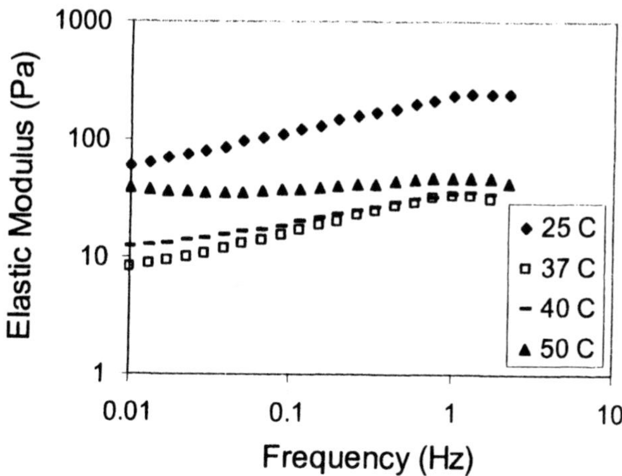

*Figure 6. Elastic modulus G' versus frequency for a gel with 20 wt% of sample **3** as a function of temperature. Note the non-monotonic dependence of G' on temperature for this sample. (Reproducd with permission from reference (29). Copyright 2004.)*

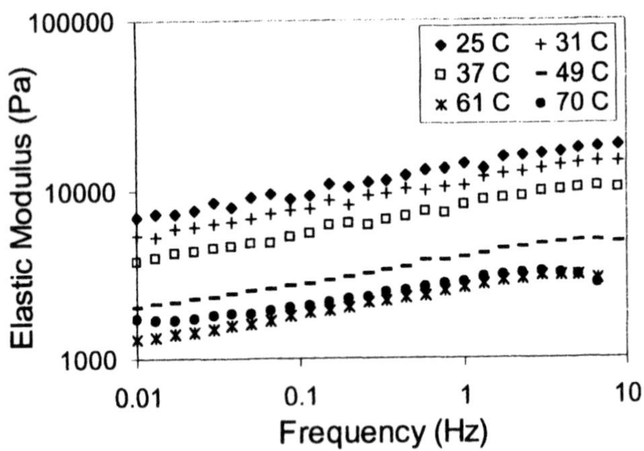

Figure 7. Elastic modulus G' versus frequency for a gel with 16 wt% of sample **5** *as a function of temperature. (Reproducd with permission from reference (29). Copyright 2004.)*

value at the gel point can be predicted to fall within the range of 0.67-1.0 for networked gels, applying classical mean-field considerations (*47*), electrical network analogies (*48*), and the percolation theory (*49,50*). Modified percolation models that account for elasticity of the network backbone (*51,52*) and excluded volume effects (*53*) have yielded predictions in the range 0-1.0. Fitting the data shown in Figure 8 to a power law gives $\Delta = 0.36$. Although it is impossible to draw any definite conclusions about the structure of the gel from this value, similar values have been reported for nonstoichiometric, chemically crosslinked polymeric gels with an excess of crosslinker present, and for chemically crosslinked gels with chains above the entanglement molecular weight (*54*). By contrast, experiments on physically associated biopolymer gels (*55,56*) and transient networks based on triblock copolymers (*57*) have yielded higher values of Δ in the range 0.5-0.7. Interestingly, gels with semicrystalline domains have lower exponents, including bacterially-derived poly(*b*-hydroxyoctanoate) *(58)* and crystallizing polypropylene (*59*). Thus, the low value of Δ may be a consequence of either entanglements of the PEO chains or rigidity of the hydrophobic PLLA domains.

Figure 9 shows the elastic modulus at fixed frequency as a function of strain for **3**, **4**, and **5** at 25 C. These polymers exhibit a linear viscoelastic response over a wide range of strain and applied stress, and no evidence of strain-hardening was observed. Thus, we expect these gels will have a uniform and predictable rheological response *in vivo*, regardless of the mechanical environment to which they are subjected.

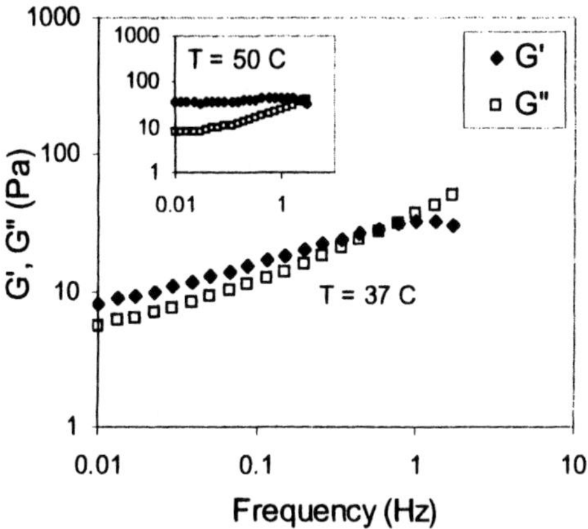

*Figure 8. Elastic modulus G' (filled diamonds) and viscous modulus G" (open squares) versus frequency for a 20 wt% gel of sample **3** at 37 °C and (inset) 50 °C. (Reproducd with permission from reference (29). Copyright 2004.)*

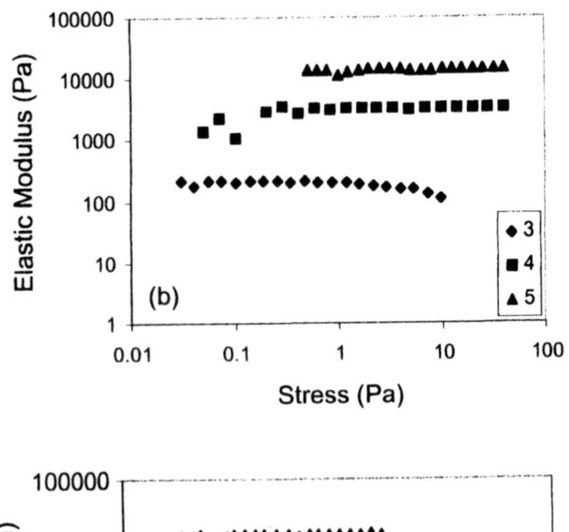

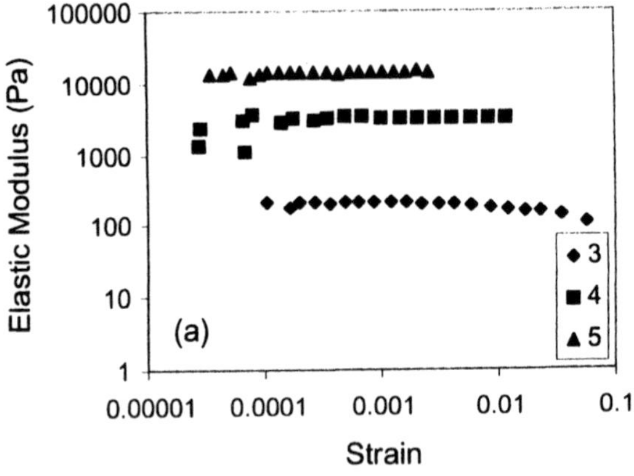

*Figure 9. Elastic modulus G' at a fixed frequency of 1.0 Hz as a function of stress (above) and strain (below) for samples **3**, **4**, and **5** at concentrations of 20, 16, and 16 wt%, respectively (25 ℃). (Reproducd with permission from reference (29). Copyright 2004.)*

Conclusions

We have been able to synthesize and characterize physically associated gels of PLLA-PEO-PLLA that have elastic moduli in the range of soft human tissues. The elastic modulus of the gels is strongly dependent on the PLLA block length, thus offering a means to control their mechanical strength, which in turn can be related to their degradation behavior. Questions still remain about the network structure and influence of PLLA block length. Work is ongoing to obtain predictable release rates of bioactive molecules from these gels, thereby establishing the usefulness of these materials for producing tailior-made matrices, loaded with drugs and growth factors for use as wound dressings, surgical implants and in other biomedical applications.

Acknowledgements

This work was partially funded by and utilized central facilities of the NSF-sponsored UMass MRSEC on Polymeric Materials (DMR-0213695). G.N.T. and S.R.B. both thank Dupont for Dupont Young Professor Awards and 3M for Nontenured Faculty Awards.

References

1. Domb, A.J.; Amselem, S.; Maniar, M. *Polymeric Biomaterials*; Marcel Dekker Inc., New York, NY, 1994, pp 339-433.
2. Langer, R. Biomaterials in drug delivery and tissue engineering: One laboratory's experience. *Acc. Chem. Res.* **2000**, *33*, 94-101.
3. Lee, K.Y.; Mooney, D.J. Hydrogels for tissue engineering. *Chem. Rev.* **2001**, *101*, 1869-1879.
4. Hubbell, J.A. Biomaterials in tissue engineering. *Biotechnology* **1995**, *13*, 565-576.
5. Hoffman, A.S. Hydrogels for biomedical applications. *Adv. Drug Deliv. Rev.* **2002**, *43*, 3-12.
6. Gref, R.; Minamitake, Y.; Peracchia, M.; Trubetskoy, Y.; Torchilin, V.; Langer, R. Biodegradable long-circulating polymeric nanospheres. *Science* **1994**, *263*, 1600.
7. Lunt, J. Large-scale production, properties and commercial applications of polylactic acid polymers. *Polym. Degrad. Stab.* **1998**, *59*, 145-152.
8. Hakkarainen, M.; Karlsson, S.; Alberlsson, A.C. Influence of low molecular weight lactic acid derivatives on degradability of polylactide. *Appl. Polym. Sci.* **2000**, *76*, 228-239.
9. Merrill, E.W.; Dennison, K.A.; Sung, C. Partitioning and diffusion of solutes in hydrogels of poly(ethylene oxide). *Biomaterials* **1993**, *14*, 1117-26.

10. Graham, N.B.; McNeill, M.E. Morphology of poly(ethylene oxide)-based hydrogels in relation to controlled drug delivery. *Makromol. Chem. Macromol. Symp.* **1988,** *19*, 255-73.
11. Kricheldorf, H.R.; Meierhaack, J. Polylactones 22. ABA triblock copolymers of L-lactide and poly(ethylene glycol). *Makromol. Chem.-Macromol. Chem. Phys.* **1993,** *194*, 715-725.
12. Kubies, D.; Rypacek, F.; Kovarova, J.; Lednicky, F. Microdomain structure in polylactide-block-poly(ethylene oxide) copolymer films. *Biomaterials* **2000**, *21*, 529-536.
13. Kissel, T.; Li, Y.X.; Unger, F. ABA-triblock copolymers from biodegradable polyester A-blocks and hydrophilic poly(ethylene oxide) B-blocks as a candidate for in situ forming hydrogel delivery systems for proteins. *Adv. Drug Deliv. Rev.* **2002,** *54*, 99-134.
14. Li, S.M.; Rashkov, I.; Espartero, J.L.; Manolova, N.; Vert, M. Synthesis, characterization, and hydrolytic degradation of PLA/PEO/PLA triblock copolymers with long poly(L-lactic acid) blocks. *Macromolecules* **1996**, *29*, 57-62.
15. Rashkov, I.; Manolova, N.; Li, S.M.; Espartero, J.L.; Vert, M. Synthesis, characterization, and hydrolytic degradation of PLA/PEO/PLA triblock copolymers with short poly(L-lactic acid) chains. *Macromolecules* **1996**, *29*, 50-56.
16. Li, Y.X.; Kissel, T. Synthesis and properties of biodegradable ABA triblock copolymers consisting of poly(L-lactic acid) or poly(L-lactic-*co*-glycolic acid) a-blocks attached to central poly(oxyethylene) B-blocks. *J. Control. Release* **1993**, *27*, 247-257.
17. Li, Y.X.; Volland, C.; Kissel, T. In-vitro degradation and bovine serum albumin release of the ABA triblock copolymers consisting of poly(L(+)-lactic acid), or poly(L(+)-lactic acid-*co*-glycolic acid) A-blocks attached to central poly(oxyethylene) B-blocks. *J. Control. Release* **1994**, *32*, 121-128.
18. Saito, N.; Okada, T.; Horiuchi, H.; Murakami, N.; Takahashi, J.; Nawata, M.; Ota, H.; Nozaki, K.; Takaoka, K. A biodegradable polymer as a cytokine delivery system for inducing bone formation. *Nat. Biotechnol.* **2001**, *19*, 332-335.
19. Molina, I.; Li, S.M.; Martinez, M.B.; Vert, M. Protein release from physically crosslinked hydrogels of the PLA/PEO/PLA triblock copolymer-type. *Biomaterials* **2001**, *22*, 363-369.
20. Jeong, B.; Bae, Y.H.; Lee, D.S.; Kim, S.W. Biodegradable block copolymers as injectable drug-delivery systems. *Nature* **1997**, *388*, 860-862.

21. Kwon, K.W.; Park, M.J.; Bae, Y.H.; Kim, H.D.; Char, K. Gelation behavior of PEO-PLGA-PEO triblock copolymers in water. *Polymer* **2002**, *43*, 3353-3358.
22. Chen, X.H.; McCarthy, S.P.; Gross, R.A. Synthesis and characterization of L-lactide-ethylene oxide multiblock copolymers. *Macromolecules* **1997**, *30*, 4295-4301.
23. Liu, L.; Li, C.X.; Liu, X.H.; He, B.L. Micellar formation in aqueous milieu from biodegradable triblock copolymer polylactide/poly (ethylene glycol)/polylactide. *Polym. J.* **1999**, *31*, 845-850.
24. Metters, A.T.; Anseth, K.S.; Bowman, C.N. Fundamental studies of a novel, biodegradable PEG-b-PLA hydrogel. *Polymer* **2000**, *41*, 3993-4004.
25. Jeong, B.; Kibbey, M.R.; Birnbaum, J.C.; Won, Y.Y.; Gutowska, A. Thermogelling biodegradable polymers with hydrophilic backbones: PEG-g-PLGA. *Macromolecules* **2000**, *33*, 8317-8322.
26. Sawhney, A.S.; Pathak, C.P.; Hubbell, J.A. Bioerodible hydrogels based on photopolymerized poly(ethylene glycol)-co-poly(α-hydroxy acid) diacrylate macromers. *Macromolecules* **1993**, *26*, 581-587.
27. West, J.L.; Hubbell, J.A. Photopolymerized hydrogel materials for drug delivery applications. *Reactive Polym.* **1995**, *25*, 139-147.
28. Peppas, N.A. Fundamentals. Hydrogels in Medicine and Pharmacy, Vol. 1; CRC Press, Boca Raton, FL, 1986.
29. Aamer, K.A.; Sardinha, H.; Bhatia, S.R.; Tew, G.N. Rheological studies of PLLA-PEO-PLLA triblock copolymer hydrogels. *Biomaterials* **2004**, *25*, 1087-1093.
30. Fujiwara, T.; Mukose, T.; Yamaoka, T.; Yamane, H.; Sakurai, S.; Kimura, Y. Novel thermo-responsive formation of a hydrogel by stereo- complexation between PLLA-PEG-PLLA and PDLA-PEG-PDLA block copolymers. *Macromol. Biosci.* **2001**, *1*, 204-208.
31. Li, S.; Vert, M. Synthesis, characterization, and stereocomplex-induced gelation of block copolymers prepared by ring-opening polymerization of L(D)-lactide in the presence of poly(ethylene glycol). *Macromolecules* **2003**, *36*, 8008-8014.
32. Stockwell, R; Meachim, G. In *Adult Articular Cartilage*; Medical, P., Ed.: London, 1979.
33. Frank, E.H.; Grodzinsky, A.J. Cartilage electromechanics-II. A continuum model of cartilage electrokinetics and correlation with experiments. *J Biomech Eng.* **1987**, *20*, 629-639.
34. Yu, Q.L.; Zhou, J.B.; Fung, Y.C. Neutral Axis Location in Bending and Young's Modulus of Different Layers of Arterial Wall. *Am. J. Physio.* **1993**, *265*, H52-H60.

35. Carter, FJ; Frank, TG; Davies, PJ; McLean, D; Cuschieri, A. Measurements and modelling of the compliance of human and porcine organs. *Med. Image Analysis* **2001**, *5*, 231-236.
36. Erkamp, R.Q.; Wiggins, P.; Skovoroda, A.R.; Emelianov, S.Y.; O'Donnell, M. Measuring the Elastic Modulus of Small Tissue Samples. *Ultrasonic Imaging* **1998**, *20*, 17-28.
37. Hutmacher, D.W. Scaffold design and fabrication technologies for engineering tissues: State of the art and future perspectives. *J Biomater. Sci. Polym. Ed.* **2001**, *12*, 107-124.
38. Pham, Q.T.; Russel, W.B.; Thibeault, J.C.; Lau, W. Micellar solutions of associative triblock copolymers: The relationship between structure and rheology. *Macromolecules* **1999**, *32*, 5139-5146.
39. Tae, G.; Kornfield, J.A., Hubbell, J.A., Lal, J. Ordering Transitions of Fluoroalkyl-Ended Poly(ethylene glycol): Rheology and SANS. *Macromolecules* **2002**, *35*, 4448-4457.
40. Tae, G.; Kornfield, J.A.; Hubbell, JA; Johannsmann, D.; Hogen-Esch, T.E. Hydrogels with controlled, surface erosion characteristics from self-assembly of fluoroalkyl-ended poly(ethylene glycol). *Macromolecules* **2001**, *34*, 6409-6419.
41. Lee, H.T., Lee, D.S. Thermoresponsive phase transitions of PLA-block-PEO-block-PLA triblock stereo-copolymers in aqueous solution. *Macromol. Res.* **2002**, *10*, 359-364.
42. Wanka, G.; Hoffmann, H.; Ulbricht, W. Phase diagrams and aggregation behavior of poly(oxyethylene)-poly(oxypropylene)-poly (oxyethylene) triblock copolymers in aqueous solutions. *Macromolecules* **1994**, *27*, 4145-4159.
43. Jeong, B.; Lee, D.S.; Shon, J.; Bae, Y.H.; Kim, S.W. Thermoreversible gelation of poly(ethylene oxide) biodegradable polyester block copolymers. *J. Poly. Sci.: Part A: Poly. Chem.* **1999**, *37*, 751-760.
44. Choi, S.W.; Choi, S.Y.; Jeong, B.; Kim, S.W.; Lee, D.S. Thermoreversible gelation of poly(ethylene oxide) biodegradable polyester block copolymers II. *J. Poly. Sci.: Part A: Poly. Chem.* **1999**, *37*, 2207-2218.
45. Zentner, G.M.; Rathi, R.; Shih, C.; McRea, J.C.; Seo, M.H.; Oh, H.; Rhee, B.G.; Mestecky, J.; Moldoveanu, Z.; Morgan, M.; Weitman, S. Biodegradable block copolymers for delivery of proteins and water-insoluble drugs. *J. Control. Release* **2001**, *72*, 203-215.
46. Winter, H.H.; Chambon, F. Analysis of the linear viscoelasticity of a cross-linking polymer at the gel point. *J. Rheol.* **1986**, *30*, 367-382.
47. Flory, P.J. Molecular size distribution in three dimensional polymers. I. Gelation. *J. Am. Chem. Soc.* **1941**, *63*, 3083-3090.

48. de Gennes, P.G. Relation Between Percolation Theory and Elasticity of Gels. *J. Phys. Lett.* **1976**, *37*, L1-L2.
49. Rubinstein, M.; Colby, R.H.; Gilmor, J.R. Dynamic scaling for polymer gelation. *Polymer Preprints* **1989**, *30*, 81-82.
50. Martin, J.E.; Adolf, D. The sol-gel transition in chemical gels. *Ann. Rev. Phys. Chem.* **1991**, *42*, 311-339.
51. Kantor, Y.; Webman, I. Elastic properties of random percolating systems. *Phys. Rev. Lett.* **1984**, *52*, 1891-1894.
52. Daoud, M. Viscoelasticity near the sol-gel transition. *Macromolecules* **2000**, *33*, 3019-3022.
53. Muthukumar, M. Screening effect on viscoelasticity near the gel point. *Macromolecules* **1989**, *22*, 4656-4658.
54. Scanlan, J.A.; Winter, H.H. Composition dependence of the viscoelasticity of end-linked poly(dimethylsiloxane) at the gel point. *Macromolecules* **1991**, *24*, 47-54.
55. Axelos, M.A.V.; Kolb, M. Experimental evidence for scalar percolation theory. *Phys. Rev. Lett.* **1990**, *64*, 1457-1460.
56. Matsumoto, T.; Kawai, M.; Masuda, T. Viscoelastic and SAXS investigation of fractal structure near the gel point in alginate aqueous systems. *Macromolecules* **1992**, *25*, 5430-5433.
57. Yu, J.M.; Dubois, P.; Teyssie, P.; Jerome, R.; Blacher, S.; Brouers, F.; L'Homme, G. Triblock copolymer based thermoreversible gels. 2. Analysis of the sol-gel transition. *Macromolecules* **1996**, *29*, 5384-5391.
58. Richtering, H..W.; Gagnon, K.D.; Lenz, R.W.; Fuller, R.C.; Winter, H.H. Physical gelation of a bacterial thermoplastic elastomer. *Macromolecules* **1992**, *25*, 2429-2433.
59. Lin, Y.G.; Mallin, D.T.; Chien, J.C.W.; Winter, H.H. Dynamic mechanical measurement of crystallization-induced gelation in thermoplastic elastomeric poly(propylene). *Macromolecules* **1991**, *24*, 850-854.

Chapter 8

Equilibrium and Kinetics of Drying and Swelling of Poloxamer Hydrogels

Zhiyong Gu and Paschalis Alexandridis*

Department of Chemical and Biological Engineering, University at Buffalo, The State University of New York, Buffalo, NY 14260–4200
***Corresponding author: palexand@eng.buffalo.edu**

Both equilibrium and transport properties are important for the characterization and utilization of Poloxamer, poly(ethylene oxide)-poly(propylene oxide), (PEO-PPO) block copolymers. We report here on equilibrium properties such as drying/swelling isotherm, osmotic pressure and water activity for representative Poloxamers in the presence of water (selective solvent for PEO). We have found the osmotic pressure to increase exponentially with increasing Poloxamer concentration in water, and the intermolecular interactions in Poloxamer-water systems to occur at different hydration levels: Poloxamer micelles, PEO coil, and PEO segment. The interaction parameter (χ_{12} between Poloxamer and water (obtained by fitting the water activity data to the Flory-Huggins equation) is above ½ and increases further with Poloxamer concentration. We also report here on the kinetics of drying and swelling of Poloxamer hydrogels when exposed to air of fixed relative humidity. We have found swelling to be a diffusion-limited process, while drying is mainly evaporation-limited. We have used a diffusion model to fit the water loss or water gain as a function of time and to extract useful parameters such as water diffusion coefficients in the hydrogel.

Introduction

Poloxamers (or Pluronics) are nonionic amphiphilic block copolymers that consist of poly(ethylene oxide) and poly(propylene oxide) and can form via self-assembly a variety of ordered, lyotropic liquid crystalline structures, e.g., cubic, hexagonal, or lamellar, in the presence of selective solvents such as water *(1-3)*. Poloxamers find numerous applications in the pharmaceutical and bio-medical fields, for example, in drug formulation, drug delivery, and biomaterials *(4,5)*.

Both equilibrium and transport properties are important factors for the characterization and utilization of Poloxamers *(6,7)*. The solvent activity is important in the modeling (and prediction) of thermodynamic and transport properties of Poloxamer–solvent systems. The osmotic pressure can reveal interactions at the molecular level and aid in the understanding of the stability of the ordered structures formed in Poloxamer hydrogels. Water sorption/desorption and transport in Poloxamer hydrogels are very relevant to the development and design of drug delivery vehicles *(7,8)*, and to the understanding of block copolymer dissolution *(9,10)*. The transport of water (due to hydration or dehydration) and/or solutes in Poloxamers may cause changes in the ordered structures formed. At the same time, these ordered structures may affect the solvent and/or solute transport in Poloxamer hydrogels *(9,11)*. We have an ongoing interest and research activity in our group on both the equilibrium and transport properties of block copolymers.

We recently studied the osmotic pressure, water activity, and intermolecular interactions in Poloxamer hydrogels *(6,12)*. We also examined the sorption (swelling) and desorption (drying) of water vapor and the transport of water in Poloxamers *(13-15)*. Here we highlight the salient information that we have obtained on the drying/swelling isotherm, osmotic pressure and water activity in Poloxamer solutions and hydrogels at different hydration levels. We also report on the kinetics of drying and swelling of representative Poloxamer hydrogels when exposed to various air relative humidity conditions in the range between 11-97%. A water transport model is presented that fits satisfactorily the experimental data and allows us to extract useful parameters.

Materials and Methods

Poloxamer 184 (Pluronic L64), Poloxamer 335 (Pluronic P105), and Poloxamer 407 (Pluronic F127) poly(ethylene oxide)–*block*–poly(propylene oxide)–*block*–poly(ethylene oxide) copolymers were obtained from BASF Corporation. They have nominal molecular weights of 2,900; 6,500 and 12,600; and PEO contents of 40, 50 and 70 wt%, respectively. Dextran T500 was purchased from Amersham Pharmacia Biotech AB. Poly(ethylene glycol)s (PEG) 4,000 and 20,000 were obtained from Fluka. All salts used are analytical grade. Millipore-filtered water was used for all sample preparations.

Constant osmotic pressure was generated either by polymer (PEG20,000 or dextran T500) aqueous solutions, or by saturated aqueous salt solutions (constant water vapor pressure, corresponding to very high osmotic pressure) *(12)*. Detailed information about the polymers and salts used, the methodology to generate and maintain constant vapor pressure, and sample preparation and handling, can be found elsewhere *(12,13)*. The measurements of water loss or gain as a function of time in Poloxamers *(13-15)*, and the osmotic stress method *(16)* and its application on Poloxamer hydrogels have been reported recently *(12)*.

Results and Discussion

Drying/Swelling Isotherm for Poloxamers and Polyethylene Glycols

Poloxamers can form gels with a variety of ordered structures (micellar cubic, hexagonal, or lamellar) in water *(1-4,17)*. In our study, the Poloxamer hydrogel films were equilibrated with air of different relative humidity (RH) (correspondingly, different water activities). At equilibrium, the chemical potential of water in the Poloxamer hydrogel is equal to that in the air. The equilibrium concentrations of Poloxamers 184, 335, and 407, and PEGs 4,000 and 20,000 homopolymers have been determined at various air RH conditions in the range between 11-97%. Because the PPO blocks in Poloxamer–water systems will try to minimize their contact with water, it is reasonable to assume that the interactions between the PEO blocks and water are the ones that primarily contribute to the intermolecular interactions between Poloxamer and water *(12)*. Based on this supposition, the equilibrium PEO concentrations (PEO wt%*, defined by Equation 1) for all the Poloxamers and PEGs examined are plotted in Figure 1 as a function of the RH:

$$PEO\ wt\%^* = \frac{(100 - H_2O\ wt\%) \times PEO\%}{(100 - H_2O\ wt\%) \times PEO\% + H_2Owt\%} \qquad (1)$$

where PEO% is the PEO content in Poloxamers and PEGs, e.g., 50% for Poloxamer 335, 100% for PEGs.

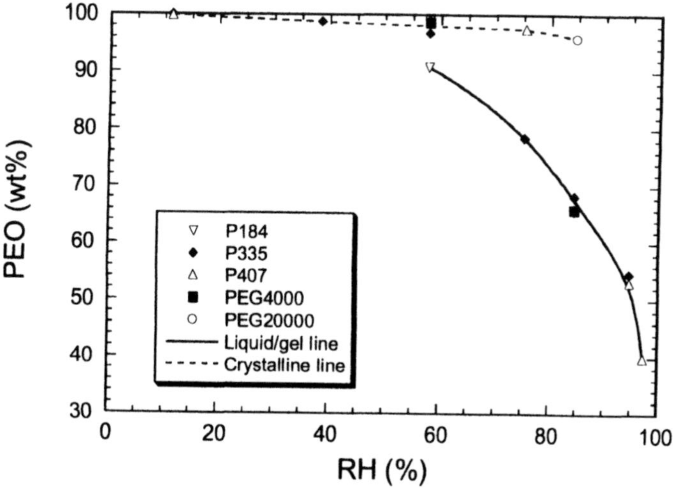

Figure 1. Drying/swelling isotherm for Poloxamer block copolymers and PEG homopolymers at 24 °C. The PEO concentration is defined by Equation 1.

Two different lines (trends) can be discerned in Figure 1: one is the liquid or gel line, where the equilibrium concentrations that fall on this line correspond to solution or lyotropic liquid crystalline phases. The other line is the crystalline (or semicrystalline) line, where the equilibrium concentrations correspond to semicrystalline states formed by PEO *(12)*. For example, Poloxamer 184 is liquid in the neat form and forms either a solution or gel when mixed with water, thus the equilibrium concentrations of Poloxamer 184 will fall on the liquid/gel line, no matter what the RH is. For Poloxamer 407, when the RH is high (≥85%), the equilibrium state is a lyotropic liquid crystalline hydrogel *(18,19)* and the equilibrium concentrations fall on the liquid/gel line; however, for RH <75%, Poloxamer 407 contains semicrystalline domains *(18,19)*, therefore the equilibrium concentrations will be located on the crystalline line. These two lines capture the PEO coil interactions in the lyotropic liquid crystalline region and the PEO segment interactions in the semicrystalline region, as discussed in the following section. The features of Figure 1 can be used to correlate qualitatively the water content of Poloxamer or PEG systems in equilibrium with water vapor with the polymer microstructure at these conditions. A quantitative model *(20)* is needed in order to predict the equilibrium self-assembled structure of Poloxamers at different water activities.

Osmotic Pressure of Poloxamer Solutions and Hydrogels

Osmotic pressure is an intrinsic property of solutions and gels and is a manifestation of intermolecular interactions. In our study, the osmotic pressure of Poloxamer solutions and hydrogels was obtained by the osmotic stress method, where the Poloxamer samples were equilibrated with either air of constant RH (as discussed in the previous section) or aqueous polymer solutions (such as dextran or PEG) through a semipermeable membrane *(12)*. Osmotic pressure (π), solvent activity (a), and solvent chemical potential ($\Delta\mu$) are related by Equation 2 *(16)*:

$$\Delta\mu = -\pi v = RT \ln a \tag{2}$$

where v is the molar volume of solvent (water), R is the gas constant, and T is the absolute temperature. In the case of water vapor in the air, the water activity is described by Equation 3:

$$a = \frac{p}{p_0} = \frac{RH}{100} \tag{3}$$

where p is the water vapor pressure in the air and p_0 is the saturated water vapor pressure at the same temperature *(16)*.

The osmotic pressure values obtained range from 5×10^3 to 3×10^8 Pa (5 orders of magnitude) in the Poloxamer concentration range between 6–99.9 wt%, and increase exponentially with an increase of Poloxamer concentration (see Figure 2). As discussed in the previous section, since PPO blocks try to avoid contact with water, it is reasonable to assume that the interactions between the PEO block and water primarily contribute to the osmotic pressure of Poloxamer solutions and hydrogels. The osmotic pressure is thus plotted in Figure 2 against the PEO concentration (defined by Equation 1).

The osmotic pressure of Poloxamer 335 solutions and hydrogels is almost the same as that of the Poloxamer 407 system in the PEO concentration range between 30–80 wt%. This observation supports the notion that the molecular interactions originate mainly from the interaction between PEO and water. Below about 30 wt% PEO, the osmotic pressure of the Poloxamer 407 system is lower than that of the Poloxamer 335 system, which points to the importance of particles (micelles) formed by block copolymers, rather than the number of PEO segments. Above about 95 wt% PEO, the osmotic pressure increases much more compared to the lower concentration region, which reveals stronger interactions.

The different slopes in Figure 2 indicate that different molecular interactions in Poloxamer–water systems occur at different hydration levels. The osmotic stress measurements obtained above give the osmotic pressure as a function of Poloxamer concentration. Small-angle x-ray scattering (SAXS) measurements give information on the separation distance of the ordered (self-assembled) structures as a function of Poloxamer concentration. By combining the osmotic pressure and separation distance data at a given Poloxamer concentration, we generated a force (pressure) *versus* distance curve that reveals the interactions at different ordered structure levels. In the concentration range between 40–80% PEO (lyotropic liquid crystalline region), the decay length (the parameter that describes how the osmotic pressure changes over distance) is comparable to the radius of gyration of the PEO coil, indicating that the interactions occur at the PEO coil level. In the concentration range above 95% PEO (semicrystalline region), the decay length is comparable to the length of a PEO segment, indicating that the interactions occur at the polymer segment level *(12)*.

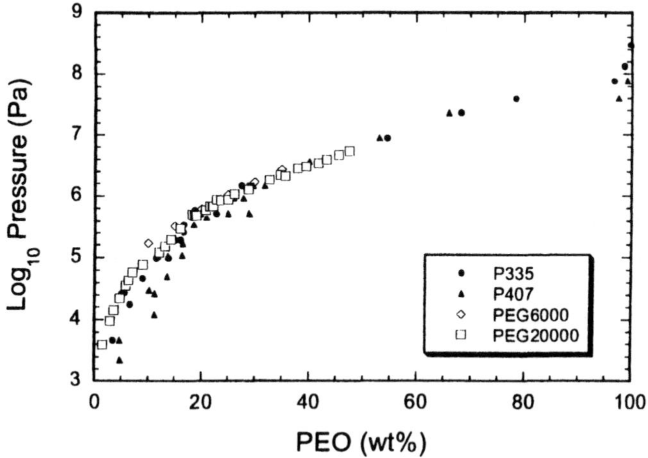

Figure 2. Osmotic pressure of Poloxamer solutions and hydrogels plotted as a function of PEO concentration (defined by Equation 1) at 24 °C. Also shown in the figure are osmotic pressure data for PEG 6,000 and 20,000 homopolymers.

Water Activity and Flory-Huggins Interaction Parameter (χ)

Solvent activity is important in the modeling of the thermodynamic properties of polymer–solvent systems *(21)*. Water activity in Poloxamer solutions and hydrogels can be obtained by Equations 2 and 3 from the osmotic pressure data discussed in the previous sections. In the context of the Flory-Huggins theory that is widely used for describing the thermodynamics of polymer solutions, the activity of the solvent (water) in a polymer solution can be obtained from the following equation:

$$\ln(a_1) = \ln(1-\varphi_2) + (1-\frac{1}{r_2})\varphi_2 + \chi_{12}\varphi_2^{\ 2} \quad (4)$$

where φ_2 is the polymer volume fraction, r_2 is the number of polymer segments, and χ_{12} the polymer–solvent interaction parameter.

The χ_{12} parameter describes interactions between the polymer and the solvent. When $\chi_{12} < ½$, the solvent is called "good" for the polymer; when $\chi_{12} >$ ½, the solvent is called "bad" and the polymer is relatively insoluble. Typically only one parameter (χ_{12}) is used to describe the solvent activity in the polymer solutions over the whole concentration range. Figure 3 shows the experimental data of water activity (left axis) in Poloxamer-water systems fitted with Equation 4 using $\chi_{12} = 0.55$. The fit compares well with the experimental data up to $\varphi_{PEO} = 0.65$. The use of the PEO volume fraction (φ_{PEO}, similar definition as that of Equation 1) instead of the Poloxamer volume fraction improved the fitting ability of the Flory-Huggins equation.

Since a constant χ_{12} was not adequate to fit the water activity in Poloxamer–water systems over the whole concentration range *(6,12)*, the interaction parameter may be better represented as a function of Poloxamer concentration. The interaction parameter values calculated on the basis of Equation 4 are plotted *versus* φ_{PEO} in Figure 3 (right axis). The χ_{12} values from the two Poloxamers considered here overlap when plotted against the PEO volume fraction. χ_{12} is higher than 1/2 over the concentration range examined, indicating that water is not a good solvent for Poloxamers. When the PEO volume fraction is above 0.6, χ_{12} increases rapidly. This indicates that water becomes progressively a worse solvent for Poloxamers. However, the assumptions of Flory-Huggins theory are not valid at high Poloxamer concentration, where a variety of ordered structures form, thus the interaction parameters reported in Figure 3 at high Poloxamer concentration are to be considered as apparent values. The increase of χ_{12} with an increase of Poloxamer concentration is similar to the trend of χ_{12} values reported between PPO-PEO-PPO block copolymers and water *(22)*.

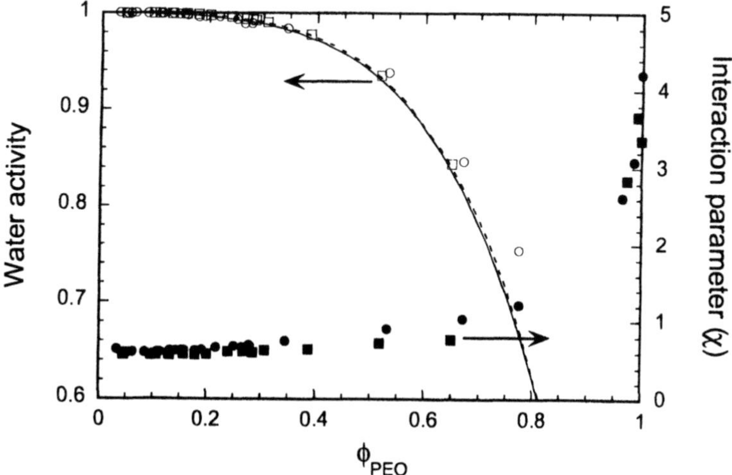

Figure 3. Water activity (left axis) and Flory-Huggins interaction parameter (right axis) as a function of PEO volume fraction for Poloxamer–water systems at 24 °C. ○ and ● – Poloxamer 335; □ and ■ – Poloxamer 407. The lines represent the fit to Equation 4 with χ_{12} = 0.55: solid line – Poloxamer 335; dashed line – Poloxamer 407.

Kinetics of Drying and Swelling of Poloxamer Hydrogels

As discussed in the section "Drying/Swelling Isotherm", when the water chemical potential in the Poloxamer hydrogel is equal to that in the air, equilibrium is achieved. However, when the water chemical potential in the Poloxamer hydrogels is different than that in the air, either drying or swelling will occur. The kinetics of drying and swelling of Poloxamer hydrogels exposed to air of fixed RH values have been determined. The initial water content in the Poloxamer hydrogel is 70 wt% in the drying studies, and 0 in the swelling studies. Figure 4 shows the evolution of Poloxamer 335 concentration as a function of time at 38% and 85% RH. The overlap of the equilibrium concentrations obtained from both the drying and swelling experiments confirms the attainment of equilibrium. In the drying process the percentage of water loss in Poloxamers (see Figure 5a) is initially a linear function of time, indicating a evaporation-limited process. In the swelling process, the percentage of water gain in Poloxamers (see Figure 5b) is initially a linear function of square root of time, indicating a diffusion-limited process *(13,14,23)*. Drying and swelling results for Poloxamer hydrogels have also been obtained at other RH conditions *(13-15)*.

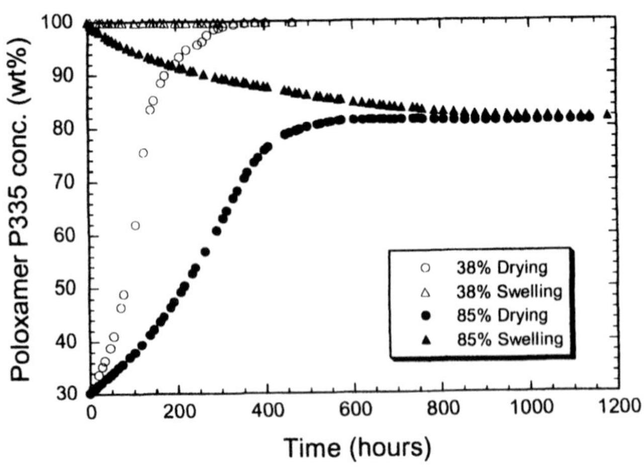

Figure 4. Poloxamer 335 concentration (wt%) versus time at 38% and 85% RH at 24 °C, when the initial film thickness is 5 mm. The initial water concentration was 70 wt% in the drying experiments, while neat Poloxamer was used for the swelling experiments.

The diffusion of water in the Poloxamer hydrogel film can be described by Fick's second law:

$$\frac{\partial C}{\partial t} = \frac{\partial}{\partial x}\left(D\frac{\partial C}{\partial x}\right) \qquad (5)$$

where C is the water concentration; t is time; x is the water diffusion direction; and D is the water diffusion coefficient, normally a function of water concentration *(24)*. The initial condition (C_0) is either 70 wt% water (in the case of drying) or 0% water (in the case of swelling):

$$C = C_0 \quad x = 0 \rightarrow L_0 \qquad (6)$$

where L_0 is the initial film thickness.

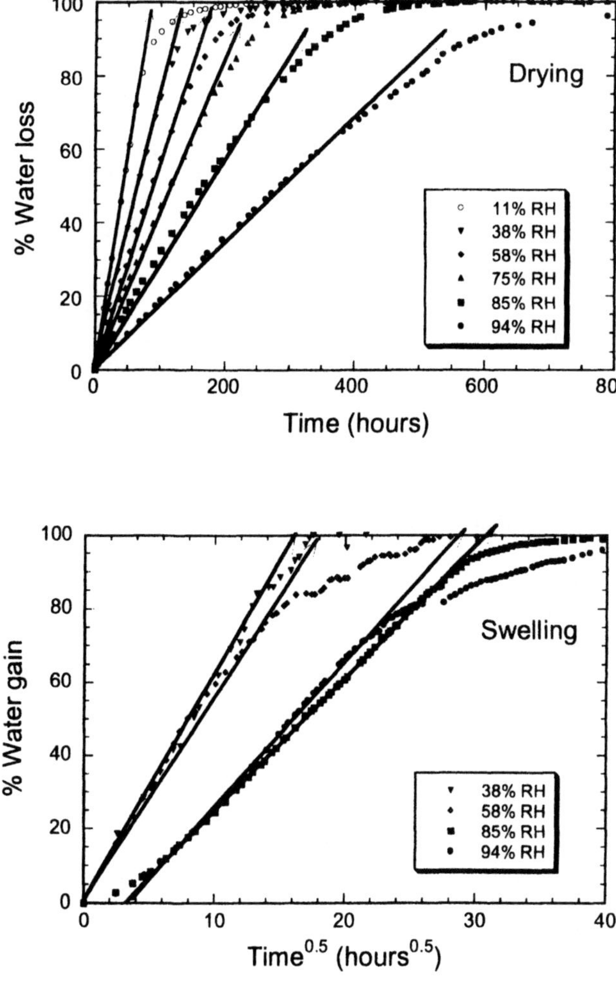

Figure 5. Percentage of water loss (above) and water gain (below) in Poloxamer 335 hydrogels at different air RH conditions at 24 °C. The lines are meant to guide the eye.

The boundary conditions at the film bottom and surface are described by Equations 7 and 8, respectively:

$$D\frac{\partial C}{\partial x} = 0, \quad x = 0 \tag{7}$$

$$-D\frac{\partial C}{\partial x} - C\frac{dh(t)}{dt} = \alpha(C - C_{\infty}), \quad x = h(t) \tag{8}$$

where h(t) is the Poloxamer film thickness as a function of time; C_{∞} is the water concentration that is in equilibrium with the air RH; α is a proportionality constant that represents the product of k_G and m. k_G is the mass transfer coefficient of water between the film surface and the bulk air; m is a proportionality constant that describes the relationship between the water concentration at the film surface and the water vapor pressure in the bulk air *(14,15)*.

The water loss or gain and the film thickness can be obtained by a polymer mass balance in the hydrogel film. The water diffusion coefficient is described by an exponential function. $D=D_0\exp(-k/C)$ (D_0 and k are constants) was used in the fits for the drying experiments and $D=D_0\exp(kC)$ for the swelling experiments. The above governing equation (Equation 5), with initial and boundary conditions, have been solved numerically and used to fit the experimental data *(14,15,25)*. The fitted values for water loss or gain, together with the experimental results, are shown in Figures 6a and 6b for Poloxamer 335 at 24 °C. The fitted values compare very well with the experimental water loss or water gain values. The model can also fit well the film thickness change during the drying or sweeling processes. Useful information, such as water diffusion coefficient (D) values in Poloxamer hydrogels, and proportionality constant α values, can be extracted from the fits *(14,15,24)*.

The α values used in the fittings are very similar to the evaporation rates of pure water at the same conditions *(14)*. At high water content (i.e., >30 wt%, corresponding to the micellar cubic or hexagonal phase of Poloxamer 335), the D values obtained are in the same order of magnitude as values reported in the literature (in the order of 10^{-10} m^2/s) *(14)*, while larger differences are observed between the values obtained here (in the order of 10^{-11} m^2/s for the lamellar or semicrystalline phase formed by Poloxamer 335) and those reported in the literature (in the order of 10^{-12} m^2/s for the lamellar phase formed by Poloxamer 335) when the water content becomes low *(15)*. Further parametric analysis confirmed that in the conditions examined here, the swelling of Poloxamers is a diffusion-limited process, while the drying of Poloxamer hydrogels is mainly evaporation-limited *(14,15)*.

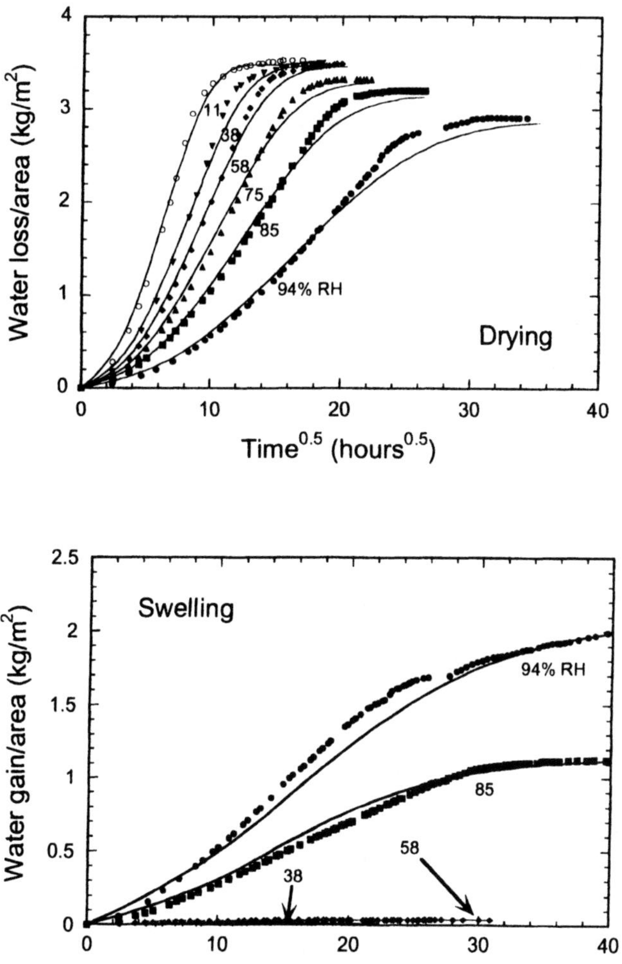

Figure 6. Water loss (above) and water gain (below) per surface area versus time for Poloxamer 335 at different air RH conditions (indicated next to each curve) at 24 °C. The lines represent the fitted values from the diffusion model.

Conclusions

Both equilibrium properties (osmotic pressure and water activity) and transport properties of Poloxamer hydrogels are reported here. Water vapor sorption (swelling) or desorption (drying) in Poloxamers involves the successive transformation of ordered structures (e.g., micellar cubic, hexagonal, lamellar). The ordered structures formed at low water contents decrease greatly the water diffusion in Poloxamer hydrogels.

The drying/swelling isotherm has been obtained for Poloxamers exposed to air of known relative humidity and two distinct regions can be discerned: liquid/gel and crystalline. The osmotic pressure and water activity of Poloxamers have been determined by the osmotic stress method. The osmotic pressure increases exponentially with an increase of Poloxamer concentration, and the intermolecular interactions in Poloxamer-water systems occur at different hydration levels. Below 30 wt% PEO, Poloxamer micelles play an important role in determining the osmotic pressure. In the concentration range between 30– 80% PEO, the interactions are at PEO coil level. At low hydration (>95% PEO), the interactions are stronger and occur at the polymer segment level. The Flory-Huggins interaction parameter χ_{12} between Poloxamers and water has been determined as a function of Poloxamer volume fraction and is above ½. χ_{12} increases rapidly when the PEO volume fraction is above 0.6.

The kinetics of drying and swelling of Poloxamer hydrogels have been investigated at various air relative humidity conditions ranging from 11-97%. Swelling of Poloxamers is a diffusion-limited process, while drying is mainly evaporation-limited in the conditions examined. A model for one-dimensional water diffusion, which accounts for a variable film thickness and water diffusion coefficient, is used to successfully fit the experimental results at different air conditions and to extract useful parameters such as the water diffusion coefficient (D).

Acknowledgements

We gratefully acknowledge financial support from the National Science Foundation (CTS-9875848/CAREER and CTS-0124848/TSE).

References

1. Alexandridis, P. Poly(ethylene oxide)-pPoly(propylene oxide) Block Copolymer Surfactants. *Curr. Opin. Colloid Interface Sci.* **1997**, *2*, 478-489.

2. Alexandridis, P.; Spontak, R.J. Solvent-Regulated Ordering in Block Copolymers. *Curr. Opin. Colloid Interface Sci.* **1999**, *4*, 130-139.
3. Alexandridis, P. Small-Angle Scattering Characterization of Block Copolymer Micelles and Lyotropic Liquid Crystals. *ACS Symp. Ser.* **2003**, *861*, 60-80.
4. Alexandridis, P.; Lindman, B. (Editors) *Amphiphilic Block Copolymers: Self-Assembly and Applications*, Elsevier Science B.V.: Amsterdam, 2000.
5. Yang, L.; Alexandridis, P. Physicochemical Aspects of Drug Delivery and Release from Polymer-Based Colloids. *Curr. Opin. Colloid Interface Sci.* **2000**, *5*, 132-143.
6. Gu, Z.; Alexandridis, P. Water Activity in Poly(oxyethylene)-*b*-Poly(oxypropylene) (Poloxamer) Aqueous Solutions and Gels. *Polym. Mat. Sci. Eng.* **2003**, *89*, 238-239.
7. Alexandridis, P.; Gu, Z. Equilibrium Swelling and Drying Kinetics of Poloxamer Hydrogel Films. *Polym. Prepr. (Am. Chem. Soc., Div. Polym. Chem.)* **2002**, *43*, 627-628.
8. Yang, L.; Alexandridis, P. Controlled Release from Ordered Microstructures Formed by Poloxamer Block Copolymers. *ACS Symp. Ser.* **2000**, *752*, 364-374.
9. Yang, L.; Alexandridis, P. Mass Transport in Ordered Microstructures Formed by Block Copolymers: Ramifications for Controlled Release Applications. *Polym. Prepr. (Am. Chem. Soc., Div. Polym. Chem.)* **1999**, *40*, 349-350.
10. Miller-Chou, B.A.; Koenig, J.L. A Review of Polymer Dissolution. *Prog. Polym. Sci.* **2003**, *28*, 1223-1270.
11. Walderhaug, H. PFG-NMR Study of Polymer and Solubilizate Dynamics in Aqueous Isotropic Mesophases of Some Poloxamers. *J. Phys. Chem. B* **1999**, *103*, 3352-3357.
12. Gu, Z.; Alexandridis, P. Osmotic Stress Measurements of Intermolecular Forces in Ordered Assemblies Formed by Solvated Block Copolymers. *Macromolecules* **2004**, *37*, 912-924.
13. Gu, Z.; Alexandridis, P. Drying of Poloxamer Hydrogel Films. *J. Pharm. Sci.* **2004**, *93*, 1454-1470.
14. Gu, Z.; Alexandridis, P. Drying of Films Formed by Ordered Poly(ethylene oxide)-Poly(propylene oxide) Block Copolymer Films. *Langmuir* **2005**, 21, 1806-1817.
15. Gu, Z.; Alexandridis, P. Sorption and Transport of Water Vapor in Amphiphilic Block Copolymer Films. *J. Disp. Sci. Tech.* **2004**, *25*, 619-629.
16. Parsegian, V.A.; Rand, R.P.; Fuller, N.L.; Rau, D.C. Osmotic Stress for the Direct Measurement of Intermolecular Forces. *Methods Enzymol.* **1986**, *127*, 400-416.

17. Alexandridis, P.; Zhou, D.L.; Khan, A. Lyotropic Liquid Crystallinity in Amphiphilic Block Copolymers: Temperature Effects on Phase Behavior and Structure for Poly(ethylene oxide)-b-Poly(propylene oxide)-*b*-Poly (ethylene oxide) Copolymers of Different Composition. *Langmuir* **1996**, *12*, 2690-2700.
18. Ivanova, R.; Lindman, B.; Alexandridis, P. Effect of Pharmaceutically Acceptable Glycols on the Stability of the Liquid Crystalline Gels Formed by Poloxamer 407 in Water. *J. Colloid Interface Sci.* **2002**, *252*, 226-235.
19. Ivanova, R.; Lindman, B.; Alexandridis, P. Evolution of the Structural Polymorphism of Pluronic F127 Poly(Ethylene Oxide)-Poly(Propylene Oxide) Block Copolymer in Ternary Systems with Water and Pharmaceutically Acceptable Organic Solvents: From 'Glycols' to 'Oils'. *Langmuir* **2000**, *16*, 9058-9069.
20. Svensson, M.; Alexandridis, P.; Linse, P. Phase Behavior and Microstructure in Binary Block Copolymer-Selective Solvent Systems: Experiments and Theory. *Macromolecules* **1999**, *32*, 637-645.
21. Zhu, H.J.; Yuen, C.M.; Grant, D.J.W. Influence of Water Activity in Organic Solvent Plus Water Mixtures on the Nature of the Crystallizing Drug Phase .1. Theophylline. *Int. J. Pharm.* **1996**, *135*, 151-160.
22. Petrik, S.; Bohdanecky, M.; Hadobas, F.; Simek, L. A Comparative Study of Sorption of Water Vapor by Oligomeric Triblock Copolymers of Ethylene and Propylene Oxides. *J. Appl. Polym. Sci.* **1991**, *42*, 1759-1765.
23. Balik, C.M. On the Extraction of Diffusion Coefficients from Gravimetric Data for Sorption of Small Molecules by Polymer Thin Films. *Macromolecules* **1996**, *29*, 3025-3029.
24. Crank, J. *The Mathematics of Diffusion*, 2nd ed., Clarendon Press: Oxford, U. K., 1975.
25. Vrentas, J.S.; Vrentas, C.M. Drying of Solvent-Coated Polymer Films. *J. Polym. Sci. B Polym. Phys.* **1994**, *32*, 187-194.

Chapter 9

Poly(Ethylene oxide) Hydrogels: Valuable Candidates for an Artificial Pancreas?

E. Alexandre[1], K. Boudjema[1,3], B. Schmitt[1,2], J. Cinqualbre[1], D. Jaeck[1], and P. J. Lutz[2,*]

[1]Fondation Transplantation, 5 Avenue Molière, F–67200 Strasbourg, France
[2]Institute Charles Sadron, 6 rue Boussingault, F–67083 Strasbourg, France
[3]Current address: CHU, F–25033 Rennes, France
[*]Corresponding author: lutz@cerbere.u-strasbg.fr

Poly(ethylene oxide) (PEO) hydrogels were synthesized directly in water or physiological medium by free radical homopolymerization of telechelic PEO macromonomers. Their ability to serve as semi-permeable, biocompatible membranes for an artificial pancreas was examined. *In vitro* tests confirmed their good biocompatibility. Both glucose and insulin diffuse through these hydrogels but the behavior of the latter is more complex. The crosslinking reaction could be extended to include direct encapsulation of biologically active materials such as hepatocytes.

Introduction

Diabetes affects at least 5% of the population in the Western industrialized countries and constitutes one of the most important chronic diseases (*1*). Type 1 diabetes mellitus, the insulin-dependent form (Insulin Dependent Diabetes Mellitus - IDDM), is caused by the immunological environment and genetic factors, leading to a destruction of the Langerhans islets in the pancreas. The four times more frequent type 2 diabetes mellitus, the insulin-independent form

(Non Insulin Dependent Diabetes Mellitus - NIDDM), is often related to diet, obesity or genetic factors *(1)*. In both cases, the metabolic effect of insulin is insufficient or absent, causing severe hyperglycemia without treatment, soon leading to death. Oral hypoglycemic drugs and a low carbohydrate diet are often sufficient to treat type 2 diabetes. However, life-long insulin substitution is necessary to treat type 1 diabetes, where insulin secretion is completely absent. Conventional insulin therapy, i.e., the administration of exogenous insulin, is efficient for treating the acute metabolic complications of IDDM, but has failed to prevent or even stabilize long-term complications. The use of transplanted exogeneic Langerhans islets, protected by a biocompatible, semi-permeable and non-degradable membrane that allows permanent and pulsed insulin secretion, would ensure prevention and even regression of the degenerative lesions, which now constitute the severity of this disease. In addition, the membrane would have to act as an efficient barrier to antibodies and lymphocytes. Several types of membranes have been studied during the last years.

The main goal of the work described here was to design implantable, biocompatible, and semi-permeable membranes based on hydrogels, and to evaluate their capacity to serve in the design of a bioartificial pancreas. First, the advantages and limits with respect to biocompatility and permeability of existing approaches will be discussed. In particular, the synthesis and some properties of macromonomer-based PEO hydrogels aimed to be used as a semi-permeable membrane will be presented. Then, selected hydrogels will be examined with respect to biocompatibility and their efficiency in glucose and insulin diffusion and compared to PEO star-hydrogels. The use of dispersed cells rather than islets, possibly improving the access of nutrients to the core, will be evaluated. Therefore, crosslinking of hydrogels directly in the presence of cells was attempted to achieve a better dispersion and accessibility of the cells.

Results and Discussion

General Remarks on Artificial Pancreas

Langerhans islets (average size from 0.05 to 0.6 mm) can be isolated from human pancreas and implanted in the liver or spleen *(2)*. However, this source of grafts is limited and implantation of allografts requires the use of anti-rejection immunosuppressant therapy. The isolation of Langerhans islets from porcine pancreas represents a promising alternative to the shortage of human grafts, but the high immunoreactivity of xenoantigens prevents the use of free grafts *(3)*. Therefore, cell encapsulation by semi-permeable, biocompatible membranes has gained increasing interest. Several approaches have been developed to design a biocompatible reservoir for Langerhans islets, aimed to be implanted as an artificial pancreas *(4)*. Microbeads, hollow fibers and vascular devices represent the most studied devices.

In the first case, islets are encapsulated into micro or macro devices (Figure 1). For example, microencapsulation of Langerhans islets into alginate-poly(L-lysine) microbeads provides efficient protection against cell-mediated immune destruction, and should allow transplantation of these capsules without immunosuppressive treatment *(5)*. While the capsules allow good oxygenation of the Langerhans islets, their poor biocompatibility as well as low chemical and mechanical stability poses problems for long-term implantation. In addition, these capsules cannot be removed easily once implanted. Various modifications of these alginate-based microbeads were synthesized, involving the inclusion of poly(ethylene glycol), (PEG), the use of crosslinkers such as carbodiimide (EDC) and glutaraldehyde (GA) in the core and onto the microcapsule membrane surface *(6)*, and the utilization of macroporous gelatin microcarriers *(7)*. Macroencapsulation was achieved with immunoisolation devices derived from the microdialysis technique *(8)*.

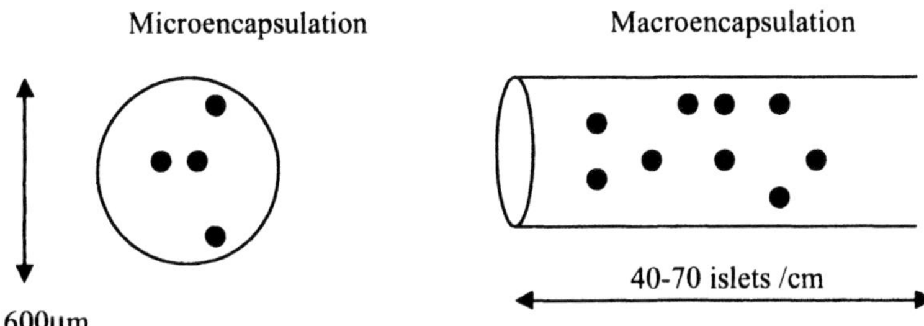

Figure 1. Schematic representation of devices used for micro and macro-encapsulation of Langerhans islets (black dots).

Hollow fibers are easy to implant and are characterized by high diffusion rates. However, as they are implanted into the peritoneal cavity, in most cases diffusion is restricted, and therefore, oxygen diffusion to the islets is poor. The artificial membrane AN69 (Hospal), for example, a synthetic copolymer of acrylonitrile and sodium methallyl sulphonate suitable for pancreatic islet encapsulation, was submitted to corona discharge to improve its insulin permeability *(9)*. Its porous structure provided good diffusion of glucose but revealed restricted insulin diffusion. However, the insulin diffusion could be improved after treatment with a copolymer or adsorption of selected proteins. In another example, porous polyamide membranes were evaluated as semi-permeable membranes for an artificial pancreas, showing most of the characteristics necessary for efficient islets encapsulation *(10)*.

Surgically implanted vascular devices present many advantages as they are in direct contact with the blood circulation, enabling rapid exchange *(11)*. In this case, the membrane is the key parameter in the efficiency of the device and must be perfectly biocompatible (Figure 2). Otherwise, proteins will rapidly grow onto the surface of the material after implantation. Depending on conformation, concentration, type, and thickness of this protein layer, the surface will be passivated or activated. If the surface is activated, all the reactions of the host immune system will take place, and within a few hours the surface will be completely covered or entrapped in a cluster of fibrin, fibrinogens, white cells and platelets. If such a surface is used, it is obvious that deposition and aggregation would dramatically decrease its diffusion properties.

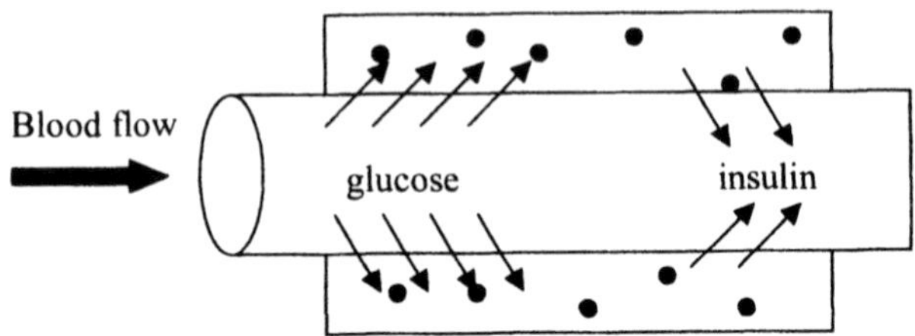

Figure 2. Schematic representation of a vascular device (black dots = Langerhans islets).

As in the case of micro or macroencapsulation, the membrane surrounding the islets has to be perfectly permeable to glucose (diameter of 3.6 Å) and to other nutrients (including oxygen) necessary for the islets. The membrane must also be permeable to insulin (diameter of 16 Å), produced by the islets after an increase in glucose blood level. The lag time between the stimulus and the flow of insulin has to be as short as possible. The islets must not be recognized by the immune system as foreign, and the membrane must be impermeable to immunoglobulins (diameter between 140 and 150 Å).

Synthesis and Characterization of Macromonomer-based PEO Hydrogels

In the present work, we have chosen to test poly(ethylene oxide), (PEO), hydrogels as possible candidates for a semi-permeable membrane. Increasing interest has been devoted over the years to the synthesis of polymeric hydrogels based on PEO chains *(12,13)*. PEO is well known to exhibit good bio-compatibility and is non-thrombogenic, and therefore, is well suited for biomedical applications *(14,15)*. Nevertheless, PEO is water-soluble and cannot be used directly in contact with blood. PEO chains for biomedical applications have either to be grafted onto surfaces or crosslinked. Among the different approaches to design PEO hydrogels, we selected those applicable directly in aqueous medium, yielding hydrogels of controllable structural parameters *(16-18)*. The polymerization of telechelic PEO macromonomers to hydrogels in water or physiological medium represents important progress in the preparation of hydrogels designed for biomedical applications *(18)*.

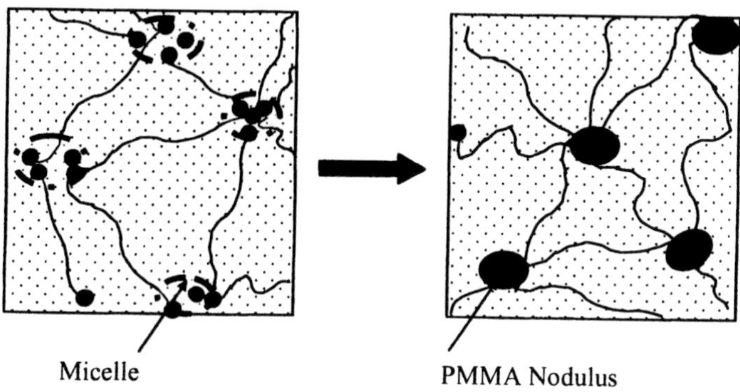

Figure 3. Schematic representation of the polymerization of methacrylate telechelic PEO macromonomers in water at 60°C before (left) and after reaction (right).

Well-defined PEO macromonomers with number average molar masses (M_n) of 3,000, 6,500, 11,500 or 15,000 gmol^{-1}, carrying polymerizable methyl-methacrylate units at both chain ends, were synthesized as described previously and directly polymerized in water to form hydrogels (Figure 3) *(19)*. Potassium peroxodisulfate was used as the initiator (1 mol% relative to the double bond content). The gel point was reached between 1 and 4 hours, dependent on the reaction temperature and the chain length of the macromonomer. After pre-paration, the gels were placed in water for swelling. Once swollen to equilibrium and free of linear unconnected chains, the gels were characterized in terms of swelling behavior and uniaxial compression modulus *(18,19)*. In most cases, gels were kept in water in the presence of 0.3 wt% sodium azide to avoid microorganism proliferation and were used as such for biomedical applications. Some characteristics of these hydrogels are given in Table I.

Table I. Preparation Conditions and Physico-chemical Characteristics of Hydrogels Obtained from PEO Macromonomers

Reference	M_n [a)] [$gmol^{-1}$]	*PEO* [b)] *[wt%]*	ε [c)] *[wt%]*	$Q_{w,\,water}$ [d)] *[%]*	E_G [e)] *[Pa]*
PEO1	3000	30	2.9	6.2	66000
PEO2	6500	30	10.0	7.2	168000
PEO3	11500	30	9.1	11.6	54500
PEO4	11500	45	11.1	11.1	80400
PEO5	11500	60	9.0	8.7	110000
PEO6	15000	30	7.1	13.6	47300

a) Number average molar mass of PEO macromonomer in [$gmol^{-1}$]
b) Concentration of PEO macromonomer crosslinked in water in [wt%]
c) Amount of extractable polymer in [wt%] after 48 hours
d) $Q_{w,\,water}$,volume equilibrium swelling degree in water in [%]
e) E_G, uniaxial compression modulus in [Pa]
Crosslinking was carried out in water at 60°C for 48 hours (initiator $K_2S_20_8$)

These results indicate that the structural parameters of macromonomer-based hydrogels can be controlled in a fashion similar to networks prepared by chain end-linking. Their swelling capacity, porosity and mechanical properties are directly related to molar mass and concentration of the precursor. In addition, these hydrogels are accessible directly in water, and therefore, no solvent exchange is necessary, which is inevitable when using the classical end-linking procedure involving isocyanate as crosslinking agent. It has to be mentioned that this strategy could be extended to the crosslinking of PEO macromonomers with PEO stars, partially functionalized with polymerizable units at the end of the branches *(20)*.

Synthesis and Characterization of Star-shaped PEO Hydrogels

In several publications, Merrill *et al.* discussed the advantages of PEO star-shaped structures compared to linear PEO chains *(21)*. These structures offer a combination of (i) high primary molar mass, (ii) non-binding character (contrary to linear PEO) and (iii) high concentration of terminal hydroxy functions. Once crosslinked via irradiation or even grafted onto surfaces, these materials revealed particularly interesting properties as blood contacting devices or as anchors to hold biologically active materials *(17,21)*. Biocompatible and porous, they should also be suitable as a semi-permeable membrane for an artificial pancreas.

For these reasons, we synthesized PEO stars *(22)* and exposed them to irradiation according to procedures discussed elsewhere *(21,23)* at different concentrations. The resulting hydrogels were characterized when swollen to equilibrium. The amount of polymer incorporated into the network increased with the initial PEO concentration and, for a given initial PEO concentration, increased with the radiation dose. Both observations can be explained easily. With increasing concentration of PEO chains, the concentration of potential chains in the network increases. A similar situation occurs with increased irradiation dose, which enhances the probability of connecting chains and increasing the density of junction points, leading to more compact materials with lower swelling degrees. It can also be anticipated that in such networks the crosslinking density should be much higher than for networks prepared from linear precursors with the same radiation dose. The star-shaped precursor contains nodules, which may contribute to the structure of the resulting network.

Biocompatibility of PEO Hydrogels

If cells adsorb onto a hydrogel surface, the adhesion will strongly reduce the flow of solutes through the membrane as well as the molecular weight cut-off of the membrane. Therefore, the *in vitro* and *in vivo* biocompatibility of our hydrogels had to be verified before tests regarding their diffusion properties could be conducted. Fibroblast adhesion experiments were performed on PEO macromonomer and star hydrogels. Human skin fibroblasts (gift from the Inserm U381 Strasbourg) were cultured on PEO hydrogel surfaces for 24 hours at 37°C in DMEM, containing 15% fetal calf serum, with a polystyrene (PS) surface as a reference. As expected, these experiments revealed strong fibroblast adsorption onto the reference PS surface. On the contrary, only a few fibroblasts were adsorbed onto the hydrogel surface. To test the *in vivo* biocompatibility, hydrogel samples were implanted into the intraperitoneal cavity or under the back of rats. Neither adsorbed cells nor fibrin were found on the PEO membranes one month and four months after implantation. Only a few macrophages were detected on the PEO hydrogel surfaces. These *in vitro* and *in vivo* experiments demonstrated good biocompatibility and hemocompatibility for this new class of PEO hydrogels. In addition, these hydrogels do not induce inflammatory reactions after intraperitoneal or subcutaneous implantation in rats, and they are non-thrombogenic after vascular implantation in pigs. Details have been reported elsewhere *(19,23)*.

Glucose and Insulin Diffusion Experiments

To reach fast diffusion of glucose and insulin through the hydrogel membranes, the pores should be interconnected and their volume fraction should be high (corresponding to high water content). In addition, the hydrophilic-lipophilic balance (HLB) of the hydrogel itself plays a key role in diffusion characteristics, especially for insulin diffusion, whose hydrophobic domains may strongly interact. Moreover, the membrane thickness should be at a minimum but, at the same time, its strength should be sufficient to resist the blood pressure. For diffusion experiments, we first had to prepare film-shaped hydrogel samples. The experimental set-up is presented in Figure 4.

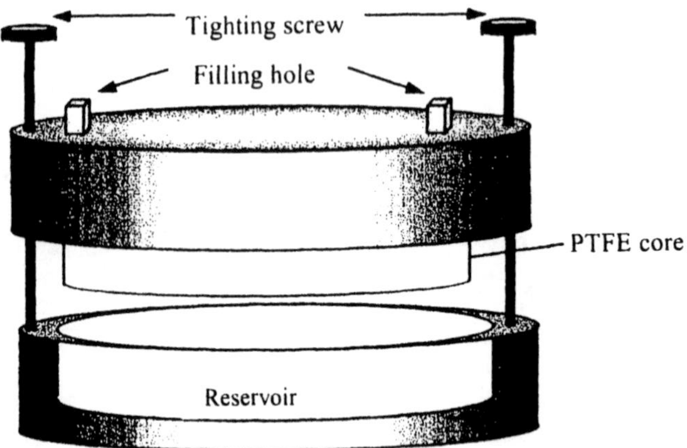

Figure 4. Experimental set-up for the preparation of PEO hydrogel films aimed to be used for diffusion studies. The distance between the reservoir and the poly(tetrafluoroethylene) (PTFE) core was 0.6 mm.

Glucose diffusion

Glucose diffusion was determined using the lag time method described by Hannoun *(24)*. From the plot of the total flux of glucose *versus* time, the lag time t_0 (i.e., the time needed by the membrane to be saturated in glucose solution) can be extrapolated. The resolution of Fick's law by Fourier analysis gives a simple relation between D_a and t_0, where $D_a = e^2 / 6t_0$ is the absolute diffusion coefficient given in [cm^2s^{-1}] and '*e*' is the thickness of the membrane in cm. Other required parameters are D_r, the relative diffusion coefficient of glucose in the gel, and D_w, the diffusion coefficient of glucose in pure water. The measured glucose diffusion data are plotted in Figure 5, together with values measured for hydrogels grafted onto PTFE surfaces *(23)*.

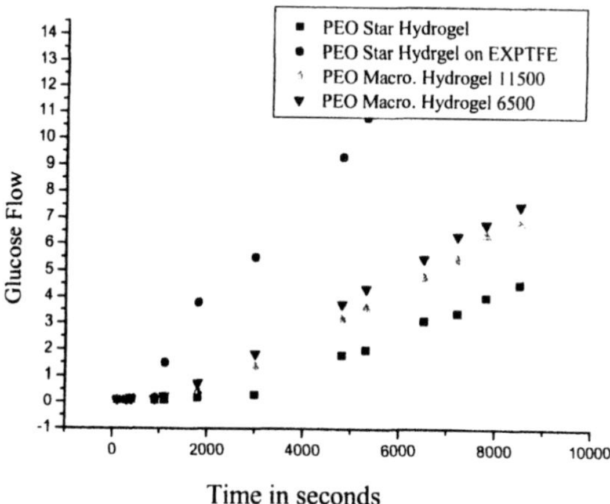

Figure 5. Comparison of glucose diffusion within PEO macromonomer-based hydrogels and PEO star-shaped hydrogels, either grafted onto EXPTFE surfaces or not grafted.

These results indicate that D_a and D_r increased with the molar mass of the macromonomer, corresponding to an improvement of the diffusion properties with increasing macromonomer molar mass. D_r was even higher for PEO star–shaped hydrogels, obtained by irradiation, which were prepared at much lower concentrations. Interestingly, the presence of a PTFE membrane, which enhanced the mechanical properties of the hydrogels, had little effect on the diffusion properties *(23)*.

Insulin diffusion

^{125}I-radiolabelled insulin was used to measure the insulin diffusion. A cylindrical-shaped macromonomer-based hydrogel, swollen to equilibrium in pure water, was immersed in a solution of ^{125}I-radiolabelled insulin of known concentration (Figure 6). The sample was periodically taken out, and the radioactivity of the gel and the solution were measured. Details of the mathematical treatment used to determine the insulin diffusion coefficent have been presented elsewhere *(26)*. The increase of radioactivity with time for a PEO macromonomer hydrogel with precursor molar mass of 11,500 $gmol^{-1}$ is shown in Figure 7.

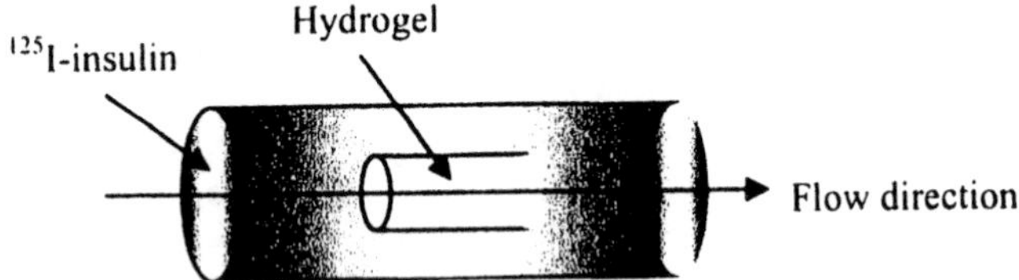

Figure 6 . Schematic representation of ^{125}I-radiolabelled insulin diffusion through a PEO macromonomer hydrogel.

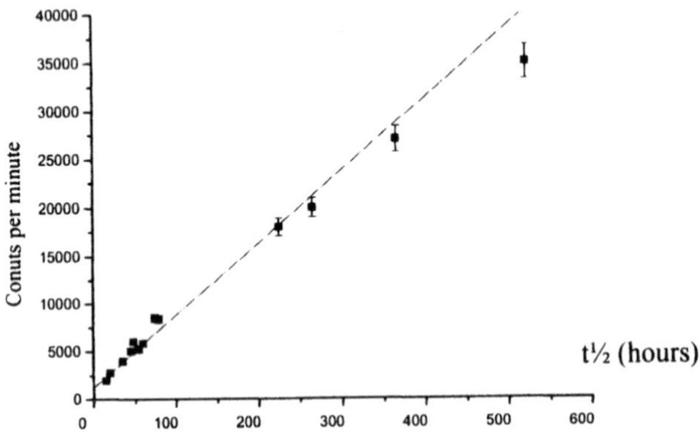

Figure 7. Increase in radioactivity with time for a PEO hydrogel with precursor molar mass of 11,500 $gmol^{-1}$, loaded with ^{125}I-radiolabelled insulin.

The resolution of Fick's law with respect to the present problem gives a relation between the slope of the curve and the diffusion:

$$S=\frac{2}{\alpha}(1+\alpha)\frac{2}{\sqrt{\pi}}\sqrt{\frac{D}{a^2}}M$$

where S is the slope of the curve; D the diffusion coefficient; $\alpha = A/\pi a^2$ and $M = AC_0/(1 + \alpha)$. The values for the shown experiment were $C_0 = 10{,}400$ cpm/mL, A = 100 mL, a = 1.92 mL and $\alpha = 8.6$, resulting in $M = 1.085\text{x}10^5$. Rearrangement of the above equation leads to the following expression:

$$D = \left[\frac{S\alpha\sqrt{\pi}a}{4(1+\alpha)M}\right]^2$$

from which the value of the insulin diffusion coefficient D_a in the hydrogel was calculated to be $1.16\text{x}10^{-6}$ cm^2s^{-1}. For comparison, the diffusion coefficient of insulin in pure water (estimated from the Stocks-Einstein relation) is equal to $D_a = 1.34\text{x}10^{-6}$ cm^2s^{-1}. The value of the relative diffusion coefficient of insulin in the gel is obtained by taking the ratio Da / $1.34\text{x}10^{-6} = 0.87$. This value of 0.87 is rather high compared to the diffusion of glucose through the same network. This observation can be explained by a partial adsorption of insulin onto the PMMA domains of the PEO hydrogel. However, these results are somewhat controversial and further work needs to be done in order to clarify this issue.

Incorporation of Cells during Hydrogel Crosslinking

Most of the devices developed for an artificial pancreas are based on the incorporation or encapsulation of Langerhans islets into micro or macrocapsules or their placement behind a semi-permeable membrane. However, insufficient nutrient flow to the cores of the islets in theses devices represents a major drawback for their long-term implantation.

The possibility of producing hydrogels directly in water by free radical polymerization of macromonomers offers the interesting perspective of direct encapsulation of living cells. In most cases, biological actives are incorporated into the dry gel. The following crosslinking reaction in these cases usually has to be conducted at least at 60°C, a temperature incompatible with cell survival. Provided the crosslinking process of macromonomers could be achieved at a temperature compatible with the survival of cells, the cells could be incorporated during the crosslinking process and thus distributed homogeneously in the three-dimensional scaffold. Therefore, we attempted in preliminary experiments to polymerize bifunctional PEO macromonomers at 37°C, with potassium peroxo-disulfate or redox compounds as the initiator. Whereas no crosslinking was observed with potassium peroxodisulfate, the use of redox initiators resulted in successful crosslinking. These results confirm that rapid crosslinking is possible even at 37°C, provided redox initiators are used for the polymerization. Once crosslinked, no change of the mechanical properties has been observed over a period of three hours *(26)*.

These PEO hydrogels were also tested with respect to their ability to serve as a template for the survival and growth of hepatocytes. Two systems were considered, (i) surfaces of pre-existing hydrogels with selected structural parameters were seeded with isolated rat hepatocytes, or (ii) hepatocytes were dispersed in physiological medium containing the macromonomer and initiator and heated to 37°C (Figure 8) *(25)*. In the first case, cells were examined at given times over two days after spreading. These results were compared to those obtained from the crosslinked dispersion of fibroblasts. The effect of pre-existing hydrogel surface structure on the extent of hepatocyte attachment and their morphology were investigated. Almost identical results were observed for both systems, prepared from PEO macromonomer precursors of molar mass of 15,000 gmol^{-1}.

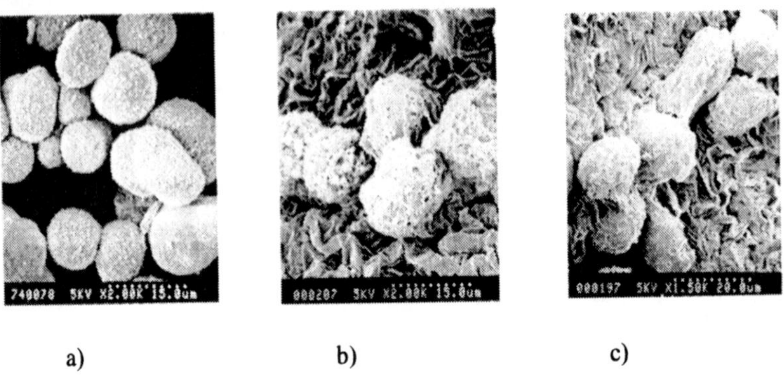

a) b) c)

Figure 8. Rat hepatocytes encapsulated into a PEO hydrogel (15,000 gmol^{-1}) after 3 hours (7b) and 24 hours (7c) in culture. Figure 7a shows freshly isolated rat hepatocytes and serves as a control. After 3 hours in culture, the cells appeared shrinkled and in some cases showed holes in their membrane. This phenomenon was increased after 24 hours in culture.

Conclusions

It has been shown that hydrogels based on either linear PEO macromonomers or star-shaped PEO are well-suited biocompatible supports for the conception of artificial organs. When implanted, they are both well tolerated by the host immune system and allow efficient diffusion of glucose. The presence of an EXPTFE membrane reinforced the mechanical properties of the PEO membrane without affecting their diffusion properties. Insulin diffusion still poses some problems, which can be attributed to the presence of the hydrophobic poly(methylmethacrylate) core in the macromonomer-based hydrogels. Therefore, copolymerization experiments of PEO macromonomers with hydrophilic monomers are now under progress in order to access hydrogels with increasing content of hydrophilic monomers in the core. This derivatization should reduce the adsorption of insulin. Lastly, and of particular significance, the possibility of crosslinking PEO macromonomers directly in the presence of hepatocytes opens many new perspectives in this field.

Acknowledgements

The authors thank Prof. Ph. Wolf (CHU, F-Strasbourg), Prof. L. Richert (Fondation Transplantation, F-Strasbourg/Faculté de Médecine et de Pharmacie, F-Besançon) and Prof E. W. Merrill (MIT, Boston) for helpful and stimulating discussions, and the CNRS, the French Ministry of Education (ACI Technology pour la Santé) and the Fondation Transplantation for financial support. Special thanks are addressed to Mrs. C. Royer (Faculté de Médecine, F-Strasbourg) for the microscopy experiments, and to Dr. P. Baxter and F. Isel for their contributions.

References

1. Ross, S.A.; Gulve, E.A.; Wang, M. Chemistry and biochemistry of Type 2 Diabetes. *Chem. Rev.* **2004**, *104*, 1255-1282.
2. Boker, A.; Rothenberg, L.; Hernandez, C.; Kenyon, N.K.; Ricordi, C.; Alejandro, M. Human Islet Transplantation: Update. *World J. Surg.* **2001**, *25*, 481-486.
3. Boudjema, K.; Alexandre, E. Islet of Langerhans Transplantation. *Ann. Chir.* **1995**, *49*, 902-908.
4. Calafiore, R. In: "Bioartificial Pancreas" Severian, D. (Ed), Polymeric Biomaterials, Marcel Dekker, New-York (NY), 2002, 983-1005.
5. Kulseng, B.; Skjak-Braek, G.; Ryan, L.; Andersson, A.; King, A.; Faxvaag, A.; Espevik, T. Transplantation of alginate microcapsules: generation of antibodies against alginates and encapsulated porcine islet-like cell clusters. *Transplantation* **1999**, *67*, 978-984.
6 Chandy, T.; Mooradian, D.L.; Rao, G.H. Evaluation of modified alginate-chitosan-polyethylene glycol microcapsules for cell encapsulation. *Artif. Organs* **1999**, *23*, 894-903.

7. Del Guerra, S.; Bracci, C.; Nilsson, K.; Belcourt, A.; Kessler, L.; Lupi, R.; Marselli, L.; De Vos, P.; Marchetti, P. Entrapment of dispersed pancreatic islet cells in Cultispher-S macroporeous gelatin microcarriers : Preparation, in vitro characterization, and microencapsulation *Biotechnol. Bioengin.* **2001**, *75*, 741-744.
8. Ehab, R.; Wernerson, A.; Arner, P. and Tibell, A. In vivo studies on insulin permeability of an immunoisolation device intended for islet transplantation using the microdialysis technique. *Eur. Surg. Res.* **1999**, *31*, 249-258.
9. Kessler, L.; Legeay, G.; West, R.; Belcourt, A.; Pinget, M. Physicochemical and biological studies of corona-treated artificial membranes used for pancreatic islets encapsulation: mechanism of diffusion and interface modification. *J. Biomed. Mater. Res.* **1997**, *34*, 235-245.
10. Lhommeau, C.; Toillon, S.; Pith, T.; Kessler, L.; Jesser, C.; Pinguet, M. Polyamide 4,6-membranes for the encapsulation of Langerhans islets: Preparation, physico-chemical properties and biocompatibility studies. *J. Mater. Sci.: Mater. Med.* **1997**, *8*, 163-174.
11. Monaco, A.P. Transplantation of pancreactic islets with immunoexclusion membranes. *Transplant. Proc.* **1993**, *25*, 2234-2236.
12 Peppas, N.A.; Bures, P.; Leobandung, W.; Ichikawa, H. Hydrogels in pharmaceutical formulations. *Eur. J. Pharmac. Biopharmac.* **2000**, *50*, 27-46.
13. Hubbell, J.A. Bioactive Biomaterials. *Curr. Opin. Biotechnol.* **1999**, *10*, 123-129.
14. Merrill, E.W.; Salzman, E.W. Poly(ethylene oxide) as biomaterial. *Am. Soc. Artif. Int. Org.* **1983**, *6*, 60-64.
15. Harris, M.J in "Poly(ethylene Glycol) Chemistry: Biotechnical and biomedical Applications" , Plenum Press, New-York, 1992.
16. Gnanou, Y.; Hild, G.; Rempp, P. Molecular structure and elastic behavior of poly(ethylene oxide) networks swollen to equilibrium. *Macromolecules* **1987**, *20*, 1662.
17. Lopina, S.T.; Wu, G.; Merrill, E.W.; Cima, L.G. Hepatocyte Culture on Carbohydrate-Modified Star Polyethylene Oxide Hydrogels. *Biomaterials* **1996**, *17*, 559-569.
18. Schmitt, B.; Alexandre, E.; Boudjema, K.; Lutz, P.J. Synthesis and study of poly(ethylene oxide) membranes obtained from homopolymerization of PEO macromonomers. *Macromol. Symp.* **1995**, *93*, 117-124.
19. Schmitt, B.; Alexandre, E.; Boudjema, K.; Lutz, P.J. Poly(ethylene oxide) hydrogels as a semi-permeable membrane for an artificial pancreas. *Macromol. Biosci.* **2002**, *93*, 341-351.
20. Knischka, R.; Lutz, P.J.; Sunder, A.; Frey, H. Structured Hydrogels based on poly(ethylene oxide) multiarm stars with hyperbranched polyglycerol cores. *Polymer. Mater. Sc. Engin.* **2001**, *84*, 945-946.

21. Merrill, E.W.; Wright, K.A.; Sagar, A.; Pekala, R.W.; Dennison, K.A.; Tay, S.W.; Sung, C.; Chaikof, E.; Rempp, P.; Lutz, P.J.; Callow, A.D.; Connolly, R.; Ramber, K.; Verdon, S. "Versions of Immobilized Poly(ethylene oxide) for Medical Applications" In: *Polymers in Medicine: Biomedical and Pharmaceutical Applications*, Ottenbrite, R.M. and Chiellini, L.E. (Eds.), Technonic Publishing, 1992.
22. Lutz, P.J. Free radical homopolymerization, in heterogeneous medium, of linear and star-shaped polymerizable amphiphilic poly(ethers) : A new way to design hydrogels well suited for biomedical applications. *Macromol. Symp.* **2001**, *164*, 277-292.
23. Alexandre, E.; Schmitt, B.; Boudjema, K.; Merrill, E.W.; Lutz, P.J. Hydrogel networks of poly(ethylene oxide) star molecules supported by expanded poly(tetrafluoroethylene) membranes: Characterization, biocompatibility evaluation and glucose diffusion characteristics. *Macromol. Biosci.* **2004**, *4*, 639-648.
24 Betty, L.M; Hannoun, G.; Stephanopoulos, G. Diffusion coefficient of glucose and ethanol in cell-free and cell-occupied calcium alginate membranes. *Biotch. and Bioeng.* **1986**, *28*, 824.
25. Alexandre, E.; Cinqualbre, J.; Jaeck, D.; Richert, L.; Isel, F.; Lutz, P.J. Poly(ethylene oxide) Macromonomer based Hydrogels as a Template for the Culture of Hepatocytes. *Macromol. Symp.* **2004**, *210*, 475-481.
26 Alexandre, E.; Boudjema, K.; Schmitt, B.; Cinqualbre, J.; Jaeck, D.; Lux, C.; Isel, F.; Lutz, P.J.; Poly(Ethylene Oxide) Based Hydrogels Designed for Artificial Organs. *Polymer. Mater. Sci. Engin.* **2003**, *89*, 240-241.
27 Schmitt, B. PhD. Thesis, University Louis Pasteur, Strasbourg, France, 1995.

Chapter 10

Silk–Elastinlike Hydrogels: Thermal Characterization and Gene Delivery

Ramesh Dandu[1], Zaki Megeed[1,5], Mohamed Haider[1], Joseph Cappello[2], and Hamidreza Ghandehari[1,3,4,*]

[1]Department of Pharmaceutical Sciences, [3]Program in Bioengineering, and [4]Greenebaum Cancer Center, University of Maryland, Baltimore, MD 21201
[2]Protein Polymer Technologies, Inc., 10655 Sorrento Valley Road, San Diego, CA 92121
[5]Current address: The Center for Engineering in Medicine, Massachusetts General Hospital, Shriners Burns Hospital, and Harvard Medical School, Boston, MA 02114
***Corresponding author: hghandeh@rx.umaryland.edu**

Silk-elastinlike protein polymers (SELP) are a class of genetically engineered block copolymers composed of tandemly repeated silk-like (GAGAGS) and elastin-like (GVGVP) peptide blocks. One of these polymers, SELP-47K, undergoes self-assembly in aqueous medium and has been extensively characterized for controlled drug and gene delivery applications requiring *in situ* gelation. This chapter provides a review of the thermal characterization of SELP-47K hydrogels, *in vitro* and *in vivo* delivery of plasmid DNA, and the potential of controlled adenoviral gene delivery from these systems.

Introduction

The significant interest and resources that the scientific community has dedicated to gene therapy research have rapidly led to an interesting and somewhat conflicting state of development. Specifically, the potential of gene therapy in treating diseases and the significant challenges facing the field now occupy almost equal status in the minds of many scientists. The resources dedicated to gene therapy have been justified by the fact that it has the potential to treat a number of diseases, including those of congenital origin, such as cystic fibrosis, and those that arise from genetics gone awry, like cancer. More recently, researchers have recognized the potential of gene therapy to treat infectious diseases, adding additional emphasis to the "potential" side of the equation, and providing even more motivation for addressing the "challenges". One of the greatest challenges facing the field of gene therapy is the effective delivery of genes. This fact has stimulated tremendous interest in understanding the barriers to gene delivery, and in developing delivery systems (vectors) that overcome them. Gene delivery vectors can be broadly categorized as viral or nonviral *(1)*. Nonviral vectors, predominantly composed of cationic polymers, lipids, or peptides, are generally considered to be relatively safe, but their effectiveness is limited by low transfection efficiencies *(2-4)*. Viral vectors generally exhibit higher transfection efficiencies *(5)*, but their usage has been limited by their actual and potential toxicities and immunogenicity *(6)*.

In cancer gene therapy, two primary methods have been used for the administration of viral and nonviral vectors: *systemic* - usually intravenous, or *localized* - often intratumoral. The major obstacles facing therapeutic application of systemically administered gene vectors to solid tumors are the degradation of nucleic acids in the circulation, transient transgene expression after transfection, low transfection efficiency, a large volume of distribution, and inefficient penetration of most vectors beyond the tumor periphery *(1,2,7,8)*.

Intratumorally administered, matrix-mediated controlled delivery of genes and viruses can potentially address some of these challenges. Encapsulating naked DNA, nonviral vectors, or viral vectors in polymeric matrices or hydrogels provides several advantages. These include (i) protection of the vector from enzymatic degradation, (ii) prolonged vector delivery, potentially increasing the duration of transgene expression and decreasing the frequency of administration, and (iii) efficient localization of transgene expression to the tumor, possibly decreasing the systemic toxicity of some transgenes (e.g., interferons, interleukins) *(8,9)*. Various chemically synthesized and natural polymers (e.g., poly(D,L-lactide-*co*-glycolide), (PLGA), gelatin-alginate coacervates) have been used as matrices for controlled gene delivery *(9-11)*. While some degree of success has been achieved, the limitations of these polymers must also be considered. Chemically synthesized polymers, generally produced by random copolymerization, have a distribution of molecular weights (polydisperse). There is little or no control over the sequence of the monomer repeats in the polymer backbone, making the prediction of their biological fate(s) such as biodistribution and biodegradation complex from a drug delivery perspective. Moreover, residual organic solvents from polymer synthesis and/or matrix fabrication may negatively impact the stability of bioactive agents incorporated in the matrix *(12)*. Due to their complex and sometimes variable structure, most matrices composed of natural polymers are not easily customized for optimizing gene delivery for specific clinical needs.

Recombinant DNA technology has enabled the synthesis of genetically engineered protein-based polymers, incorporating peptide blocks from naturally occurring proteins such as elastin, silk, and collagen, to produce materials not found in nature, such as silk-elastin and silk-collagen *(13,14)*. The sequence of these polymers is encoded at the DNA level, which is subsequently translated to protein polymer. DNA-directed polymer synthesis can be carried out with unprecedented fidelity using the cellular machinery. This approach allows exquisite control over the sequence and molecular weights of these polymers, enabling the optimization of the macromolecular architecture and thereby their physicochemical properties. These polymers are stereoregular, monodisperse, biocompatible, biodegradable, and their synthesis does not require the use of organic solvents. The bioprocessing technologies used for the scale-up and purification of conventional recombinant proteins have been adapted for protein-based polymers to produce large quantities of endotoxin-free materials for use *in vivo*. A number of classes of recombinant polymers have now been synthesized, including the elastin-like polymers (ELP) *(15,16)*, silk-like polymers (SLP) *(17,18)*, silk-elastinlike polymers (SELP) *(13,19)*, poly(glutamic acid)s *(20)*, silk-collagen polymers *(14)*, and others *(21)*. Several of these polymers have been characterized for biomedical applications, including controlled release *(22, 23)*, tissue repair *(24)*, and targeted drug delivery *(25)*. In our laboratory, we use genetically engineered protein polymers both as a research tool, to understand structure/property relationships, and as a therapeutic tool, to deliver DNA to solid tumors. This chapter reviews our efforts to use thermal methods to understand polymer properties of SELP as they relate to drug delivery as well as their evaluation as matrices for controlled delivery of plasmid DNA and adenoviral vectors for cancer gene therapy.

Results and Discussion

Silk-Elastinlike Protein Polymers

Silk-elastinlike polymers (SELP) are a class of genetically engineered block copolymers composed of tandemly repeated silk-like (with the amino acid sequence GAGAGS) and elastin-like (with the amino acid sequence GVGVP) peptide blocks. Incorporation of silk-like blocks imparts crystallinity and enhances the crosslinking density, while the incorporation of elastin-like blocks improves aqueous solubility. The structural and mechanical properties of these polymers can be optimized by carefully altering the ratio(s) and sequence(s) of the silk-like or elastin-like units present in the monomer repeat and/or the number of monomer repeats in the polymer. This range of variations allows the use of SELP for a variety of biomedical applications, including dermal augmentation, structural materials for control of urinary incontinence, and drug delivery. Recombinant techniques enable customization of the physicochemical properties of SELP by incorporation of peptide motifs that induce gel formation, control biodegradation, and enable stimuli-sensitivity or biorecognition to suit specific drug delivery applications *(26,27)*. The synthesis, characterization, biocompatibility, and biodegradation of SELP have been well-established *(26-29)*. SELP-47K, a copolymer with four silk-like blocks and seven elastin-like blocks in its primary repetitive sequence (Figure 1, bold) is soluble in aqueous medium, and undergoes an irreversible sol-to-gel transition to form hydrogels. This transition is slow at room temperature (several hours), yet occurs within minutes at 37 °C, making SELP-47K potentially useful for controlled delivery applications requiring *in situ* gelation.

MDPVVLQRRDWENPGVTQLNRLAAHPPFASDP
MGAGSGAGAGS[**(GVGVP)$_4$GKGVP(GVGVP)$_3$**
(GAGAGS)$_4$]$_{12}$(GVGVP)$_4$GKGVP(GVGVP)$_3$(GAGA

Figure 1. SELP-47K amino acid sequence. (M-methionine, D-aspartate, P-proline, V-valine, L-leucine, Q-glutamine, R-arginine, W-tryptophan, E-glutamate, N-aspargine, G-glycine, T-threonine, H-histidine, F-phenylalanine, K-lysine, Y-tyrosine, S-serine).

Analysis of SELP-47K Hydrogels by Differential Scanning Calorimetry

Hydrogel swelling, the release of solutes, phase transitions, and the chemical stability of solutes incorporated into hydrogels are potentially influenced by the nature and extent of water imbibed by the hydrogel and its interaction with the polymer matrix *(30-34)*. For many years, a discrepancy has been noted between the (total) amount of water present in a hydrogel (determined gravimetrically) and the amount detectable by calorimetric methods during freezing. This "non-freezable" fraction of water has often been attributed to hydrogen bonding between polymeric chains and water in their vicinity *(35,36)*. However, recent reports indicate that other factors play a role, including polymer hydrophobicity, polymer glass transition temperature, hydrogel pore size, and the diffusivity of water in the hydrogel matrix *(37,30)*. A certain fraction of the non-freezable water could be "bound", with an inherently restricted molecular mobility, hindering solute diffusion through the hydrogel network *(30)*. In an effort to better understand the interaction of SELP-47K with water, we used DSC to probe the nature and the existence of non-freezable water, and its influence on solute diffusion in these hydrogel systems *(38)*.

Amount of Non-freezable Water in SELP-47K Hydrogels

Calorimetric measurements indicated that SELP-47K hydrogels contain up to 27 wt% of non-freezable water (Figure 2A). This trend was qualitatively similar to that observed with poly(methylmethacrylate) (PMMA) hydrogels, though SELP-47K hydrogels contained 4-5 times as much non-freezable water at equivalent levels of hydration.

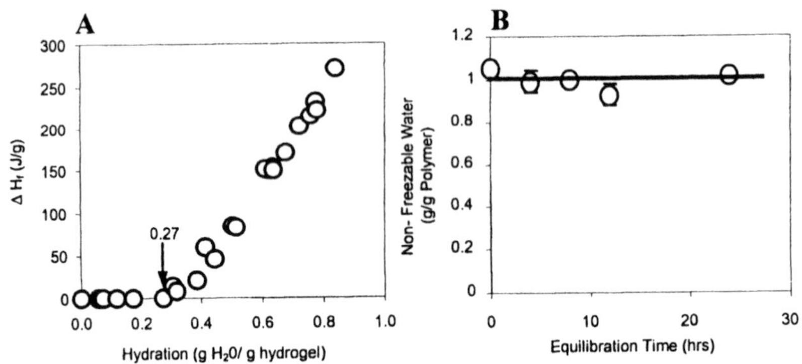

Figure 2. (A) Heat of fusion of water in hydrogels as a function of hydration (g H_2O/g hydrogel). DSC did not detect any water in the hydrogels until approximately 27 wt%. Above this water content, the heat of fusion increased linearly (r2=0.9831). (B) Effect of equilibration time at -15 °C on the amount of non-freezable water in hydrogels with 87 wt% water. Each data point is the mean ± SD of triplicates. (Reproduced with permission from ref. (38). Copyright 2004 American Chemical Society.)

This discrepancy may be attributed to the relatively high hydrophobicity of SELP-47K (22.4% valine, 12.2% alanine). Previous studies have shown that increasing the fraction of hydrophobic polymer in an interpenetrating polymer network resulted in an increase in the non-freezable water content *(37)*. Equilibration of PMMA hydrogels at -15 °C for up to 14 hours resulted in a significant decrease in the amount of non-freezable water *(39)*. This observation was explained by considering the restricted diffusion of water molecules during freezing and the accompanying transition from a rubbery to a glassy state. By contrast, equilibration of SELP-47K hydrogels for up to 24 hours did not decrease the amount of non-freezable water (Figure 2B), indicating that the intrinsic properties of the polymer (i.e., hydrophobicity) may be more important than kinetic factors.

Heat Capacity of Non-Freezable Water

Truly bound water can be considered to be in a different thermodynamic state than bulk water, with inherent restriction in its molecular mobility due to hydrogen bonding between water molecules themselves and/or between water molecules and polymer chains *(40,41)*. To evaluate the molecular mobility of non-freezable water, the heat capacities of SELP-47K hydrogels were measured at different temperatures and levels of hydration, and compared to those calculated from Equation 1:

$$C_{p(\text{hydrogel})} = f_{\text{water}}\, C_{p\,(\text{water})} + f_{\text{polymer}}\, C_{p(\text{ polymer})} \qquad \textit{Eq. (1)}$$

where C_p is the heat capacity of the hydrogel, water, or polymer, and f is the fraction (g/g hydrogel) of the water or polymer. Equation 1 is applicable to a binary system composed of a water phase and a polymer phase. The measured hydrogel heat capacities fit well with the predictions of Equation 1 (Figures 3A-C). Deviations from the predictions of Equation 1 were observed at levels of hydration between 0.2 and 0.4, and were attributed to a previously observed elastin glass transition *(42)*. Calculated values for the heat capacity of water in the hydrogels closely correspond with literature values for bulk water (Figure 3D), indicating that it is unlikely that all of the non-freezable water in SELP-47K hydrogels is "bound."

The DSC studies demonstrate the presence of up to 27 wt% non-freezable water in SELP-47K hydrogels, which is not substantially affected by equilibration of the hydrogels at -15 °C. The presence of non-freezable water in SELP-47K hydrogels may be influenced by the hydrophobicity of the polymer and/or the presence of nanocavities in the hydrogel network *(36,37)*, as observed in other protein-based systems. However, the indirect nature of calorimetric measurements precludes conclusions about the primary origins of the non-

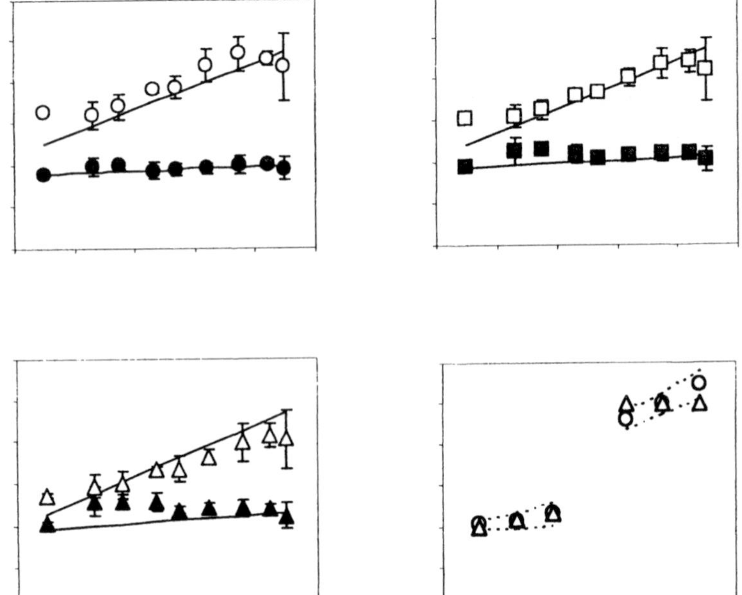

Figure 3. Effect of hydrogel hydration (g H_2O/ g polymer) on Cp at (A) 60 °C (open circle) and -60 °C(filled circle); (B) 40°C (open box) and -40 °C (filled box); (C) 20 °C (open triangle) and -20 °C (filled triangle); and (D) Calculated (○) partial heat capacity of water in SELP-47K hydrogels, at various temperatures, versus accepted values in the literature (△). Dotted lines are 95% confidence interval of data set with 45 data points for each temperature. (Reproduced with permission from ref. (38). Copyright 2003 American Chemical Society.)

freezable water and its influence on the macromolecular diffusion in these hydrogels. Further structural characterization by pulse field gradient nuclear magnetic resonance (PFG-NMR) experiments could provide insight into the nature of water in these systems. This could aid in the understanding of the diffusion of solutes (genes) through their matrices, since non-freezable water is not available for solute diffusion through the pores of the hydrogel.

Gene Delivery from SELP-47K Hydrogels

As previously described, SELP-47K undergoes spontaneous self-assembly in aqueous solution to form hydrogels. This self-assembly occurs through physical crosslinking of the silk-like units by hydrogen bonding, and it is accelerated when the polymer is transferred from room temperature to body temperature. The kinetics of this sol-to-gel transition allow solutions of polymer and bioactive agents with aqueous solubility to be prepared at room temperature and injected through a small gauge needle, with hydrogel matrices formed *in situ* within minutes. The hydrogels release the incorporated bioactive agents through diffusion and/or matrix degradation *(22,23)*. While the applications of SELP-mediated controlled gene delivery are numerous, we have focused our efforts on the controlled delivery of plasmid DNA and adenoviral vectors to solid tumors.

Controlled Delivery of Plasmid DNA from SELP-47K

As mentioned previously, the ability to obtain sufficient transfection efficiency and duration are two major challenges facing viral and nonviral gene delivery. Matrix-mediated gene delivery can address these issues by controlling the delivery of nucleic acids or gene vectors over prolonged periods of time. An additional benefit that arises from this approach is the precise spatial localization of the delivered vectors. The matrix also has the potential to protect nucleic acids or gene vectors from degradation. In an effort to evaluate the potential of SELP as matrices for the delivery of plasmid DNA, we have evaluated the release of DNA from SELP-47K matrices *in vitro*, and the delivery of DNA from SELP-47K matrices to solid tumors *in vivo (43)*.

In vitro Release of Plasmid DNA from SELP-47K Hydrogels

SELP hydrogel systems undergo *in situ* gelation and can be molded to a variety of shapes, including cylinders, disks and films. We chose to use a cylindrical shape for our hydrogel systems for *in vitro* evaluation(s), as our experience with these polymers has been that cylindrical hydrogels could be made with minimal intersample variability (size and shape) and waste. This shape also allowed us to adapt established diffusion models from literature for the evaluation of their release properties. The influence of plasmid DNA conformation and size, polymer concentration, hydrogel geometry and cure time, and DNA concentration, on DNA release from cylindrical SELP-47K hydrogels were evaluated over a period of 28 days *(23,43)*. Hydrogels were prepared by mixing aqueous solutions of DNA and SELP-47K at room temperature, and gel formation was induced by incubation at 37 °C. Hydrogel were cut and the release studies were carried out. The diffusivity of DNA from these hydrogels occurring in two dimensions (axial and radial) was calculated using Equation 2 *(44,45)*:

$$\frac{M_t}{M_\infty} = 1 - \frac{32}{\pi^2}\sum_{i=1}^{\infty}\frac{1}{\alpha^2_i}\exp\left(-\frac{\alpha^2_i}{r^2}D_e t\right)\sum_{j=0}^{\infty}\frac{1}{(2j+1)^2}\exp\left(-\frac{(2j+1)^2\pi^2}{h^2}D_e t\right) \quad Eq.\ (2)$$

where r is the radius and h the height of a cylinder, M_t is the cumulative amount of solute released at time t, M_∞ is the amount released as $t \to \infty$, D_e is the average effective intragel diffusivity of the solute, and α_i are the roots of the zero-order Bessel function, $J_0(\alpha_i) = 0$. D_e was estimated from a nonlinear fit of Equation 2 to the experimental release data, using *Mathematica* software.

Effect of Plasmid DNA Conformation and Size on In Vitro Release

Restriction enzymes were employed to produce DNA predominantly in the linear, open-circular, or supercoiled forms. The cumulative percentage of each form of DNA released from SELP-47K hydrogels was in the order of linear > supercoiled > open-circular. The open-circular form was practically not released, probably due to its impalement on the polymer chains. Plasmids were released in a size-dependent manner, from 2.6 to 11 kilobases (kb) (Table I) *(43)*.

Previous studies have shown that increasing the polymer concentration and hydrogel cure time decreases the rate of DNA release, while increasing the DNA concentration from 50 to 250 μg/ml did not influence the rate of release *(23)*.

Effect of Hydrogel Geometry on In Vitro Release

Information on the influence of hydrogel geometry on the release rate may be useful in determining where and how to inject SELP matrices. The effect of hydrogel geometry on DNA release was studied by preparing cylinder-like and disc-like hydrogels both having equivalent volumes. Disc-like hydrogels released DNA faster than their cylindrical analogues (Table 1) *(43)*. This observation was attributed to the larger surface-to-volume ratio of discs and was accurately described by fitting the release data to Equation 2.

Table I. Influence of size and conformation of plasmid DNA and geometry of hydrogel on release

Plasmid	Geometry	Plasmid size (kbp)[a]	Hydrogel surface area (cm^2)	De[b] (cm^2/sec)
pUC18	Disc	2.60	0.74	$2.55 \pm 0.51 \times 10^{-10}$
l-pRL-CMV[c]	Disc	4.08	0.74	$1.94 \pm 0.27 \times 10^{-10}$
sc-pRL-CMV[d]	Disc	4.08	0.74	$8.90 \pm 0.12 \times 10^{-11}$
oc-pRL-CMV[e]	Disc	4.08	0.74	$1.96 \pm 0.83 \times 10^{-13}$
pRL-CMV	Large Disc	4.08	1.70	$1.76 \pm 0.28 \times 10^{-10}$
pRL-CMV	Cylinder	4.08	1.07	$9.23 \pm 0.15 \times 10^{-11}$
pCFB-EGSH-Luc	Disc	8.50	0.74	$3.09 \pm 0.43 \times 10^{-11}$
pFB-ERV	Disc	11.00	0.74	$1.70 \pm 0.52 \times 10^{-12}$

[a] *kbp* = kilobase pairs, [b] D_e = average effective diffusivity of plasmid DNA in hydrogel determined by nonlinear fit of release data to Equation 2, [c] *l* = linearized, [d] *sc* = supercoiled, [e] *oc* = open-circular. Samples in which the conformation is not denoted were a mixture of the three conformations, primarily supercoiled (Reproduced with permission from ref. *(43)*. Copyright 2004 Elsevier Science).

In Vitro Bioactivity of Encapsulated Plasmid DNA.

Plasmid DNA incorporated in SELP-47K hydrogels retained bioactivity, even after incubation in PBS, at 37 °C for 28 days *(43)*. Effective diffusivities of plasmid DNA described above can be correlated with the amount released from SELP-47K hydrogels. These results demonstrate that release is dependent on the size and conformation of plasmid DNA, the geometry and cure time of the hydrogel, and the concentration of the polymer.

From a clinical perspective, optimal delivery rates for different cancer gene therapy applications vary, dependent on factors such as tumor stage, its aggressiveness (metastatic or non-metastatic), the type and location of the tumor, among others. The *in vitro* studies described above demonstrate the potential of optimizing the release kinetics of plasmid DNA of different sizes from SELP-based systems by carefully altering their polymer concentration and cure time to suit specific gene delivery applications.

Delivery of Plasmid DNA from SELP-47K Hydrogels to Solid Tumors

In an effort to establish the feasibility of using SELP-47K hydrogels to deliver plasmid DNA to solid tumors, we investigated the intratumoral delivery of a reporter plasmid (*Renilla* luciferase) to subcutaneous tumors in a murine (athymic *nu/nu*) model of human breast cancer (MDA-MB-435 cell line) *(43)*. SELP-47K/DNA solutions containing 4, 8, or 12 wt% polymer and 70 μg DNA per 100 μl were injected directly into the tumors. At predetermined time points, animals were euthanized, tumors were resected and homogenized, and luciferase expression was assayed. To evaluate the potential of SELP-47K to localize delivery to the tumor, we also evaluated the transfection of the skin approx. 1 cm around each tumor.

Intratumoral injection of 4 or 8 wt% hydrogels resulted in significantly enhanced tumor transfection for up to 21 days when compared to naked DNA, while 12 wt% matrices enhanced transfection up to 3 days (Figure 4A). The levels of tumor transfection mediated by the three concentrations of polymer were statistically equivalent until 7 days, when the 4 and 8 wt% matrices were both more effective than 12 wt% and naked DNA (Figure 4A). The high levels of tumor transfection observed initially (up to 7 days) in case of 12 wt% hydrogels are probably due to burst release upon administration. The irregular (atypical) pore distribution in the perimeter of the hydrogel could lead to a faster release of the "loosely entrapped" DNA from their surfaces irrespective of polymer concentration and potentially obscure any actual differences in their release rates occurring initially.

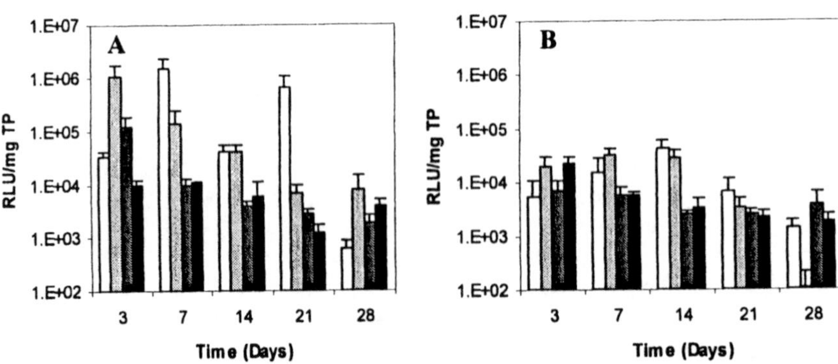

Figure 4. Expression of Renilla luciferase in tumors (A) and in skin directly surrounding the tumors (B), after intratumoral injection. Bars represent 4 wt% polymer (white), 8 wt% polymer (light gray), 12 wt% polymer (dark gray), and naked DNA without polymer (black). Each bar represents the mean +SEM for n=4 or n = 5 samples. (Reproduced with permission from ref. (43). Copyright 2004 Elsevier Science)

The higher transfection observed in case of the 4 and 8 wt% hydrogels up to day 21 is consistent with *in vitro* release data *(23)*, indicating lower release rates at higher polymer concentrations due DNA entrapment within the matrices. The delivery of DNA from 4, 8, and 12 wt% polymer hydrogels resulted in a mean 142.4-fold, 28.7-fold, and 3.5-fold increase in tumor transfection, respectively, compared with naked DNA over the entire 28-day period.

Overall, the mean tumor transfection was 42.0, 27.2, and 4.6 times greater than skin transfection for 4, 8, and 12 wt% hydrogels, respectively, over the entire 28-day period, compared to a 1.3 fold difference between tumor and skin transfection for naked DNA (Figure 4A and 4B). This indicates that the enhancement of transfection was largely restricted to the tumor, suggesting a spatial localization benefit from the matrix. The decrease in transfection over time can be attributed to a corresponding decrease in the release of plasmid due to entrapment, or depletion due to release and/or degradation.

Controlled Delivery of Adenoviral Vectors from SELP-47K

As previously mentioned, viral vectors are generally associated with relatively high transfection efficiencies, but their clinical application is limited by safety concerns *(1,6)*. To reduce the risk of oncogene disruption by integrating viral vectors such as retroviruses, non-integrating viral vectors such as adenoviruses have been widely explored *(5)*. However, non-integrating viral vectors are similar to nonviral vectors in that they suffer from limitations in the duration of transgene expression, which may be exacerbated by their rapid clearance. Furthermore, viral vectors face the additional challenge of escaping, or minimizing, the body's immune response, which can increase the rate of clearance, further reducing the duration of transgene expression, and lead to serious systemic or local inflammation.

Encapsulation of viral vectors in hydrogels may begin to address some of these challenges. The controlled release of small amounts of viral vectors from a matrix reservoir may increase the duration of transfection, while simultaneously reducing the amount of antigen available for sampling and detection by the immune system. Furthermore, encapsulation of a viral vector in a matrix may protect it from degradation, though the potential to include stabilizers in the matrix remains to be explored. Based on the promising results obtained when delivering plasmid DNA from SELP-47K, we sought to investigate the potential of using this polymer as a matrix for the controlled delivery of adenoviral vectors.

In vitro Release of Adenoviral Vectors from SELP-47K Hydrogels

Adenoviral DNA is encapsulated in a protein coat that is essential for its protection. These viral coat proteins may potentially interact with the amino-acid backbone of the SELP-47K polymer, leading to an altered viability and transfection efficiency. In an effort to evaluate the long-term viability of adenoviruses incorporated in SELP-47K hydrogels, we have carried out an *in vitro* release and bioactivity study with an adenovirus containing the gene for green fluorescent protein (AdGFP) *(43)*. SELP-47K/AdGFP solutions were prepared at 4, 8, and 11.3 wt% polymer. The mixtures were allowed to gel, and hydrogel discs were placed in a phosphate buffered saline (PBS) release medium. At predetermined time points, release medium was collected and used to transfect HEK-293 cells. Transfection was observed up to 22 days with AdGFP released from the 4 wt% hydrogels, indicating that viable AdGFP continued to be released in this time frame. The number of transfected cells obtained with the virus released from the 8 wt% hydrogel was less than that obtained from the 4 wt% hydrogel, and the 11.3 wt% hydrogel did not release any detectable adenovirus after the first day. These results demonstrate that adenoviral release can be controlled over a continuum by controlling polymer composition, and that the bioactivity of the released adenoviruses decreases with time.

Influence of Adenovirus on the Swelling Ratio of SELP-47K Hydrogels

Adenoviral vectors are typically delivered in large amounts (about 10^6 to 10^{11} pfu) in the clinic to produce therapeutic effects. Given the protein nature of adenoviral coats, they could potentially interact with the polymeric chains and influence the degree of swelling of the hydrogel system. The equilibrium ratio of swelling (q = weight of wet hydrogel/weight of dry hydrogel) is an indication of the pore size and mechanical properties of a hydrogel, and can often be used to predict the rate of release of solutes. Swelling studies were performed to evaluate the influence of adenoviral incorporation and polymer concentrations on the swelling properties of SELP-47K hydrogels over time *(46)*. Adenovirus-loaded hydrogels were fabricated, and the swelling ratio was measured up to 15 days by previously described methods (Figure 5) *(47)*.

Results indicate that the incorporation of adenovirus does not significantly alter the swelling properties over a period of 15 days (Figure 6A-B). Consistent with previous observations (in the absence of adenovirus), the swelling ratio of the hydrogels with and without adenovirus decreased as the polymer concentration increased (Figure 6A) *(46)*. At given polymer concentrations (4 , 8, and 11.7 wt%), the degrees of swelling of the hydrogels were not influenced by the incubation time (Figure 6B). In the range studied (10^4, 10^6 and 10^8 pfu/50μl hydrogels), an increase in the amount of adenoviruses did not significantly change the degree of swelling of the hydrogels up to 15 days (Figure 6B).

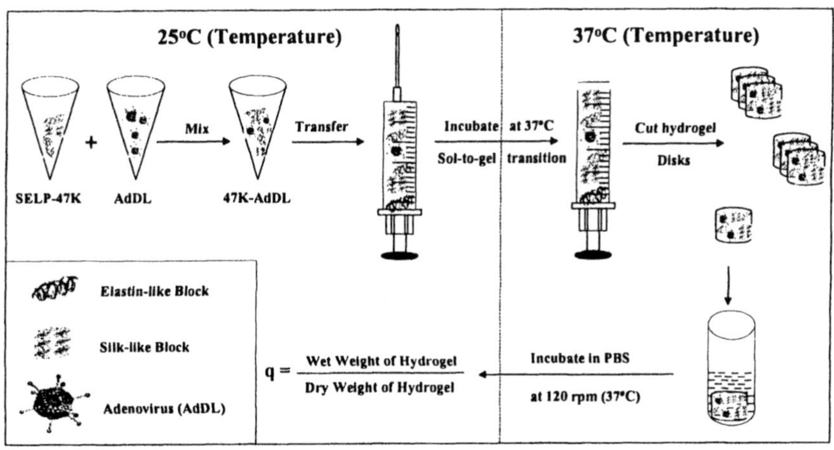

Figure 5. SELP-47K and the adenovirus DL312 (AdDL) were mixed to obtain desired compositions, and were drawn in 1-ml syringes at 25 °C. Syringes were incubated at 37 °C for 4 hours to induce gel formation. Hydrogel discs were cut using a sterile razor blade, placed in 1 ml PBS media, and mildly agitated at 120 rpm at 37 °C. Hydrogels were retrieved and the swelling ratio (q) was determined in triplicate at predetermined time points.

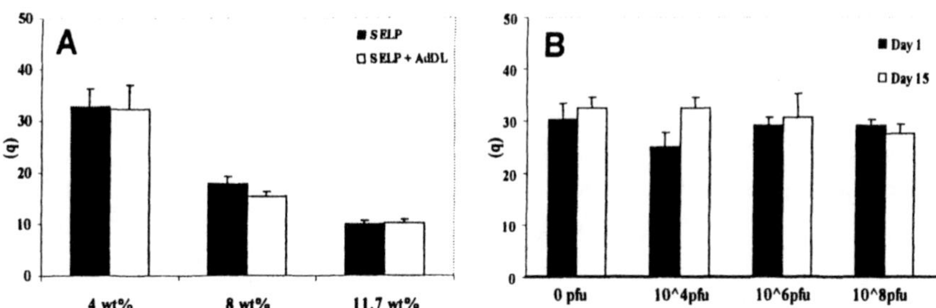

Figure 6. (A) Influence of polymer concentration (wt%) and the presence of adenovirus (10^6 pfu/50 µl hydrogel) on the degree of swelling (q, mean ± SD) at day 15. (B) Influence of adenoviral concentration on the degree of swelling (q, mean ± SD) of 4 wt% hydrogels over time (Reproduced with permission from ref. (46). Copyright 2004 Controlled Release Society.)

While the potential interactions of SELP copolymers with adenoviral particles need to be further characterized, these studies indicate that the network properties of SELP-47K are largely unaffected by the presence of adenovirus. Release and bioactivity studies demonstrate that bioactive adenovirus is released from SELP-47K hydrogels for up to 22 days. The swelling studies suggest a predictable release of adenovirus from these hydrogels, due to an absence of substantial changes in network properties. The next logical steps are quantifying the amount of adenovirus released, the proportion that is bioactive, and the feasibility of using SELP-47K hydrogels for adenoviral gene delivery *in vivo*.

Conclusions

Thermal characterization of SELP-47K hydrogels indicates that a fraction of the water in the hydrogel is non-freezable in nature, but with a heat capacity that is similar to bulk water. The nature of this water and the pore sizes of the hydrogels need to be further investigated to allow correlation of these parameters with DNA release rates. *In vitro* and *in vivo* studies of the release of plasmid DNA suggest that by carefully altering the polymer concentration and hydrogel cure time, one may optimize the DNA release kinetics. The *in vitro* bioactivity and swelling studies with adenoviruses demonstrate the feasibility of using SELP-47K hydrogels for controlled adenoviral delivery. The biodegradability and biocompatibility of the SELPs, and the ability to control their structure by genetic engineering techniques show promise for the design and development of novel matrices for localized and controlled nonviral and viral gene delivery.

Acknowledgements

We would like to thank Drs. Bert. W. O'Malley Jr. and Daqing Li for their valuable discussion and support on adenoviral gene delivery. Financial support from DOD (DAMD17-03-0237) (HG), National Cancer Center Predoctoral Fellowship (ZM), and Predoctoral Fellowship from the Egyptian Ministry of Higher Education (MH) is appreciated.

References

1. Verma, I.M.; Somia, N. Gene therapy - Promises, problems and prospects. *Nature* **1997**, *389*, 239-242.
2. Luo, D.; Saltzman, W.M. Synthetic DNA delivery systems. *Nat. Biotechnol.* **2000**, *18*, 33-37.
3. Liu, F.; Huang, L. Development of non-viral vectors for systemic gene delivery. *J. Controlled Rel.* **2002**, *78*, 259-266.

4. Guo, X.; Szoka Jr., F.C. Chemical approaches to triggerable lipid vesicles for drug and gene delivery. *Acc. Chem. Res.* **2003**, *36*, 335-341.
5. Rots, M.G.; Curiel, D.T.; Gerritsen, W.R.; Haisma, H.J. Targeted cancer gene therapy: The flexibility of adenoviral gene therapy vectors. *J. Controlled Rel.* **2003**, *87*, 159-165.
6. Worgall, S.; Wolff, G.; Falck-Pedersen, E.; Crystal, R.G. Innate immune mechanisms dominate elimination of adenoviral vectors following in vivo administration. *Hum. Gene Ther.* **1997**, *8*, 37-44.
7. Davidson, B.L.; Hilfinger, J.M.; Beer, S.J. Extended release of adenovirus from polymer microspheres: Potential use in gene therapy for brain tumors. *Adv. Drug Deliv. Rev.* **1997**, *27*, 59-66.
8. Jain, R.K. Delivery of molecular and cellular medicine to tumors. *Adv. Drug Deliv. Rev.* **1997**, *26*, 79-90.
9. Doukas, J.; Chandler, L.A.; Gonzalez, A.M.; Gu, D.; Hoganson, D.K.; Ma, C.; Nguyen, T.; Printz, M.A.; Nesbit, M.; Herlyn, M.; Crombleholme, T.M.; Aukerman, S.L.; Sosnowski, B.A.; Pierce, G.F. Matrix immobilization enhances the tissue repair activity of growth factor gene therapy vectors. *Hum. Gene Ther.* **2001**, *12*, 783-798.
10. Beer, S.J.; Matthews, C.B.; Stein, C.S.; Ross, B.D.; Hilfinger, J.M.; Davidson, B.L. Poly (lactic-glycolic) acid copolymer encapsulation of recombinant adenovirus reduces immunogenicity in vivo. *Gene Ther.* **1998**, *5*, 740-746.
11. Kalyanasundaram, S.; Feisnstein, S.; Nicholson, J.P.; Leong, K.; Garver Jr., R.I. Coacervate microspheres as carriers of recombinant adenovirus. *Cancer Gene Therapy* **1999**, *6*, 107-112.
12. van de Weert, M.; Hennink, W.E.; Jiskoot, W. Protein instability in poly(lactic-co-glycolic acid) microparticles. *Pharm. Res.* **2000**, *17*, 1159-1167.
13. Cappello, J.; Crissman, J.; Dorman, M.; Mikolajczak, M.; Textor, G.; Marquet, M.; Ferrari, F. Genetic engineering of structural protein polymers. *Biotechnol. Prog.* **1990**, *6*, 198-202.
14. Teule, F; Aube, C.; Abbott, A.G; Ellison, M.S. Production of customized novel fiber proteins in yeast (Pichia pastoris) for specialized applications: in Proceedings of the 3rd International Conference on Silk, Montreal, Canada, June 16-19, **2003**.
15. McPherson, D.T.; Morrow, C; Minehan, D.S.; Wu, J; Hunter, E; Urry, D.W. Production and purification of a recombinant elastomeric polypeptide, G-(VPGVG)19-VPGV, from Escherichia coli. *Biotechnol. Prog.* **1992**, *8*, 347-52.
16. Meyer, D.E.; Chilkoti, A. Genetically encoded synthesis of protein-based polymers with precisely specified molecular weight and sequence by

recursive directional ligation: Examples from the elastin-like polypeptide system. *Biomacromol.* **2002**, *3*, 357-367.

17. Arcidiacono, S.; Mello, C.; Kaplan, D.; Cheley, S.; Bayley, H. Purification and characterization of recombinant spider silk expressed in Escherichia coli. *Appl. Microbiol. Biotechnol.* **1998**, *49*, 31– 38.
18. Prince, J.T.; McGrath, K.P.; DiGirolamo, C.M.; Kaplan, D.L. Construction, cloning, and expression of synthetic genes encoding spider dragline silk. *Biochemist* **1995**, *34* , 10879– 10885.
19. Nagarsekar, A.; Crissman, J.; Crissman, M.; Ferrari, F.; Cappello, J.; Ghandehari, H. Genetic engineering of stimuli-sensitive silk-elastinlike protein block copolymers. *Biomacromol.* **2003**, *4*, 602-607.
20. Zhang, G.; Fournier, M.J.; Mason, T.L.; Tirrell, D.A. Biological synthesis of monodisperse derivatives of poly(α,L-glutamic acid): model rodlike polymers. *Macromolecules* **1992**, *25*, 3601–3603.
21. Haider, M.; Megeed, Z.; Ghandehari, H. Genetically engineered polymers: Status and prospects for controlled delivery. *J. Controlled Rel.* **2004**, *95*, 1-26.
22. Dinerman, A.A.; Cappello, J.; Ghandehari, H.; Hoag, S.W. Solute diffusion in genetically engineered silk-elastinlike protein polymer hydrogels. *J. Controlled Rel.* **2002**, *82*, 277-287.
23. Megeed, Z.E.; Cappello, J.; Ghandehari, H. Controlled release of plasmid DNA from a genetically engineered silk-elastinlike hydrogel. *Pharm. Res.* **2002**, *19*, 954-959.
24. Betre, H.; Setton, L.A.; Meyer, D.E.; Chilkoti, A. Characterization of a genetically engineered elastin-like polypeptide for cartilaginous tissue repair. *Biomacromol.* **2002**, *3*, 910-916.
25. Meyer, D.E.; Shin, B.C.; Kong, G.A.; Dewhirst, M.W.; Chilkoti, A. Drug targeting using thermally responsive polymers and local hyperthermia. *J. Controlled Rel.* **2001**, *74*, 213–224.
26. Megeed, Z.; Cappello, J.; Ghandehari, H. Genetically engineered silk-elastinlike protein polymers for controlled drug delivery. *Adv. Drug Deliv. Rev.* **2002**, *54*, 1075–1091.
27. Ferrari, F.; Richardson, C.; Chambers, J.; Causey, S.C.; Pollock, T.J.; Cappello, J.; Crissman, J.W. U.S. Patent 5,243,038, USA, Protein Polymer Technologies, Inc. 1993.
28. Nagarsekar, A.; Crissman, J.; Crissman, M.; Ferrari, F.; Cappello, J.; Ghandehari, H. Genetic synthesis and characterization of pH- and temperature-sensitive silk-elastinlike protein block copolymers. *J. Biomed. Mater. Res.* **2002**, *62*, 195-203.
29. Cappello, J.; Crissman, J.W.; Crissman, M.; Ferrari, F.A.; Textor, G.; Wallis, O.; Whitledge, J.R.; Zhou, X.; Burman, D.; Aukerman, L.; Stedronsky, E.R. In-situ self-assembling protein polymer gel systems for

administration, delivery, and release of drugs. *J. Controlled Rel.* **1998**, *53*, 105-117.

30. Hoffman, A.S. Hydrogels for biomedical applications. *Adv. Drug Deliv. Rev.* **2002**, *43*, 3-12.
31. Lele, A.K.; Hirve, M.M.; Badiger, M.V.; Mashelkar, R.A. Predictions of bound water content in poly(*N*-isopropylacrylamide) gel. *Macromolecules* **1997**, *30*, 157-159.
32. Netz, P.A.; Dorfmuller, T. Computer simulation studies of anomalous diffusion in gels: Structural properties and probe-size dependence *J. Chem. Phys.* **1995**, *103*, 9074-9082.
33. Yoshioka, S.; Aso, Y.; Terao, T. Effect of water mobility on drug hydrolysis rates in gelatin gels. *Pharm. Res.* **1992**, *9*, 607-612.
34. Shibukawa, M.; Aoyagi, K.; Sakamoto, R.; Oguma, K. Liquid chromatography and differential scanning calorimetry studies on the states of water in hydrophilic polymer gel packings in relation to retention selectivity. *J. Chromatography* **1999**, *832*, 17-27.
35. Higuchi, A.; Iijima, T. DSC investigation of the states of water in poly (vinyl alcohol-co-itaconic acid) membranes. *Polymer* **1985**, *26*, 1833-1837.
36. Liu, W.G.; De Yao, K. What causes the unfrozen water in polymers: Hydrogen bonds between water and polymer chains? *Polymer* **2001**, *42*, 3943- 3947.
37. Tian, Q.; Zhao, X.; Tang, X.; Zhang, Y. Fluorocarbon-containing hydrophobically modified poly(acrylic acid) gels: Gel structure and water state. *J. Appl. Polym. Sci.* **2003**, *89*, 1258.
38. Megeed, Z.; Cappello, J.; Ghandehari, H. Thermal analysis of water in silk-elastinlike hydrogels by differential scanning calorimetry. *Biomacromol.* **2004**, *5*, 793-797.
39. Bouwstra, J.A.; Salomonsdevries, M.A.; Vanmiltenburg, J.C. The thermal behavior of water in hydrogels. *Thermochim Acta* **1995**, *248*, 319-327.
40. Ishikiriyama, K.; Todoki, M. Heat capacity of water in poly(methyl methacrylate) hydrogel membrane for an artificial kidney. *J. Polym. Sci. Polym. Phys.* **1995**, *33*, 791-800.
41. Hoeve, C.A.; Tata, A.S. The structure of water absorbed in collagen. *J. Phys. Chem.* **1978**, *82*, 1660-1663.
42. Kakivaya, S.R.; Hoeve, C.A. The glass point of elastin. *Proc. Natl. Acad. Sci.* **1975**, *72*, 3505-3507.
43. Megeed, Z.; Haider, M.; Li, D.; O'Malley Jr., B.W.; Cappello, J.; Ghandehari, H. *In vitro* and *in vivo* evaluation of recombinant silk-elastin-like hydrogels for cancer gene therapy. *J. Controlled Rel.* **2004**, *94*, 433-445.

44. Fu, J. C.; Hagemeir, C.; Moyer, D. L. A unified mathematical model for diffusion from drug-polymer composite tablets. *J. Biomed. Mater. Res.* **1976,** *10*, 743-758.
45. Siepmann, J.; Ainaoui, A.; Vergnaud, J. M.; Bodmeier, R. Calculation of the dimensions of drug-polymer devices based on diffusion parameters. *J. Pharm. Sci.* **1998**, *87*, 827-832.
46. Dandu, R.; Li, D.; O'Malley Jr., B.W.; Cappello, J.; Ghandehari, H. Matrix-mediated adenoviral gene delivery with recombinant silk-elastinlike hydrogels: in Proceedings of the 31st Annual Meeting and Exposition of the Controlled Release Society, Honolulu, Hawaii, June 12-16, **2004**.
47. Dinerman, A.; Cappello, J.; Ghandehari, H.; Hoag, S.W. Swelling characteristics of silk-elastinlike hydrogels. *Biomaterials* **2002**, *23*, 4203-4210.

Chapter 11

Development and Characterization of Dual-Release Poly(D,L-lactide-*co*-glycolide) Millirods for Tumor Treatment

Brent Weinberg, Feng Qian, and Jinming Gao*

Department of Biomedical Engineering, Case Western Reserve University, Cleveland, OH 44106
***Corresponding author: jinming.gao@case.edu**

In recent years, minimally invasive treatments of solid tumors, such as image-guided radiofrequency (RF) ablation, have emerged as a powerful alternative therapy to surgery for patients with unresectable tumors. One major limitation of RF ablation is frequent recurrence of tumors due to incomplete destruction of cancerous cells at the tumor boundary. Biodegradable polymer millirods, composed of poly(D,L-lactide-*co*-glycolide) (PLGA) and impregnated with anti-cancer agents, have been designed to be implanted in tumors after RF ablation to deliver drugs to the surrounding tissue and kill the remaining tumor cells. By tailoring device design, it is possible to create dual-release polymer implants that incorporate both a burst release to rapidly raise the surrounding tissue drug concentrations as well as a sustained release to maintain those drug concentrations for an extended period of time. Combination of RF ablation with local drug therapy will provide a minimally invasive paradigm for the effective treatment of solid tumors.

Introduction

Cancer is one of the major health issues plaguing modern society, with more than one million new cases and a half million cancer-related deaths expected in the United States in 2004 alone *(1)*. Much of the morbidity and mortality arising from cancer is because of solid tumors in a variety of organs. Over the past several decades, the main method for treatment of tumors has been through surgical resection of the main tumor mass in conjunction with either systemic chemotherapy or local radiation therapy. However, not all cancer patients are good candidates for surgery for several reasons, such as inaccessibility of the tumor to surgical approaches, poor overall general health, or poor expected outcome after surgical resection. As an example, resection of hepatocellular carcinoma has often had limited success with a 5-year survival rate after curative surgery around 30% *(2)*. Such difficulties in treating liver cancer has led to the development of a variety of alternative treatments for hepatic cancer, which include novel drug treatments, and minimally invasive therapies *(3)*.

Nonconventional treatments for liver cancer have grown rapidly in recent years to include such options as intra-arterial chemotherapy, chemo-embolization, ablative techniques, and even orthotopic liver transplantation *(3, 4)*. Because of the high fraction of non-resectable liver cancers, methods for tumor ablation such as ethanol injection, cryoablation, and heat treatment have gained particular favor because they can often be performed in a less invasive manner than full scale surgeries, often laparoscopically or even percutaneously *(5)*. A large effort of current research is directed towards the development of minimally invasive radiofrequency (RF) ablation as a treatment for liver tumors *(6)*. In this technique, an electrode is directly inserted percutaneously into the diseased tissue through the guidance of an imaging technique such as ultrasound *(7)*, magnetic resonance imaging (MRI) *(8)*, or computed tomography (CT) *(9)*. After correct positioning of the electrode, radiofrequency voltage is applied between the electrode tip and a reference electrode, which is most often a large conducting pad placed on the patient's body. Ionic oscillation due to the applied field leads to localized heating to 70-90°C and eventually coagulative necrosis in the region of the electrode tip *(10)*. Major advantages of RF treatment include the limited invasiveness of the procedure (requiring only local anesthesia) and the ability to treat tumors in a variety of regions. However, the procedure has had some limitations, most notably tumor recurrence around the boundary of the ablated region *(7,11-13)*. Many attempts to improve RF ablation have focused on making improvements in electrode design, including the addition of water-cooled tips and multi-tipped electrodes to widen the region of ablation, and therefore, the size of tumors that can be effectively treated with this approach *(10,14)*. One potential option for improving the outcome from RF ablation is to combine the current procedure with an additional treatment, such as local drug delivery to the treated region.

Targeted drug delivery to tumors promises to reduce systemic toxicity from drug side effects while at the same time maximizing treatment efficacy. A variety of strategies have been proposed to deliver drugs more effectively to tumors, one of which uses intravenously injected delivery devices such as liposomes *(15)* and micelles *(16)* that can selectively accumulate in tumors. Other local delivery mechanisms are more direct and involve implantation or injection of a drug delivery device directly into the treatment region. One clinically successful example of an implantable polymer device is the BCNU-containing Gliadel wafer used in the treatment of brain cancer *(17-21)*. Other local delivery methods, such as the injection of polymer microspheres containing anti-cancer drugs are also under development *(22)*. These approaches have inspired the development of implantable polymer devices to be used in conjunction with RF ablation to limit the recurrence of tumors after the procedure.

In our laboratory, biodegradable polymer drug delivery devices impregnated with anti-cancer drugs have been developed for use with RF ablation in the treatment of liver cancer. These implants, also named as polymer millirods, are small, cylindrical (1.6 mm in diameter, 10 mm in length) poly(D,L-lactide-*co*-glycolide), (PLGA), devices that are designed to be implanted in a minimally invasive fashion through a modified 14-gauge biopsy needle. This procedure can be performed percutaneously with image guidance immediately after RF ablation of the tumor region. Because tumor recurrence at the boundary between ablated and normal tissue is the primary problem these implants seek to address, their main function is to deliver a therapeutic level of drug to this tissue boundary. RF ablation of the liver grossly changes the physiological properties of the surrounding tissue. The vascularity of the region is largely destroyed and drug diffusion rates change drastically, and any implant design must take account of these parameters *(23)*. Optimally, these millirods would deliver a quantity of drug that would allow the targeted tissue to reach a therapeutic concentration quickly and stay at that level for as long a time as possible *(24)*. The simplest millirod design is a monolithic system that releases the majority of loaded drug over a short period of time *(25)*. By making improvements to this monolithic design, however, it is possible to create dual-release millirods with optimized release kinetics. These implants release an initial burst of drug to rapidly bring the treatment region to a therapeutic concentration, followed by a sustained dose to maintain this concentration throughout several days *(24)*. In order to successfully develop a drug implant to be used alongside RF ablation, it is critically important to understand the drug delivery properties of both types of devices *in vivo* to facilitate the intelligent design of future drug delivery implants.

Materials and Methods

Materials

Poly(D,L-lactide) (PLA; inherent viscosity 0.67 dL/g) and poly(D,L-lactide-*co*-glycolide) (PLGA 1:1 lactide:glycolide, 50,000 Da molecular weight, 0.65 dL/g inherent viscosity) were purchased from Birmingham Polymers (Birmingham, AL). PLGA microspheres with an average diameter of 4 μm were produced by a single emulsion technique *(25)*. Trypan blue, poly(ethylene glycol) (PEG; M_n 4,600), and poly(ethylene oxide) (PEO; M_w 200,000) were purchased from Aldrich (Milwaukee, WI). 5-Fluorouracil (5-FU), phosphate-buffered saline (PBS), Tris-buffered saline, sodium chloride (NaCl), and methylene chloride were obtained from Fischer Scientific (Pittsburgh, PA). Doxorubicin HCl was purchased from Bedford Laboratories (Bedford, OH), desalted by dialysis, lyophilized, and used in a fine powder form. Teflon tubing (1.6 mm ID, 2.0 mm OD) was purchased from McMaster-Carr Supply (Cleveland, OH).

Fabrication of Monolithic Millirods

Monolithic polymer millirods were fabricated by a compression heat molding technique *(25)*. Briefly, the drug, PLGA microspheres, and any desired excipient molecule were weighed separately according to the desired composition of the implants. These powder components were then placed together in a plastic tube and mixed by a vortex mixer for 10 minutes. After mixing, 50 mg of this mixture was weighed out and placed in a Teflon tube, which was then inserted into a stainless steel mold. The molds were heated to the desired temperature (60, 70, 80, or 90°C) in an oven for two hours, while compressing the powder mixture with stainless steel plungers (1.6 mm OD) with a pressure of 4.6 MPa. The annealed millirods were then pushed out of the Teflon tubing with a stainless steel plunger and allowed to cool. The resulting millirods had a diameter of 1.6 mm and were cut to approximately 8-10 mm in length.

Fabrication of Dual-Release Millirods

Dual-release millirods were manufactured by modification of monolithic millirods with two consecutive dip coating procedures *(26)*. First, monolithic millirods of desired composition were fabricated as described above. Then, a sustained release coating composed of a blend of PEG and PLA was added to this millirod. To add this coating, both polymers were dissolved in methylene chloride to a total concentration of 200 mg/mL. Millirods were then dipped at a speed of 2 mm/s into this solution using a vertically placed syringe pump. The rate of sustained drug release could be controlled by varying the PEG content of the coating from 5-20%. Millirods with this single coating were dubbed sustained release millirods. To impart a burst drug release to the millirods, a second dip coating of drug and PEO (75% drug, 25% PEO, total concentration 100 mg/mL), dissolved in methylene chloride, was applied to the sustained release rods. The amount of burst release was controlled by varying the number of coatings with this burst solution. In this manner, implants that exhibited both burst and sustained release properties were produced.

In Vitro Drug Release

In vitro release studies were performed in an orbital shaker at 37°C and a speed of 100 rpm. Millirods were placed in solution in this shaker, and at each time point were removed and placed into fresh solution. Drug concentrations of the resulting solutions were determined using a UV-Vis spectrophotometer (Perkin-Elmer, Lambda 20), and cumulative release was determined by summing the release throughout all of the study time periods. Studies with Trypan blue and 5-FU were performed in 10 mL vials of PBS buffer, and corresponding studies with doxorubicin were performed in 2 mL vials of Tris buffer. The following absorption wavelengths and extinction coefficients were used: Trypan blue, 597.2 nm, 64.6 mL/(cm mg) *(25)*; 5-FU, 266.1 nm, 46.1 mL/(cm mg) *(27)*; and doxorubicin, 480.8 nm, 16.8 mL/(cm mg) *(26)*.

SEM Analysis

Scanning electron microscopy (SEM, JEOL 840, 20 keV) was performed to gain an understanding of the implant morphology. Millirods were fractured, mounted on an aluminum stub with double sided tape, and sputter-coated with a 10-nm layer of Pd. Images were obtained of both the outer surface and cross-section of the implant.

RF Ablation of Livers and Implantation of Polymer Millirods

All animal procedures followed an approved protocol of the Institutional Animal Care and Use Committee (IACUC) at Case Western Reserve University. Studies were performed in male New Zealand white rabbits *(28)* or male Sprague-Dawley rats *(23)*. Animals were anesthetized, the abdomen shaved and prepped, and the liver exposed through a midline incision. Tissue was ablated using an RF-current generator with a 19-gauge needle electrode to a temperature of 90±2°C for 3 minutes in rabbits and 2 minutes in rats. The ablation volume typically had a radius of 4-6 mm from the electrode tip. After the ablation procedure, millirods were placed along the ablation needle track. For normal tissue studies, the liver lobe was punctured with a needle and the millirods placed into unablated tissue. Animals were sacrificed at the appropriate time points, the millirod explanted, and tissue samples removed for analysis.

Histology Analysis and Drug Distribution in Liver Tissues

After removal of the treated tissue, alternating slices of tissue were either fixed for histological analysis or maintained for concentration analysis. Histological samples were fixed in 10% formalin solution, embedded in paraffin, sliced, and stained. Doxorubicin concentrations were determined by a fluorescence imaging technique *(28)*. Liver tissue was mounted on a cryostat in embedding media and sliced perpendicular to the long axis of the millirod. Images of the slices were then obtained on a fluorescence imager (FluoroImager SI, Molecular Dynamics), and fluorescence intensities converted to doxorubicin concentrations using predetermined calibration values.

Results and Discussion

Monolithic Millirods

Drug Release

Monolithic millirods loaded with 20% (w/w) Trypan blue were fabricated to investigate the effects of annealing conditions on implant properties *(25)*. For this study, compression pressure and annealing time were held constant at 4.6 MPa and 2 hours, respectively, but four sets of implants were annealed at 60, 70, 80, and 90°C. The release of drug was then monitored *in vitro* until more than 90% of the dye loading was released (Figure 1). While all the temperatures are above the glass transition temperature of PLGA (T_g = 45°C), the drug release from the rods manufactured at 60°C differed drastically from the other sets, which suggested that the polymer did not anneal completely at this temperature, allowing for more rapid release of the drug. SEM revealed the morphology of these millirods generated at different temperatures and allowed for some insight into their release properties. The SEM of the 60°C set demonstrated particles that largely retain the original morphology of the component particles, while both the 80°C and 90°C sets reveal PLGA microparticles that are largely annealed. The intermediate temperature, 70°C, led to millirods that were largely annealed on the outer surface of the implant but much less annealed at the core of the implant. Because these millirods share similar release kinetics with the completely annealed implants, it seems that drug release was largely dictated by the outer surface of the implant, which would act as the main barrier for drug release. For the remaining studies, completely annealed implants were desired, and an annealing temperature of 90°C was chosen because it yielded highly reproducible implants.

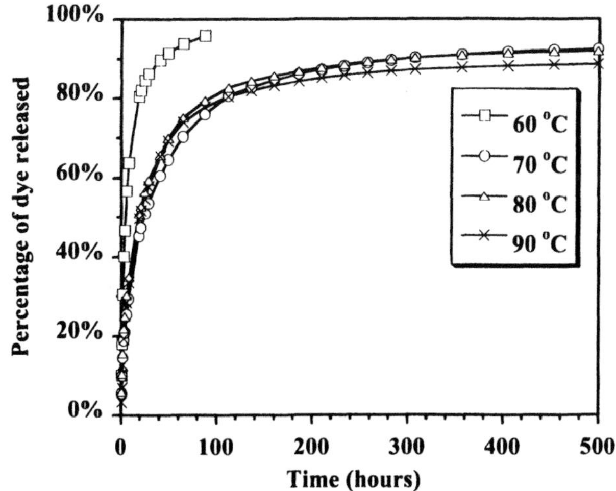

Figure 1. Release profiles of 20% Trypan-blue PLGA millirods annealed at different temperatures. (Reproduced with permission from ref. (25). Copyright 2001 John Wiley & Sons.)

Another major factor in drug release rates was the amount of drug loading. PLGA millirods with various loading densities of 5-FU demonstrated the effects of loading density on drug release *(27)*. Monolithic millirods containing 10, 20, and 30% 5-FU were used, and the release data are demonstrated in Figure 2a. This figure demonstrates a considerably higher release rate for higher drug loadings; by 80 hours, the 30% millirods had released 95% of their original drug loading, while the 10% millirods had released only 25%. Such a dramatic difference can be explained by a percolation behavior, in that enough drug must be present to create a series of interconnecting pores through which drug can be released. SEM evidence for this process is shown in Figures 2b and 2c, which show images of 30% and 10% 5-FU loaded millirod cross sections after 2 days of drug release in PBS. In the 30% millirod, a series of interconnecting channels are seen throughout the cross section, while the 10% millirod contains no channels but only pores near the surface where drug has been released. It can be concluded from these findings that greater concentrations of soluble components lead to accelerated drug release from the millirod implant.

In order to design millirods with rapid release but lower overall drug loadings, it may be desirable to use an excipient molecule, such as PEG, NaCl, or glucose, to act as an additional soluble component of the millirod. For example, an implant containing a 10% (w/w) loading of 5-FU can be created that has similar release kinetics to the 30% (w/w) loaded rod through the addition of 20% NaCl. This strategy has been used throughout these studies to accelerate drug release from the implants.

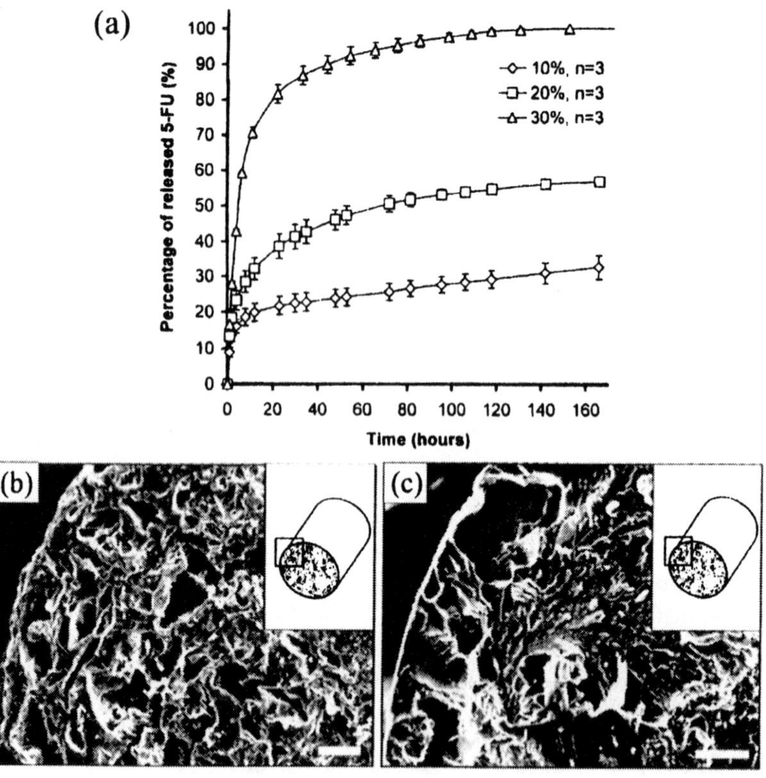

Figure 2. (a) Release profiles of monolithic millirods loaded with 10, 20, and 30% 5-FU. (b,c) SEM of 30% and 10% 5-FU millirod cross section, respectively, after 2 days of drug release in PBS (37 °C). (Adapted with permission from ref. (27). Copyright 2002 Wiley Periodicals, Inc.)

Tissue Distribution

After thorough investigation of drug release properties *in vitro*, the release of drug *in vivo* in both normal and ablated liver tissue was quantified through the use of doxorubicin containing millirods and fluorescent imaging *(28)*. Fluorescence images of drug accumulation in tissues surrounding the implant at 24 hours are shown in Figure 3 for both non-ablated and ablated liver. In normal tissue, a therapeutic concentration of doxorubicin, defined as 6.4 μg/g body weight *(29)*, is found only 1.2 mm from the implant boundary, while in the ablated liver this distance is found to be 5.2 mm. This result supports the conclusion that RF ablation can provide enhanced drug retention and penetration by minimizing drug loss due to perfusion.

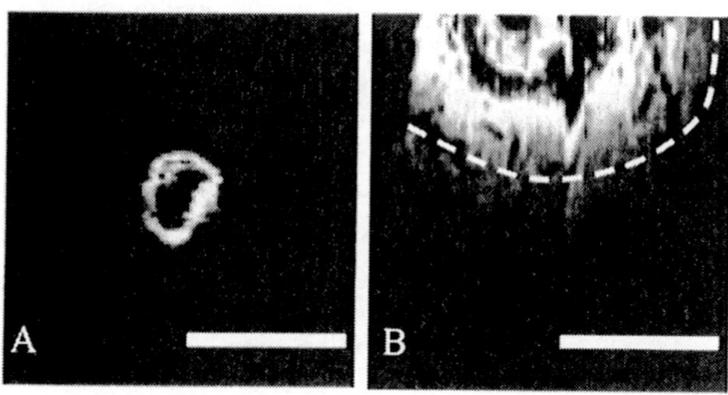

Figure 3. Fluorescence imaging of doxorubicin distribution after 24 hours. (A) Normal liver. (B) Ablated liver, with a white dotted line showing the normal/ablated tissue boundary. Only half of the ablated tissue distribution is shown so that the extensive distribution area can be seen. Scale bars are 3 mm. (Adapted with permission from ref. (28). Copyright 2002 Wiley Periodicals, Inc.)

To further examine the time course of drug distribution from 1-96 hours, doxorubicin-containing millirods were implanted into the livers of rats *(23)*. Again, normal tissue showed only a small therapeutic region, while in the ablated tissue drug accumulated and remained at high levels throughout the course of the entire study. In these experiments, fitting drug concentration levels to a model of drug distribution allowed for calculation of apparent diffusion coefficients for both normal and ablated liver tissue, and an apparent drug elimination rate from normal tissue. The diffusivity in ablated tissue (D_a^*), 1.1×10^{-7} cm^2s^{-1}, was considerably smaller than that of normal tissue (D_n^*), 6.7×10^{-7} cm^2s^{-1}. Reduction of tissue barriers by RF ablation would be expected to increase the value of D_a^*. However, the observed lower apparent diffusivity may reflect drug binding to the cellular debris (e.g. DNA, protein) in the ablated tissue, which slows down drug transport in the surrounding area *(23)*. Moreover, increased accumulation of drug in the ablated region is likely due to drug binding and lack of drug elimination from perfusion in ablated tissue, which was found to occur at a rate of 9.6×10^{-4} s^{-1} in normal liver tissue. Development of this model allowed considerable insight into the mechanism for increased drug accumulation in ablated liver tissue.

Discussion

Monolithic PLGA millirods allowed for the successful release of therapeutic drugs into both normal and ablated liver tissues with the major limitation being the rapid speed of drug delivery. Both *in vitro* and *in vivo*, the majority of loaded drug was delivered within the first 24 hours after implantation, which led to drug concentrations in the surrounding tissue peaking during this same time frame. Ablated tissue had the ability to retain drug for at least 96 hours, but in perfused tissue the drug was quickly washed away. This finding underscores the value of using an intratumoral drug delivery device in conjunction with RF ablation. In both non-ablated tissue and normal tissue surrounding the ablated region, concentrations quickly returned to levels below the therapeutic concentration. In order to have the greatest effect, it is desirable to maintain tissue concentrations in the therapeutic range for the longest period of time possible.

Dual-Release Millirods

Drug Release

Changing the release properties of millirods offers considerable opportunities to improve the tissue drug distribution. Implants releasing their loading quickly provide enough drug to rapidly reach a therapeutic concentration at the target tissue but may not maintain those elevated drug levels for long periods of time. On the other hand, implants providing sustained release often have the opposite problem: they maintain drug concentrations for long periods of time but may take a substantial amount of time to reach that concentration. Dual-release millirods seek to combine a burst release of drug with a sustained dose that is maintained over a long period of time. To accomplish this goal, monolithic millirods are supplemented by two consecutive coatings: an inner layer that acts as a diffusion barrier to provide sustained release *(27)*, and an outer layer that provides a rapid burst of drug soon after implantation.

Doxorubicin-containing dual-release millirods were manufactured using this approach and studied *in vitro (26)*. Dual-release millirods with a core of 16% doxorubicin, 24% NaCl, and 60% PLGA were dip-coated with PEG/PLA. A burst coating was added through three dip-coatings in a doxorubicin/PEO solution. The resulting drug release profiles are shown in Figure 4a. The three formulations shown contain the same burst dose of around 1.5 mg/cm, which is released quickly, but have different sustained release rates that control the total length of drug release. After the burst release, these rods average release of 0.27, 0.43, and 0.60 mg/(day cm of millirod) for B3S1, B3S2, and B3S3 formulations. Most notably, however, these rods are capable of predictable release

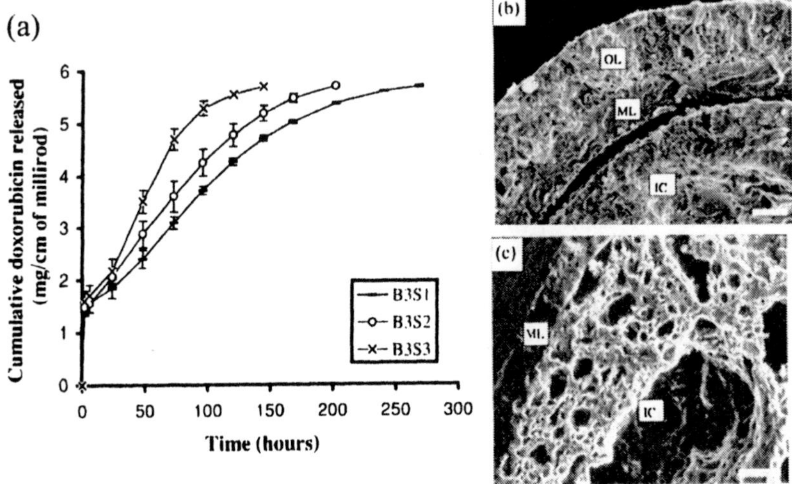

Figure 4. (a) Cumulative release from dual-millirods. The three formulations vary only in PEG content of the sustained layer: B3S1-5%, B3S2-10%, and B3S3-20%. (b) SEM of B3S2 formulation before release showing the outer layer, OL, middle layer, ML, and inner core, IC. (c) SEM of B3S2 after 7 days of release. The scale bars are 100 μm. (Adapted with permission from ref. (26). Copyright 2002 Elsevier Science B.V.)

over a period of more than one week, considerably longer than the monolithic millirods. Figure 4b and 4c show SEM images of dual-release millirods before drug release and after 7 days of release in Tris buffer. All three layers are visible in the pre-release millirod, the outer doxorubicin/PEO layer, the middle PEG/PLA layer, and the inner doxorubicin/NaCl/PLGA core. After 7 days of release, the outer layer is no longer visible due to its complete dissolution, and the middle layer appears porous, most likely due to dissolution of the PEG component of the membrane. The inner core also shows interconnecting channels similar to those in monolithic millirods after complete drug release (Figure 2b). These *in vitro* results confirmed the successful combination of burst and sustained release in these dual-release millirods.

Tissue Distribution

To determine the value of using dual-release millirods *in vivo*, tissue drug distributions using dual-release implants were compared with sustained-release millirods *(24)*. Dual-release implants were fabricated with a core of 5% doxorubicin, 25% NaCl, and 70% PLGA, a middle layer of 10% PEG and 90% PLA, and a burst outer layer of 75% doxorubicin and 25% PEO. Sustained-release implants were fabricated similarly with the exception of the outer burst layer, which was not included. The millirods were implanted into ablated rat livers, and a set of animals was sacrificed and tissue samples taken for fluorescence analysis at each time point. Fluorescence images of the doxorubicin concentrations are shown in Figure 5. The most significant difference in drug release is in the 1-day image, which shows substantially greater drug concentrations in the dual-release group, as would be expected due to the added burst dose. Throughout the remaining time points, the drug concentrations due to the sustained-release implants increase gradually from 1-8 days, eventually resembling those from the dual-release implants. In contrast, the dual-release implants rapidly approach high concentrations that do not vary significantly over the course of the study. These results underscore the value of using dual-release implants to rapidly elevate drug concentrations in the treatment region.

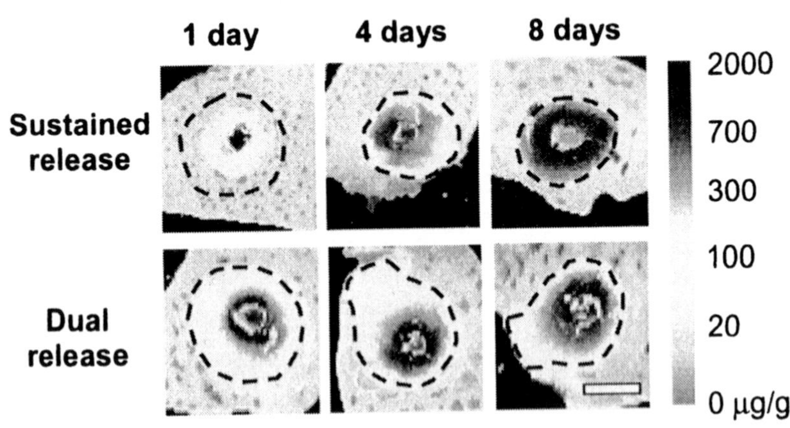

Figure 5. Fluorescence imaging of doxorubicin distribution comparing sustained-release and dual-release millirods in ablated liver over 8 days. The dotted line is the ablation boundary, and the key shows doxorubicin concentration. The scale bar is 5 mm. (Adapted with permission from ref. (24). Copyright 2002 Plenum Publishing.)

Figure 6 shows the time course of the local doxorubicin concentration at the interface between ablated and normal tissue. Similar to the results seen qualitatively in the fluorescence images, the doxorubicin concentration at the boundary increases rapidly to 43.7 µg/g after two days and remains at similar levels through 8 days. While dual-release implants reached boundary concentrations of >30 µg/g within two days, the sustained-release implants did provide these concentrations until day 6-8. The added burst release dramatically increases the ability of the millirods to deliver effective concentrations of drug to the surrounding tissue.

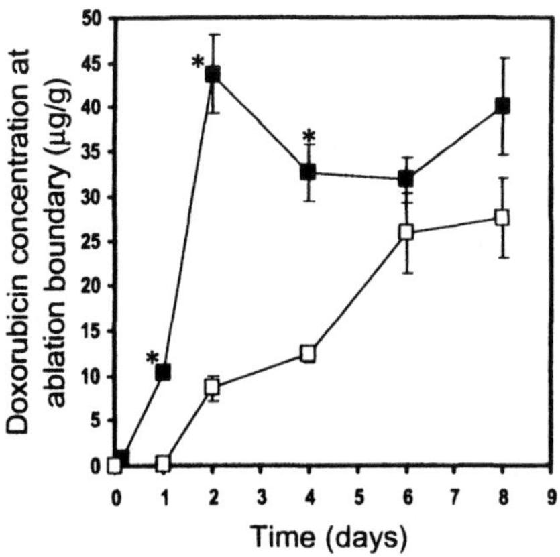

Figure 6. Time curve showing the concentration of doxorubicin at the ablated-normal tissue boundary for sustained- (□) and dual- (■) release millirods. Significant differences between the two ($p < 0.01$, $n = 6$) are marked with an asterisk. (Adapted with permission from ref. (24). Copyright 2002 Plenum Publishing.)

Although dual-release millirods were shown to successfully deliver drug to the boundary between normal and ablated tissue, later studies revealed that the host inflammatory response had a considerable effect on drug transport to the normal tissue surrounding the ablated region *(30)*. Figure 7a shows a Masson trichrome stain of an ablated liver region after 8 days, indicating extracellular collagen deposition (blue band), consistent with the formation of a fibrous capsule and granulation tissue. Figure 7b is a fluorescent micrograph from the same region. Significant doxorubicin fluorescence is visible in the entire ablated region, while little is seen in the adjacent non-ablated region. The dark area between the two regions is the same fibrous band ranging from 0.5-1.0 mm in width which does not accumulate doxorubicin. This fibrous capsule and granulation tissue act as a significant barrier to doxorubicin transport across the boundary, leading to a plateau of drug just inside the boundary.

Discussion

Dual-release millirods successfully combine the release kinetics of a burst release device with sustained release that can be maintained over a period of several days. By tailoring the properties of the individual layers, it becomes possible to control both the burst dose and the sustained release rates to achieve optimal dose kinetics. This approach represents a significant improvement over monolithic drug implants, which released most of their dose within two days. As a result, dual-release millirods were able to rapidly provide therapeutic concentrations to the surrounding tissue and maintain them for 8 days more effectively than either monolithic or sustained-release implants. However, studies of the tissue response to ablation and millirod implantation have revealed a limitation of intratumoral drug delivery in conjunction with RF ablation: the wound healing response due to ablation may significantly restrict drug delivery past the boundary between ablated and normal tissue. Since this region is of key importance in preventing tumor recurrence after RF ablation, this finding will influence the design of intratumoral drug delivery devices in the future. Current studies are investigating the incorporation of anti-inflammatory drugs within the implants to facilitate better drug delivery to the ablation boundary.

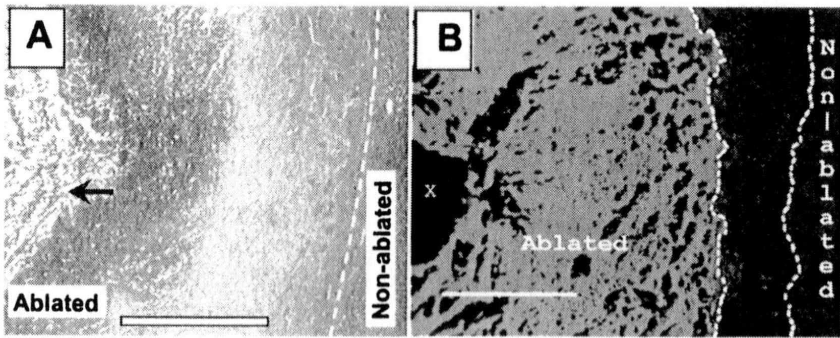

Figure 7. Ablated liver tissue 8 days after millirod implantation. (A) Masson trichrome-stained section showing the ablated-normal tissue boundary. The arrow indicates the direction of the implanted millirod and the dotted line denotes the ablated-normal tissue boundary. (B) Fluorescence micrograph of doxorubicin distribution in this tissue with an X at the millirod location. Scale bars are 1 mm.

Conclusions

In conclusion, PLGA millirods with different drug release properties have been developed for use in conjunction with RF ablation of liver tumors. RF ablation provides a novel, minimally-invasive treatment of tumors; however, tumor recurrence has limited the therapeutic outcome of this treatment. To overcome these limitations, PLGA millirods have been designed to be implanted minimally invasively after RF ablation and deliver anti-cancer drugs to the ablated region, particularly at the ablation boundary where tumor recurrence is most likely to occur. Dual-release millirods combining both a burst and sustained dose of drug were found to have optimal pharmacokinetic properties, delivering therapeutic doses of drug to the ablation boundary quickly and maintaining drug levels for 8 days. Future studies will focus on implementing the knowledge gained from pharmacokinetic studies in a tumor model and making further design improvements, such as the incorporation of anti-inflammatory drugs into the polymer matrix to reduce transport barriers and improve drug delivery efficiency.

Acknowledgements

We thank Drs. John Haaga, James Anderson, Nicholas Stowe, Agata Exner and Mr. Elvin Blanco for their assistance on this project. This work is supported by a research grant from the National Institute of Health (R01-CA-90696). Brent Weinberg was supported in part by NIH T32 GM07250.

References

1. Jemal, A.; Tiwari, R.C.; Murray, T.; Ghafoor, A.; Samuels, A.; Ward, E.; Feuer, E.J.; Thun, M.J. Cancer Statistics, 2004. *CA Cancer J Clin* **2004**, *54*, 8-29.
2. Little, S.A.; Fong, Y. Hepatocellular carcinoma: current surgical management. *Semin Oncol* **2001**, *28*, 474-486.
3. Aguayo, A.; Patt, Y.Z. Nonsurgical treatment of hepatocellular carcinoma. *Semin Oncol* **2001**, *28*, 503-513.
4. Kashef, E.; Roberts, J.P. Transplantation for hepatocellular carcinoma. *Semin Oncol* **2001**, *28*, 497-502.
5. Barnett, C.C., Jr.; Curley, S.A. Ablative techniques for hepatocellular carcinoma. *Semin Oncol* **2001**, *28*, 487-496.
6. Goldberg, S.N.; Gazelle, G.S.; Compton, C.C.; Mueller, P.R.; Tanabe, K.K. Treatment of intrahepatic malignancy with radiofrequency ablation: radiologic-pathologic correlation. *Cancer* **2000**, *88*, 2452-2463.
7. Francica, G.; Marone, G. Ultrasound-guided percutaneous treatment of hepatocellular carcinoma by radiofrequency hyperthermia with a 'cooled-tip needle'. A preliminary clinical experience. *Eur J Ultrasound* **1999**, *9*, 145-153.
8. Merkle, E.M.; Boll, D.T.; Boaz, T.; Duerk, J.L.; Chung, Y.C.; Jacobs, G.H.; Varnes, M.E.; Lewin, J.S. MRI-guided radiofrequency thermal ablation of

implanted VX2 liver tumors in a rabbit model: demonstration of feasibility at 0.2 T. *Magn Reson Med* **1999**, *42*, 141-149.

9. Hahn, P.F.; Gazelle, G.S.; Jiang, D.Y.; Compton, C.C.; Goldberg, S.N.; Mueller, P.R. Liver tumor ablation: real-time monitoring with dynamic CT. *Acad Radiol* **1997**, *4*, 634-638.
10. Goldberg, S.N. Radiofrequency tumor ablation: principles and techniques. *Eur J Ultrasound* **2001**, *13*, 129-147.
11. Dodd, G.D., 3rd; Soulen, M.C.; Kane, R.A.; Livraghi, T.; Lees, W.R.; Yamashita, Y.; Gillams, A.R.; Karahan, O.I.; Rhim, H. Minimally invasive treatment of malignant hepatic tumors: at the threshold of a major breakthrough. *Radiographics* **2000**, *20*, 9-27.
12. Lencioni, R.; Goletti, O.; Armillotta, N.; Paolicchi, A.; Moretti, M.; Cioni, D.; Donati, F.; Cicorelli, A.; Ricci, S.; Carrai, M.; Conte, P.F.; Cavina, E.; Bartolozzi, C. Radio-frequency thermal ablation of liver metastases with a cooled-tip electrode needle: results of a pilot clinical trial. *Eur Radiol* **1998**, *8*, 1205-1211.
13. Buscarini, L.; Rossi, S. Technology for Radiofrequency Thermal Ablation of Liver Tumors. *Semin Laparosc Surg* **1997**, *4*, 96-101.
14. Lorentzen, T. A cooled needle electrode for radiofrequency tissue ablation: thermodynamic aspects of improved performance compared with conventional needle design. *Acad Radiol* **1996**, *3*, 556-563.
15. Gabizon, A.; Shmeeda, H.; Barenholz, Y. Pharmacokinetics of pegylated liposomal Doxorubicin: review of animal and human studies. *Clin Pharmacokinet* **2003**, *42*, 419-436.
16. Kwon, G.S.; Kataoka, K. Block-Copolymer Micelles as Long-Circulating Drug Vehicles. *Adv Drug Deliv Rev* **1995**, *16*, 295-309.
17. Fung, L.K.; Shin, M.; Tyler, B.; Brem, H.; Saltzman, W.M. Chemotherapeutic drugs released from polymers: distribution of 1,3-bis(2-chloroethyl)-1-nitrosourea in the rat brain. *Pharm Res* **1996**, *13*, 671-682.
18. Strasser, J.F.; Fung, L.K.; Eller, S.; Grossman, S.A.; Saltzman, W.M. Distribution of 1,3-bis(2-chloroethyl)-1-nitrosourea and tracers in the rabbit brain after interstitial delivery by biodegradable polymer implants. *J Pharmacol Exp Ther* **1995**, *275*, 1647-1655.
19. Tamada, J.; Langer, R. The development of polyanhydrides for drug delivery applications. *J Biomater Sci Polym Ed* **1992**, *3*, 315-353.
20. Langer, R. Biodegradable polymers for drug delivery to the brain. *ASAIO Trans* **1988**, *34*, 945-946.
21. Domb, A.; Maniar, M.; Bogdansky, S.; Chasin, M. Drug delivery to the brain using polymers. *Crit Rev Ther Drug Carrier Syst* **1991**, *8*, 1-17.
22. Menei, P.; Jadaud, E.; Faisant, N.; Boisdron-Celle, M.L.; Michalak, S.; Fournier, D.; Delhaye, M.; Benoit, J.H.P. Stereotaxic implantation of 5-fluorouracil-releasing microspheres in malignant glioma. *Cancer* **2004**, *100*, 405-410.

23. Qian, F.; Stowe, N.; Liu, E.H.; Saidel, G.M.; Gao, J. Quantification of *in vivo* doxorubicin transport from PLGA millirods in thermoablated rat livers. *J Control Rel* **2003**, *91*, 157-166.
24. Qian, F.; Stowe, N.; Saidel, G.M.; Gao, J. Comparison of doxorubicin concentration profiles in radiofrequency-ablated rat livers from sustained- and dual-release PLGA millirods. *Pharm Res* **2004**, *21*, 394-399.
25. Qian, F.; Szymanski, A.; Gao, J. Fabrication and characterization of controlled release poly(D,L-lactide-co-glycolide) millirods. *J Biomed Mater Res* **2001,** *55*, 512-522.
26. Qian, F.; Saidel, G.M.; Sutton, D.M.; Exner, A.; Gao, J. Combined modeling and experimental approach for the development of dual-release polymer millirods. *J Control Rel* **2002,** *83*, 427-435.
27. Qian, F.; Nasongkla, N.; Gao, J. Membrane-encased polymer millirods for sustained release of 5- fluorouracil. *J Biomed Mater Res* **2002**, *61*, 203-211.
28. Gao, J.; Qian, F.; Szymanski-Exner, A.; Stowe, N.; Haaga, J. *In vivo* drug distribution dynamics in thermoablated and normal rabbit livers from biodegradable polymers. *J Biomed Mater Res* **2002**, *62*, 308-314.
29. Ridge, J.A.; Collin, C.; Bading, J.R.; Hancock, C.; Conti, P.S.; Daly, J.M.; Raaf, J.H. Increased adriamycin levels in hepatic implants of rabbit Vx-2 carcinoma from regional infusion. *Cancer Res* **1988**, *48*, 4584-4587.
30. Blanco, E.; Qian, F.; Weinberg, B.; Stowe, N.; Anderson, J.M.; Gao, J. Effect of fibrous capsule formation on doxorubicin distribution in radiofrequency ablated rat livers. *J Biomed Mat Res* **2004**, *69A*, 398-406.

Chapter 12

Prediction of Drug Solubility in Adhesive Matrix for Transdermal Drug Delivery Based on a Solvation Parameter Model

Jianwei Li[*] and Jeremy J. Masso

Transdermal Drug Delivery, 3M Drug Delivery Systems, 3M Center, Building 260–3A–05, St. Paul, MN 55144
[*]Corresponding author: jli7@mmm.com

The work presented here establishes the relationship between drug solubility in two acrylate adhesives with a previously defined drug-polymer interaction parameter and drug solubility in acetonitrile. The drug solvation parameter model decomposes a hard-to-measure quantity into two easily measurable parameters. It is concluded that there is an excellent linear relationship for the parameters involved. Thus, the model can be used to compute the solubility in the polymer for new drug candidates. Moreover, the two parameters in the relationship can be either easily measured or computed based on their molecular properties by the solvation parameter model. The methodologies presented can be applied to other adhesives.

Introduction

One of the critical parameters in the optimization of drug-in-adhesive (DIA) transdermal formulations is drug solubility within the adhesive. The desired solubility not only provides optimal permeation flux of the drug but also meets the therapeutic dose requirement (e.g., mg per day) within a reasonable patch size *(1,2)*. Although it is a critical parameter, there is no easy quantitative measurement method available for the purpose of formulation study due to the complexity of the adhesive matrix. Perhaps two of the most commonly used methods are accelerated crystal seeding and differential scanning calorimetry (DSC) *(3,4)*. The first method is essentially qualitative, while DSC is the method of choice if a quantitative study is performed. Both experiments, however, are time-consuming and require significant optimization. Accordingly, a quick methodology to estimate the drug solubility in a polymer matrix for transdermal drug delivery is highly desirable. It should be noted that either polymer or adhesive is used interchangeably to refer to the polymers used in transdermal formulations.

Several requirements have to be satisfied for the development of a reliable methodology. First, a theoretical basis is required relating drug solubility to predictable parameters that can be connected to the properties of the drug (and adhesive) by quantitative structure-property relationships (QSPR) *(5)*. Second, a quantitative method is needed to measure drug solubility within the adhesive. The DSC technique is adopted in this study because it is quantitative and relatively easy to perform. Furthermore, it has been widely used to characterize drug formulations and to study drug-excipient interaction or compatibility *(6-9)*. Third, model drugs have to be carefully selected for calibration of the theoretical relationship, covering a wide range of drug properties.

A novel approach to the characterization of drug-polymer interactions has been described previously *(10)*. In this approach, a dry adhesive is allowed to swell in dilute acetonitrile (ACN) solutions of probe compounds. After the swelling, the dissolved drugs can interact with polymer fragments or monomer functional groups, resulting in a decrease in the drug concentration due to sorption. The sorbed amount of molecules is an indication of the strength of the interaction of each compound with the polymer relative to ACN, and can be considered as an interaction parameter. This parameter is the basis of the current study.

As will be shown in the theoretical section, drug solubility within an adhesive can be thermodynamically decomposed (linearly) into the interaction parameter and the drug solubility in ACN. Both parameters have been shown in previous studies to correlate well with molecular properties (or descriptors) of the drug molecule by the solvation parameter model. In other words, the two simple parameters can be computed based on these molecular properties. Thus, if we can "calibrate" the thermodynamic relationship (eq 5) with model drugs or compounds to determine the model coefficients for an adhesive, this relationship, combined with the calculation of interaction parameter and drug solubility in ACN, can be used to predict the solubility within the adhesive for a new drug candidate.

This study is intended to validate and establish the relationships between drug solubility in two acrylate adhesives with the drug-polymer interaction parameter and drug solubility in ACN. A series of reference compounds or drugs is selected, and their solubility values in adhesives are measured by DSC technique. The data are then used to establish the relationships for the estimation of the solubility of new drug candidates.

Theoretical Consideration

The Solvation Parameter Model

The solvation parameter model is a type of general quantitative structure-activity relationship. It relates the physical properties to several types of inter-molecular interactions (dispersion, polar, hydrogen bond association, and hydrophobic interactions) *(11-12)*. A good example is the drug solubility (S_w) in water *(13)*:

$$Log\left(S_w\right) = c + eE + sS + aA + bB + vV + gAB \quad (1)$$

where c is the regression constant; E, S, A, B, and V are called solute or drug descriptors, representing their physicochemical properties; and e, s, a, b, v, and g are the coefficients. The descriptors are the excess molar refraction (E); dipolarity/polarizability (S); "overall" or "effective" hydrogen bond acidity (A) and basicity (B); and McGowan characteristic molar volume (V) *(12)*.

Equation 1 contains five product terms representing the properties (descriptors) of the drug molecules and the system (solvent) (a gAB term is used to describe solute-solute interactions, and usually absent in the model). The e-constant is the capacity of water to interact with the drug molecules through π or n electron pairs. The s-constant is the capacity of water to take part in dipole-dipole and dipole-induced dipole interactions. The a-constant is a measure of the hydrogen-bonding (HB) basicity of water, and the b-constant is a measure of its HB acidity. Finally, the v-constant is a measure of the ease of cavity formation in water. Similar equations have been obtained for retention in chromatography, octanol-water partition coefficient, skin permeation coefficient, etc. *(12,14)*.

Molecular descriptors are available for more than 4000 compounds, and further values can be obtained by parameter estimates or experiment. *(15)* A software package (*Absolv*) is available to calculate molecular properties (Sirius Analytical Instruments Ltd., East Sussex, UK).

Physicochemical Characterization of Adhesives

We have reported a methodology to characterize the physicochemical properties of an adhesive that are directly related to transdermal drug delivery *(10)*. A set of acetonitrile solutions of judiciously selected probe compounds with known physicochemical properties was brought in contact with a dry adhesive. After swelling of the adhesive, interactions of solutes with swollen polymer will occur. The strength of these interactions is indicated by the sorbed amount of molecules ($\overline{\delta n}$) onto the polymer (proportional to binding constant), and related to the properties of the solutes. As a result, the sorbed amount of probe molecules can be described by an equation similar to eq 1:

$$Ln\left(\overline{\delta n}\right) = c + eE + sS + aA + bB + vV \qquad (2)$$

The sorbed amount is a result of differential interactions between the solute (drug) and both, the polymer and ACN (both regarded as solvent), through the intermolecular interactions described above. The system constants in eq 2 are defined by the complementary (differential) interactions of both adhesive and ACN with the descriptors, and contain rich information of the properties of the polymer. The e-constant is a measure of the difference in system polarizability (or the capacity of the polymer relative to ACN to interact with the probe molecules through π or n electron pairs). The s-constant is a measure of the difference in system dipolarity/polarizability (or the capacity of the polymer relative to ACN to take part in dipole-dipole and dipole-induced dipole interactions). The a-constant is a measure of the difference in HB basicity of the polymer relative to ACN, and the b-constant is a measure of the difference of HB acidity of the polymer relative to ACN. Finally, the v-constant is a measure of system hydrophobicity of the polymer relative to ACN (or relative ease of forming a cavity for the probe in the polymer relative to ACN). ACN is a polar solvent, and essentially does not possess HB capability. The coefficients in eq 2 will mainly reflect the relative polar and absolute HB properties of the adhesive *(10)*.

Finally, the relationship between the sorbed amount and physicochemical properties of the probe compound can be regarded as a valuable model to predict the sorbed amount for a new drug entity. This sorbed amount was defined as a drug-polymer interaction parameter or index, as mentioned earlier.

Drug-Polymer Interactions

As stated, $\overline{\delta n}$ is proportional to the binding constant with the swollen adhesive. Therefore, $Ln\left(\overline{\delta n}\right)$ is the difference in interaction (energy) of drug molecules with the adhesive relative to ACN *(16)*:

$$Ln\left(\overline{\delta n}\right) = E_{Drug-Adhesive} - E_{Drug-ACN} \tag{3}$$

where $E_{Drug-Adhesive}$ and $E_{Drug-ACN}$ denote the drug-adhesive and drug-ACN interaction energy, respectively. It is then derived from eq 3 that the degree of drug-polymer interactions can be related to $Ln\left(\overline{\delta n}\right)$ and $E_{Drug-ACN}$ as follows:

$$E_{Drug-Adhesive} = Ln\left(\overline{\delta n}\right) + E_{Drug-ACN} \tag{4}$$

If the degree of drug-ACN interactions is described by the solubility ($Ln(S_{ACN})$) in CAN, and the drug solubility ($Ln(S_{Adh})$) in the adhesive is used as an indication of the absolute drug-polymer interactions, eq 4 can be further modified to:

$$Ln(S_{Adh}) = Ln(S_{Adh})_0 + pLn\left(\overline{\delta n}\right) + qLn(S_{ACN}) \tag{5}$$

where $Ln(S_{Adh})_0$ is an intercept constant, and p and q are model coefficients.

Equation 5 indicates that the logarithm of drug solubility (mol/weight) in the adhesive is linearly related to the sorbed amount of molecules onto the adhesive and the drug solubility in ACN. In fact, the measurement of drug solubility in the adhesive, usually a difficult and time-consuming experiment, has been transformed into two relatively simple and accurate measurements. If the constant and coefficients ($Ln(S_{Adh})_0$, p and q) are determined *a priori*, the results of $Ln\left(\overline{\delta n}\right)$ and $Ln(S_{ACN})$ measurements can be used to compute the drug solubility in the adhesive. Furthermore, if $Ln\left(\overline{\delta n}\right)$ and $Ln(S_{ACN})$ can be correlated with the molecular properties of drug molecules based on a set of reference compounds by the solvation parameter model, $Ln(\overline{\delta n})$ and $Ln(S_{ACN})$ can be calculated for a new drug candidate, which in turn allows computation of the drug solubility in the adhesive.

Materials and Methods

Determination of the Sorbed Amount onto Adhesives

Two adhesives (IOA/ACM/VoAC and IOA/HEA/VoAC) are evaluated in this study (IOA=isooctyl acrylate, ACM=acrylamide, HEA=2-hydroxyethyl acrylate, and VoAC=vinyl acetate). The monomer ratio for IOA/ACM/VoAC is 75/5/20, and that for IOA/HEA/VoAC is 58/20/18 with 4% Elvacite macromer. The monomer structures are shown in Figure 1. It is noted that ACM is mainly a polar component, and HEA possesses a significant HB capability due to its hydroxyl group.

The reference drugs/compounds used in this study for the two adhesives are shown in Tables I and II. All chemicals used in this study were acquired from Aldrich (Aldrich Chemical Company). The sorbed amounts of molecules onto the two adhesives have been measured previously (*10*) or estimated based on eqs 10 and 11 (see Results and Discussion section), and are also shown in Tables I and II. All experimental procedures for the preparation of adhesives and measurement of the sorbed amount of molecules have been described previously (*10*).

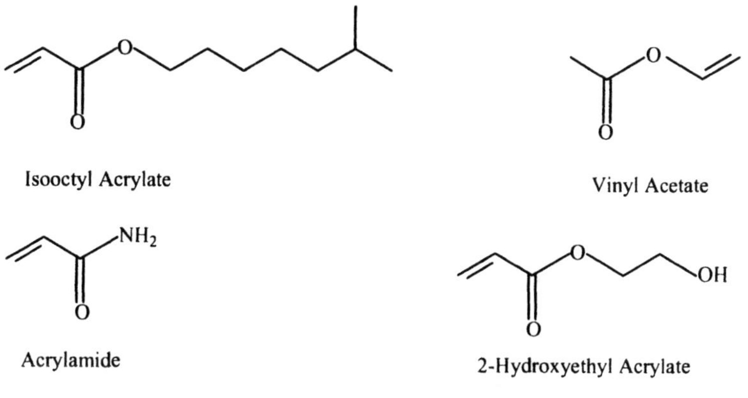

Figure 1. Monomer structures used in this study.

Table I. Drugs used for the correlation of solubility in IOA/ACM/VoAC adhesive.

Compound	$Ln(\overline{\delta n})$ *Per g Adhesive*	$Ln(S_{ACN})$ *(mol/100 g ACN)*	S_{Adh} *(%, w/w)*	$Ln(S_{Adh})$ *(mol/100 g Formulation)*
3-Cyanophenol	3.94	0.41	21.2	-1.73
4'-Aminoacetophenone	3.26	-2.93	5.1	-3.27
4-Nitroaniline	3.29	-1.74	8.9	-2.74
4-Nitrophenol	4.29	0.40	28.3	-1.59
Benzamide	3.46	-2.71	6.5	-2.92
Butyl Paraben	4.18	-0.56	21.6	-2.19
Estradiol	3.54	-6.24	7.9	-3.54
Estriol	3.47	-8.80	4.0	-4.28
Lidocaine	2.65	0.74	15.5	-2.71
Pentachlorophenol	5.15	-2.83	38.0	-1.95

Table II. Drugs used for the correlation of solubility in IOA/HEA/VoAC adhesive.

Compound	$Ln(\overline{\delta n})^{a}$ *Per g Adhesive*	$Ln(S_{ACN})^{b}$ *(mol/100 g ACN)*	S_{Adh} *(%, w/w)*	$Ln(S_{Adh})$ *(mol/100 g Formulation)*
Estradiol	3.49	-6.24	5.0	-4.00
Estriol	3.55	-8.80	2.0	-4.97
Estrone	2.77	-6.37	1.5	-5.19
Levonorgestrel	2.71	-7.80	1.5	-5.34
Cortexolone	2.59	-6.17	1.5	-5.44
2,4,5-Trichlorophenol	3.87	0.94	60.0	-1.19
Ethyl Paraben	3.64	-1.80	15.0	-2.41
Butyl Paraben	3.45	-0.56	35.0	-1.71
Pentachlorophenol	4.34	-2.83	45.0	-1.78
4'-Aminoacetophenone	3.26	-2.93	10.0	-2.60
3-Cyanophenol	3.89	0.41	35.0	-1.22
Lidocaine	3.04	0.74	25.0	-2.24

Measured in this study or calculated by eq 11[a] or eq 13[b].

Measurement of Drug Solubility in Adhesive by DSC

The drug solubility within the polymer matrix was measured by DSC. The measurement is based on the heat (enthalpy change) needed to melt a quantitative amount of solid crystals in the polymer. The calculation of the method is based on the following assumptions: (i) only solid not dissolved in the polymer matrix (i.e., free solid) contributes to the enthalpy change, and (ii) the amount of solid dissolved by a given mass of polymer matrix is the same for all concentrations greater than the solubility. The following equation is then derived to linearly relate the drug loading to the melting enthalpy change *(16-18)*:

$$\chi = S_{Adh} + k\left[\Delta H^0(\chi)\right] \tag{6}$$

where χ is the drug concentration (%, w/w) in a series of formulations, $\Delta H^0(\chi)$ is the melting enthalpy of the drug in the formulation, and k is the slope. The intercept (S_{Adh}) is the drug solubility. Figure 2 illustrates an example of the measurement using 4'-aminoacetophenone. Drug solubility values in adhesive IOA/ACM/VoAC are taken from a previous publication *(16)* and indicated in Table I. Drug solubility values in IOA/HEA/VoAC adhesive are measured in this study and shown in Table II. The preparation of drug/adhesive formulations together with DSC operating conditions have been explained previously *(16)*. It is shown in Tables I and II that the drug solubility in both adhesives covers a wide range.

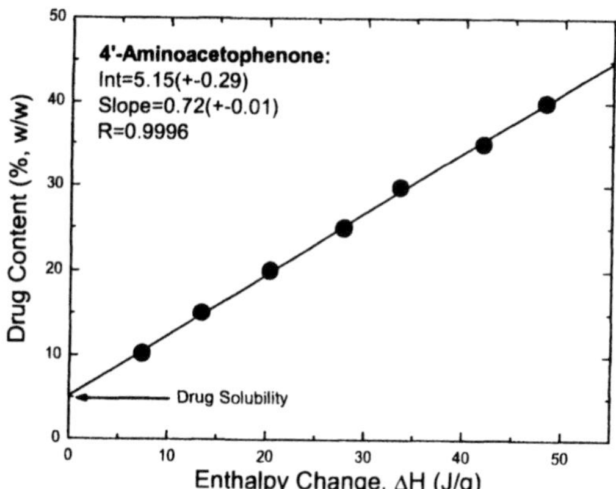

Figure 2. Determination of 4'-aminoacetophenone solubility in IOA/ACM/VoAC adhesive by DSC.

Measurement of Solubility in ACN

The measurement of drug solubility in ACN has been described previously *(16)*. Briefly, a small volume of ACN was saturated with an excess amount of each drug at ambient temperature (25°C). About 200 μL of each supernatant was transferred to a 100-mL volumetric flask by pipette, and the flask was weighed quickly and accurately (weight designated as W_T). Then the sample was diluted with additional CAN, and the concentration of the solution measured by liquid chromatography. The weights (W_D) of compounds in ACN were computed based on the volume of the dilution and their concentrations. The amount (W_{ACN}) of ACN in the samples was computed as the difference (W_T-W_D) of the total weight and drug weight. The solubility in ACN in terms of mol/100 g ACN was computed as follows:

$$S_{ACN} = \left(\frac{W_D}{W_{ACN}}\right)\left(\frac{100}{MW}\right) = \left(\frac{W_D}{W_T - W_D}\right)\left(\frac{100}{MW}\right) \quad (7)$$

Tables I and II also include the solubility values in ACN. They were either measured in a previous study *(16)* or calculated by eq 13.

Results and Discussion

Correlation of Drug Solubility in Adhesive with Drug-Polymer Interaction Parameter

To evaluate the suitability of the drug solubility model (eq 5) in adhesive systems, the data in Table I for IOA/ACM/VoAC adhesive are first analyzed in terms of the quality of the fit and the accuracy of the calculated values. If the model accurately and correctly describes the dependence of drug solubility, the correlation should be both precise and accurate. The correlation results and statistics based on the data in Table I are shown in Table III. Note that the solubility in Tables III and IV is in mol/100 g formulation.

It can be seen from Table III that the linear relationship between the drug solubility in the adhesive and two independent variables ($Ln(\overline{\delta n})$ and $Ln(S_{ACN})$) is very acceptable, as indicated by the correlation coefficient (0.989) and standard error (SE= 0.146). The adjusted correlation coefficient is greater than 0.97, and the relative standard error in $Ln(S_{Adh})$ is no more than 5%. The relative model error in S_{Adh} on average is about 10%. This level of linearity is considered excellent taking into consideration the complexity of the system. The quality of the fit is further examined in Figure 3, illustrating the calculated vs.

Table III. Coefficients and statistics for the correlation for IOA/ACM/VoAC adhesive.

Term	*Coefficient*	*Coefficient Value*[a]	*t-Statistic*	*P-Value*
$Ln(S_{Adh})_0$		-4.49(0.27)	-16.6	7.16E-07
$Ln(\overline{\delta n})$	p	0.63(0.07)	8.9	4.43E-05
$Ln(S_{ACN})$	q	0.22(0.02)	14.1	2.12E-06
Statistics:				
R=0.989; Adj. R=0.971; SE=0.146; RSE(%)[b]=4.43; RME(%)[c]=10; F=150				

[a]Values in parentheses indicate the standard errors of the coefficients. [b]Relative standard error of the fit, computed by the standard error divided by half range of $Ln(S_{Adh})$. [c]Relative model error (average) in S_{Adh}, computed by $\overline{100|\sigma_i|}$ based on the error propagation, where σ_i is the fit residual for each compound. The average value is reported.

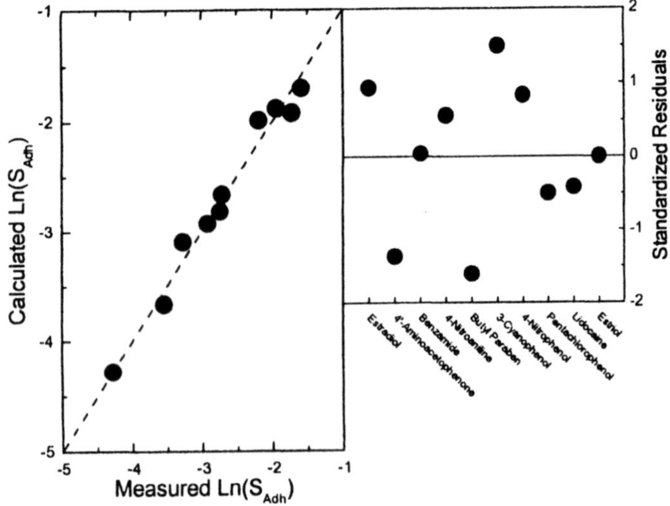

Figure 3. Relationship between computed and measured $Ln(S_{Adh})$ (left), and standardized residuals against the drug compounds for IOA/ACM/VoAC adhesive (right). (Reproduced with permission from reference (16). Copyright 2002 Elsevier Science B.V.)

measured solubility. The calculated amounts were computed by the $Ln(\overline{\delta n})$ and $Ln(S_{ACN})$ values (Table I) and regression coefficients (Table III) of eq 5.

It can be clearly seen from Figure 3 that calculated values are very well linearly correlated with measured values. The slope of the correlation is essentially unity. Thus, eq 5 gives accurate predictions of the values actually observed. It can be observed from Figure 3 (right, standardized residual plot) that the residual distribution is symmetrical, and the absolute standardized residuals are no more than 3. There is not one single unusually large residual in the data. It is noted that the standardized residuals are determined by the residuals divided by the standard error of the linear regression. Overall, the results in Table III and Figure 3 clearly validate the precision and accuracy of the relationship (eq 5). The final model can be written as follows:

$$Ln(S_{Adh}) = \underbrace{-4.49 + 0.63Log(\overline{\delta n}) + 0.22Log(S_{ACN})}_{IOA/ACM/VoAC} \quad (8)$$

$$n = 10,\ R = 0.989,\ SE = 0.146,\ F = 150$$

Table IV shows coefficients and statistics for IOA/HEA/VoAC adhesive.

Table IV. Coefficients and statistics for the correlation for IOA/HEA/VoAC adhesive.

Term	*Coefficient*	*Coefficient Value*[a]	*t-Statistic*	*P-Value*
$Ln(S_{Adh})_0$		-6.18(0.65)	-9.5	5.71E-06
$Ln(\overline{\delta n})$	p	1.25(0.18)	7.1	5.90E-05
$Ln(S_{ACN})$	q	0.36(0.03)	13.3	3.14E-07
Statistics:				
R=0.990; Adj. R=0.986; SE=0.277; RSE(%)[b]=13.0; RME(%)[c]=19.5; F=200.				

Note: Notation same as in Table III.

The correlation coefficient is 0.990 and SE is 0.277. The model error in S_{Adh} is 19.5% on average. This level of model error is acceptable in transdermal formulation study. Figure 4 shows the plot between the computed and measured solubility. It can be seen that all data points are scattered across the diagonal line and the residuals are symmetrically distributed, thereby indicating that the linear dependence model (eq 5) correctly describes the solubility data for this adhesive. The final model equation is then given as follows:

$$Ln(S_{Adh}) = \underbrace{-6.18 + 1.25Log(\overline{\delta n}) + 0.36Log(S_{ACN})}_{IOA/HEA/VoAC} \qquad (9)$$
$$n = 12,\ R = 0.990,\ SE = 0.277,\ F = 200$$

Accordingly, the drug-polymer interaction parameter defined previously and the drug solubility in ACN are experimentally confirmed to be linearly related to the drug solubility in the adhesive. Equations 8 and 9 provide the foundation for the calculation of solubility in the two adhesives for new drug candidates in a transdermal feasibility study. The two simple parameters ($Ln(\overline{\delta n})$ and $Ln(S_{ACN})$) can be easily measured experimentally or calculated based on the correlations established previously by the solvation parameter model (see next section).

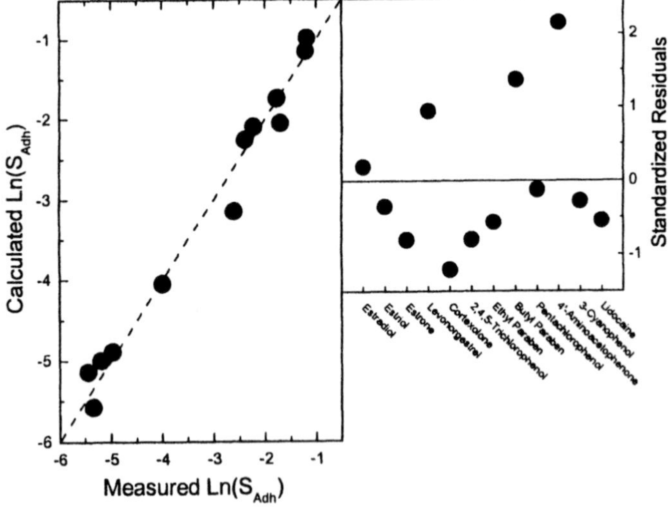

Figure 4. Relationship between computed and measured $Ln(S_{Adh})$ (left), and standardized residuals against the drug compounds for IOA/HEA/VoAC adhesive (right).

Correlation of $Ln(\overline{\delta n})$ and $Ln(S_{ACN})$ with the Molecular Descriptors by the Solvation Parameter Model

The $Ln(\overline{\delta n})$ and $Ln(S_{ACN})$ parameters have been measured for a set of reference compounds in previous studies *(10,16)*. The measured values have been correlated with the physicochemical properties or molecular descriptors of the compounds. The results are summarized in this section.

The sorbed amount of molecules for IOA/ACM/VoAC adhesive is described by the following equation:

$$Ln(\overline{\delta n}) = \underbrace{3.78 + 0.07E - 0.56S + 1.54A - 0.78B + 0.21V}_{IOA/ACM/VoAC} \quad (10)$$

$$n = 18,\ R = 0.972,\ SE = 0.167,\ F = 41$$

For IOA/HEA/VoAC adhesive, the sorbed amount is related to the molecular descriptors as follows:

$$Ln\left(\overline{\delta n}\right) = \underbrace{3.37 + 0.36E - 0.54S + 0.96A - 0.11B + 0.00V}_{IOA/HEA/VoAC} \tag{11}$$
$$n = 17,\ R = 0.976,\ SE = 0.096,\ F = 45$$

The drug solubility in ACN has been correlated by the following two equations:

$$Ln\left(S_{ACN}\right) = 4.59 - 2.03S + 1.92A + 4.00B - 2.83V - 6.47A \cdot B \tag{12}$$
$$n = 39,\ R = 0.971,\ SE = 0.771,\ F = 87$$

and

$$Ln\left(S_{ACN}\right) = 4.12 - 1.10V - 0.34\left(\frac{T_m}{10}\right) \tag{13}$$
$$n = 41,\ R = 0.978,\ SE = 0.600,\ F = 415$$

where T_m is the melting point of the drug.

For a new drug candidate, the molecular descriptors should be obtained by either software or estimation method based on the molecular structure *(12,14)*. The obtained descriptors can then be used to compute $Ln\left(\overline{\delta n}\right)$ and $Ln\left(S_{ACN}\right)$, which in turn are used to calculate the drug solubility in the two adhesives by either eq 8 or eq 9.

Comparison of Dependence of Solubility on Adhesive Properties

Because the two adhesives contain different monomer units, the drug solubility in the two adhesives should be different. In IOA/ACM/VoAC adhesive, ACM is essentially a very polar component. However, HEA in IOA/HEA/VoAC adhesive is not only moderately polar, but also possesses a considerable HB capability. The HB property of the adhesive should increase the solubility of drugs with strong HB capability. To compare the two adhesives, $Ln\left(\overline{\delta n}\right)$ and $Ln\left(S_{ACN}\right)$ in eq 8 are substituted by eqs 10 and 13, to yield the following equation:

$$Ln\left(S_{Adh}\right) = \underbrace{\begin{matrix} -1.19 + 0.04E - 0.35S + 0.97A \\ -0.49B - 0.11V - 0.08\left(\frac{T_m}{10}\right) \end{matrix}}_{IOA/ACM/VoAC} \tag{14}$$

A similar substitution of eq 9 by eqs. 11 and 13 leads to the following equation for IOA/HEA/VoAC adhesive:

$$Ln\left(S_{Adh}\right) = \underbrace{\begin{matrix} -0.47 + 0.45E - 0.68S + 1.20A \\ -0.14B - 0.40V - 0.12\left(\frac{T_m}{10}\right) \end{matrix}}_{IOA/HEA/VoAC} \tag{15}$$

A careful comparison of eqs 14 and 15 clearly indicates that the HB capability of the drug molecule will have a much more pronounced effect on the solubility in IOA/HEA/VoAC adhesive. The a-coefficient is 0.97 for IOA/ACM/VoAC adhesive compared to 1.20 for IOA/HEA/VoAC adhesive. Similarly, the b-coefficient is –0.49 for IOA/ACM/VoAC adhesive compared to –0.14 for IOA/HEA/VoAC adhesive. Both coefficients for IOA/HEA/VoAC adhesive are greater than those for IOA/ACM/VoAC adhesive, thereby indicating a higher HB acidity and basicity of IOA/HEA/VoAC adhesive. The results obtained in this study for the two adhesives are consistent with the physical properties of the monomer units.

Conclusions

This work presented here established experimentally the relationship between drug solubility in two acrylate adhesives with a previously defined drug-polymer interaction parameter and drug solubility in ACN. The model equation (eq 5) decomposes a hard-to-measure quantity into two easily measurable parameters. This model has been evaluated by drug solubility measured in two adhesives. It is concluded that there is an excellent linear relationship for the parameters involved. Thus, the model can be used to compute the solubility in the polymer for new drug candidates. Moreover, the two parameters in the relationship can be either easily measured or computed based on their molecular properties by the solvation parameter model. The methodologies presented can be applied to other adhesives.

Acknowledgements

The authors would like to acknowledge the support for this work by the Department of Transdermal Drug Delivery of 3M Drug Delivery Systems Division.

References

1. Moser K.; Kriwet K.; Naik A.; Kalia Y.N.; Guy R.H. Passive Skin Permeation Enhancement and Its Quantification in Vitro. *Eur. J. Pharm. Biopharm.* **2001**, 52, 103.
2. Barry, B.W. Novel Mechanisms and Devices to Enable Successful Transdermal Drug Delivery. *Eur. J. Pharm. Sci.* **2001**, 14, 101.
3. Canter, A.S.; Wirtanen, D.J. Novel Acrylate Adhesives for Transdermal Drug Delivery. *Pharm. Res.* **2002**, 26, 28.
4. Theeuwes, F.; Hussain, A.R.; Higuchi, T. Quantitative Analytical Method for the Determination of Drugs Dispersed in Polymers Using Differential Scanning Calorimetry. *J. Pharm. Sci.* **1974**, 63, 427.
5. Grover, M.; Singh, B.; Bakshi, M.; Singh, S. Quantitative Structure-Property Relationships in Pharmaceutical Research – Part 1. *Pharm. Sci. Tech. Today* **2000**, 3, 28.
6. Puttipipatkhachorn, S.; Nunthanid, J.; Yamamoto, K.; Peck, G.E. Drug Physical State and Drug-Polymer Interaction on Drug Release from Chitosan Matrix Films. *J. Control. Rel.* **2001**, 75, 143.
7. Sato, K.; Mitsui, N.; Hasegawa, T.; Sugibayashi, K. Potential Usefulness of the Solubility Index for the Prediction of the Skin Permeation Rate of 5-ISMN from Pressure-Sensitive Adhesive Tape. *J. Control. Rel.* **2001**, 73, 269.
8. Nair, R.; Nyamweya, N.; Gonen, S.; Martinez-Miranda, L.J.; Hoag, S.W. Influence of Various Drugs on the Glass Transition Temperature of Poly (Vinylpyrrolidone): A Thermodynamic and Spectroscopic Investigation. *Int. J. Pharm.* **2001**, 225, 83.
9. Guyot, M.; Fawaz, F. Design and In Vitro Evaluation of Adhesive Matrix for Transdermal Delivery of Propranolol. *Int. J. Pharm.* **2000**, 204,171.
10. Li, J.; Masso J.J.; Rendon S. Quantitative Evaluation of Adhesive Properties and Drug-Adhesive Interactions for Transdermal Drug Delivery Formulations Using Linear Solvation Energy Relationships. *J. Control. Rel.* **2002**,82, 1.
11. Schultz, R.M.; Liebman, M.N. In *Textbook of Biochemistry with Clinical Correlations*, Devlin, T.M., ed., Wiley & Sons, New York, 2002, 5th. Ed., p 135.
12. Abraham, M.H.; Chadha, H.S. In *Lipophilicity in Drug Action and Toxicology*, Pliskay, V.; Testar, B.; Waterbeemed, V.H., eds., VCH, Weinheim (Germany), 1996, p 311.
13. Abraham, M.H.; Le, J. The Correlation and Prediction of the Solubility Of Compounds in Water Using an Amended Solvation Energy Relationship. *J. Pharm. Sci.* **1999**, 88, 868.

14. Abraham, M.H.; Chadha, H.S.; Martins, F.; Mitchell, R.C.; Bradbury, M.W.; Gratton, J.A. Hydrogen Bonding Part 46: A Review of the Correlation and Prediction of Transport Properties by an LFER Method: Physicochemical Properties, Brain Penetration and Skin Permeability. *Pestic Sci.* **1999**, 55, 78.
15. Poole, C.F.; Gunatilleka, A.D.; Poole, S.K. In *Advances in Chromatography*, Brown, P.R.; Grushka, E., eds., Marcel Dekker, NY. 2000, Vol 40, p 159.
16. Li, J.; Masso, J.J.; Guertin, J.A. Prediction of Drug Solubility in An Acrylate Adhesive Based on the Drug–Polymer Interaction Parameter and Drug Solubility in Acetonitrile. *J. Control. Rel.* **2002**, 83, 211.
17. Smith, G.W. A Thermodynamic Model for Solubilities in Phase-Separated Systems: Polymer-Dispersed Liquid Crystals. *Mol. Cryst. Liq. Cryst.* **1993**, 225, 113.
18. Smith, G.W. A Calorimetric Determination of Fundamental Properties of Polymer-Dispersed Liquid Crystals. *Mol. Cryst. Liq. Cryst.* **1992**, 213, 11.

Chapter 13

Dissolution of Solid Dispersions of Ibuprofen Studied by Fourier Transform Infrared Imaging

K. L. Andrew Chan and Sergei G. Kazarian*

Department of Chemical Engineering and Chemical Technology, Imperial College, London SW7 2AZ, United Kingdom
***Corresponding author: s.kazarian@imperial.ac.uk**

The application of Fourier Transform infrared (FTIR) imaging using the macro attenuated total reflection (ATR)-IR methodology to formulations of ibuprofen in poly(ethylene glycol) (PEG) enabled characterization of the distribution of both, polymer and drug, in the tablet. Visualization of the dissolution process upon contact of the tablet with flowing water has been demonstrated. The mechanism of dissolution and drug release for two tablet preparation methods (mechanical mixing and melt) have been compared using this spectroscopic imaging approach.

Improvement of the dissolution of poorly water-soluble drugs is often achieved via preparation of solid dispersions of the drugs in hydrophilic polymeric carriers. However, the issues of stability, predictability, efficiency, and ultimately bioavailability of these formulations often hinder their adoption by pharmaceutical manufactures. *(1-6)* Therefore, a better understanding of the mechanism of drug release from solid dispersion is needed to facilitate the optimal design of such formulations. Despite numerous studies of solid dispersions, there is still a lack of understanding of processes that occur within these dispersions upon contact with dissolution media. *(2)* This situation is caused by the fact that conventional dissolution studies, such as described in the United States Pharmacopeia (USP), only analyze the drug concentration within the dissolution media as a function of time, without providing any insight into the complex processes that occur within the formulation. As a consequence, studies based on the conventional dissolution approach might result in erroneous conclusions and recommendations. For example, while it is generally believed that formulations of molecularly dispersed drugs within polymer carriers are required for the enhancement of dissolution rates, it was recently shown that these drugs can form a drug-enriched layer containing drug crystallites that actually hinder the overall drug dissolution. *(7,8)*

Visualization of processes occurring within formulations during dissolution, such as water uptake, drug diffusion, and polymer dissolution or erosion, are crucial for understanding the mechanism of drug release. This visualization has been achieved using magnetic resonance imaging (MRI), however, its spatial resolution and acquisition speed is limited. *(9-11)* A new experimental approach, based on Fourier Transform infrared (FTIR) spectroscopic imaging, has recently been developed to obtain more detailed insight into the processes within formulations upon contact with dissolution media. *(7)* Chemical specificity of FTIR imaging makes this a powerful method to analyze these processes. We have recently coined the term "chemical photography" to describe this imaging method, which is based on the use of the infrared array detector to simultaneously measure thousands of spectra from different locations. This approach has already been used to analyze *in situ* compacted tablets and to study pharmaceutical formulations under controlled humidity. *(12-14)* FTIR imaging in transmission mode was also applied recently to investigate drug delivery systems. *(15)*

This FTIR imaging method is limited by the strong absorption of water in the infrared range, which requires either using D_2O as the dissolution medium or preparing very thin samples. *(15)* Fortunately, the attenuated total reflection (ATR) sampling method allows overcoming this difficulty by measuring a relatively thin layer of the sample immersed in water. We have recently developed the methodology to image the dissolution of formulations in water using a diamond ATR accessory. *(13,16)* In a previous study, solid dispersions in PEG have been prepared via supercritical fluid impregnation or melting. *(7,8)*

Although supercritical fluid impregnation is a very promising method due to the lack of solvent residues and the use of near-ambient temperatures, it is still not broadly applicable. *(17)* Conventional techniques include mechanical mixing of drug and polymer or melting of polymer carrier to incorporate the drug. In this work we discuss the application of FTIR spectroscopic imaging to compare the effect of these two conventional preparation methods to image formulations in contact with flowing water.

Materials and Methods

Mechanically Prepared Sample

PEG powder (MW 8000 Da, purchased from Sigma) with particle size between 30-60 μm was prepared by grinding, followed by sieving of the PEG particles. The prepared powder was mechanically mixed with 20 wt% of ibuprofen in a beaker.

Sample Prepared by the Melt Method

20 wt% of ibuprofen was dissolved in molten PEG (MW 8000 Da) at 70 °C. The molten mixture was allowed to solidify at room temperature and then powdered with a spatula.

Tablet Preparation

1 mg of the mixture was compacted in a newly developed tablet press, which was specifically designed for studying tablet dissolution. *(13)* The tablet press is integrated into a supercritical fluid analyzer (Specac, Ltd), allowing subsequent dissolution studies of the compacted tablet. A schematic presentation of the tablet compaction cell is shown in Figure 1 (compaction mode). The tablet is compacted on one side of the diamond such that half of the imaging area will be exposed to an empty space. This arrangement allows studying the tablet disintegration into the dissolution medium. The advantage of this cell compared to our previous work *(7)* is that it allows studying the drug release under flow of water, which is more physiologically relevant. The design and the operational details of the tablet compaction cell are described elsewhere. *(13)* The powder was introduced into the compaction area and pressed by a piston, which was driven toward the compaction area by a screw anvil. The compaction force was controlled by a torque wrench with approximately 100 N of force applied (giving a hydraulic pressure of ca. 1400 bar).

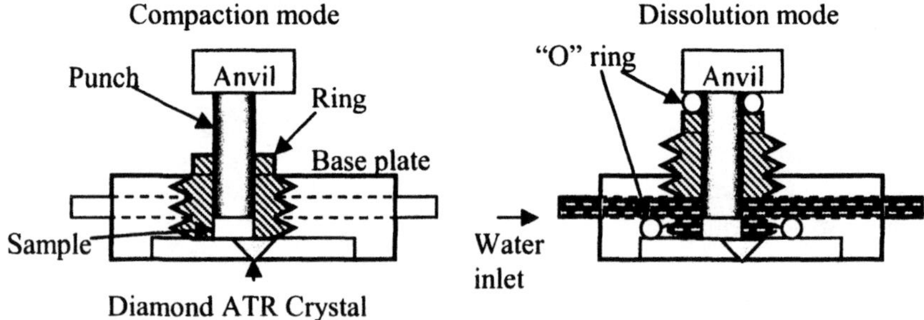

Figure 1. Schematic cross-section view of the compaction cell in compaction mode and dissolution mode. When the tablet is formed, the ring can be lifted using a spanner, allowing water flow around the compacted tablet.

FTIR Spectroscopic Imaging

All FTIR images were obtained with an IFS66/s step-scan spectrometer (Bruker Optics) with a macro-chamber (IMAC™) extension. The diamond ATR-IR accessory (Specac, Ltd.) was placed in the sample compartment of the macro-chamber and aligned as described in a previous work. *(16)* A 64 x 64 focal plane array (FPA) detector was used to measure spectra. The distribution of absorbance of the characteristic absorption bands of ibuprofen and PEG in the measured area of the sample were used to construct corresponding chemical images. A low-pass filter was employed to reduce the measured spectral range to 1800-900 cm^{-1} in order to increase the acquisition speed and improve temporal resolution. A spectral resolution of 16 cm^{-1} was used with 5 co-additions giving a total scanning time of ca. 30 s.

Dissolution Study

After compaction of the tablet, the compaction force was reduced to ca. 50 N to obtain a homogeneous contact between the ATR crystal and the tablet without deforming the tablet upon removal of the metal ring surrounding the tablet (see Figure 1, dissolution mode). Distilled water with a flow rate of 1 ml s^{-1} was pumped through the compaction cell using a Kontron 320 HPLC pump, and FTIR images were acquired at 1-minute intervals for the first 10 minutes and then at ca. 7-minute intervals afterward.

Results and Discussion

The advantages of FTIR spectroscopy for chemical analysis are based on the rich chemical information shown in IR spectra. The characteristic absorption band of a component can be used to generate an image that represents the distribution of the particular component in the measured area of the sample. Reference spectra of crystalline ibuprofen and semi-crystalline PEG are shown in Figure 2. The absorption band of the carbonyl group vibration ν(C=O) at 1710 cm^{-1} has been used to represent the distribution of ibuprofen, and the absorption band at 1100 cm^{-1} was used to plot the distribution of PEG. Chemical images are generated by plotting the integral absorbance value of these bands as a map such that the changing distribution of different components during dissolution can be monitored. The grey-scales of all images are adjusted such that direct comparison can be made between each component.

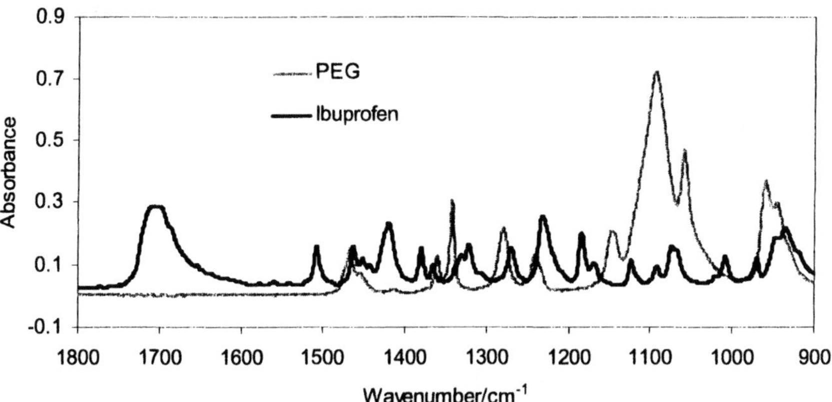

Figure 2. Reference FT-IR spectra of ibuprofen (black) and PEG (grey) in the fingerprint region measured with 4 cm^{-1} spectral resolution.

Dissolution of the Sample Prepared by Mechanical Mixing.

The image of the compacted tablet acquired before the addition of water is shown in Figure 3 and indicated as time 0 min. The image shows a heterogeneous distribution of the drug as evidenced by enhanced drug concentrations at several localized areas. The spectrum extracted from the locations in which PEG and ibuprofen were both present shows that ibuprofen in this formulation exists in its crystalline form, which is confirmed by comparing the position of the carbonyl band in Figure 4 to the reference spectrum in Figure 2. Further-more, a previous study has shown that the carbonyl band of crystalline ibu-profen absorbs at 1710 cm^{-1}, which is in agreement with the carbonyl position shown in Figure 4. *(18)* The images in Figure 3 show the distribution of PEG and ibuprofen after the flow of water had started.

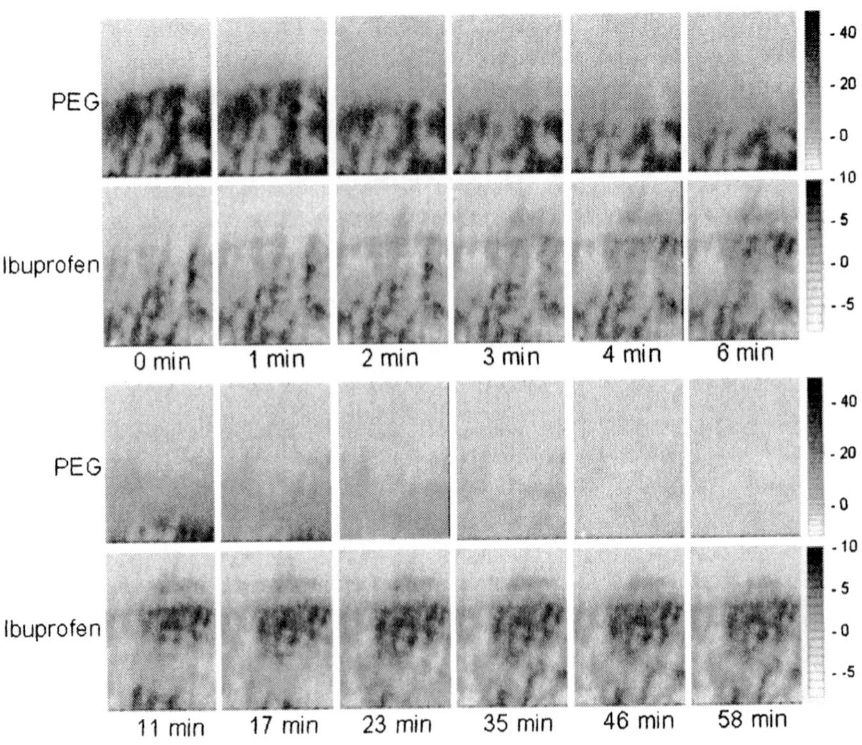

Figure 3. Grey-scale images showing the absorbance of PEG (top row) and ibuprofen (bottom row) change during the dissolution of the tablet prepared by the mechanical method. The dark area indicates a high absorbance, while the light area indicates a low absorbance. The image area is ca. 1140 x 820 μm^2.

PEG dissolves gradually, almost completely disappearing after 58 minutes. As PEG dissolves, a small amount of ibuprofen advances away from the initial position within the tablet for the first six minutes and then starts to accumulate, forming a large cluster of crystals. The movement of the ibuprofen is apparently caused by the swelling of PEG, pushing several smaller drug particles for a small distance. However, as dissolved PEG is quickly removed by the water flow, the movement of the drug ended after a short distance (ca. 300 μm). Evidently, the used water flow was not sufficient to carry away the drug particles.

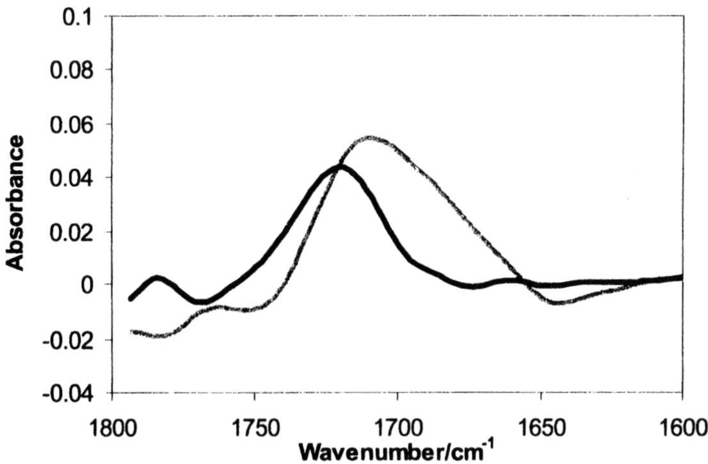

Figure 4. FTIR spectra extracted from tablet images prepared by the mechanical (grey) and melt (black) methods.

The images also show that the area where the ibuprofen concentration suddenly increases coincides with the area where PEG disappears completely. This suggests the deposition of ibuprofen particles on the ATR crystal after the removal of PEG by water. This type of dissolution mechanism has been described as "drug-controlled". *(2)* The solubility of the drug in the polymer-rich diffusion layer is low, hence the drug release rate will be controlled by the dissolution of solid drug particles.

Dissolution of the Sample Prepared by the Melt Method.

Before the dissolution, the image in Figure 5 at 0 min shows a well-defined and sharp edge of the tablet and a very uniform distribution of the polymer in the measured tablet area. Figure 5 also reveals that ibuprofen is homogeneously distributed within the sample. The black dot on the top right corner and the white dot on the bottom left corner are due to bad pixels of the FPA detector. The infrared spectrum extracted from the measured tablet area shows a shift of the carbonyl band position to 1720 cm^{-1} (see Figure 4), indicating that the drug is molecularly dispersed in the polymer matrix.

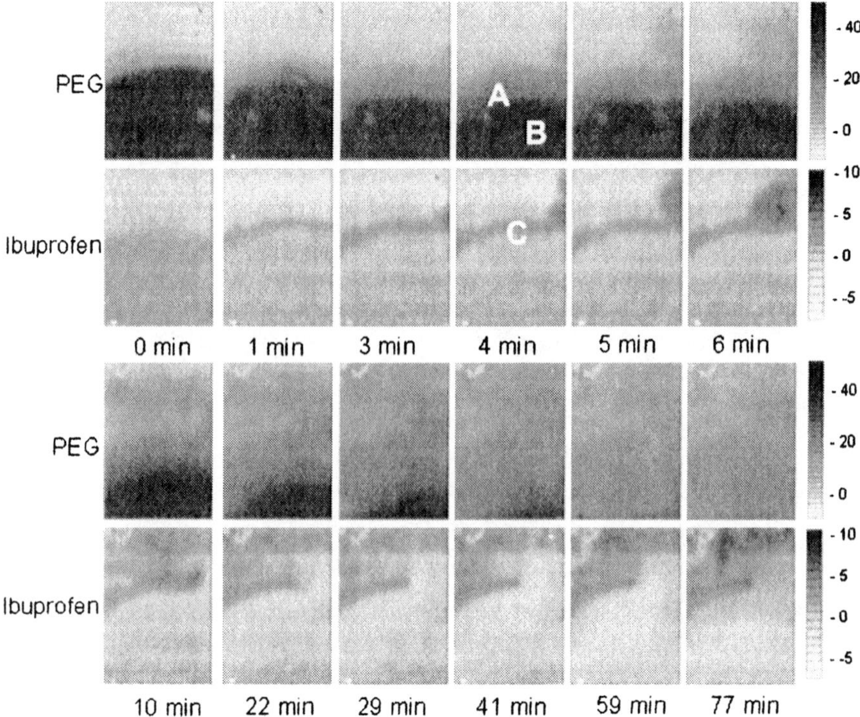

Figure 5. Grey-scale images showing the absorbance of PEG (top row) and ibuprofen (bottom row) change during the dissolution of the tablet prepared by the melt method. The dark area indicates a high absorbance, while the light area indicates a low absorbance. The image area is ca. 1140 x 820 μm^2.

The images in Figure 5 show that PEG at the edges of the tablet dissolves quickly as the flow of water begins, leading to a less-defined tablet edge. However, the images at the following time intervals reveal two distinct steps in the PEG concentration formed 3 minutes after the water addition. This behavior is different from the observations in previous studies, where PEG formed a smooth variation in the distribution. *(7,8)* The image of ibuprofen revealed the formation of a line of increased concentration at the edge of the tablet after the beginning of the water flow. This line grew between the images captured at 1 and 3 minutes. The observed increase of the ibuprofen concentration in that area was caused by crystallization of ibuprofen, which was originally molecularly dispersed within the polymer matrix. A similar phenomenon has been observed in a previous study using static water. *(7)* This result shows that flow of water does not remove the formed layer of a drug barrier.

Spectra from three different regions, A, B and C, of the image shown in Figure 5 have been extracted for direct comparison. These spectra are shown in Figure 6. The spectra measured at location B represent the non-dissolved part of the tablet, as indicated by the shoulder at 1060 cm^{-1}, a distinct feature of semi-crystalline PEG. The position of the carbonyl band at B indicates that ibuprofen is still molecularly dispersed at that point. The disappearance of the shoulder at 1060 cm^{-1} at location A together with the appearance of the absorption band due to the bending mode vibration of water at 1640 cm^{-1} is a good indication that PEG is being dissolved in this region. However, the position of the carbonyl band of ibuprofen measured in that region remains at 1720 cm^{-1}. This behavior is probable caused by the high concentration of PEG (although it is already dissolved), preventing the crystallization of ibuprofen. Further investigation is required to prove this hypothesis and will be explored in future studies. Spectra extracted from region C show the spectrum of crystalline ibuprofen, as indicated by the shift of the carbonyl band to 1710 cm^{-1}. However, the spectrum also shows the presence of PEG and water in this area. This observation could be explained by the assumption that the drug barrier is not completely impermeable so both, water and dissolved polymer, can still diffuse through this drug barrier.

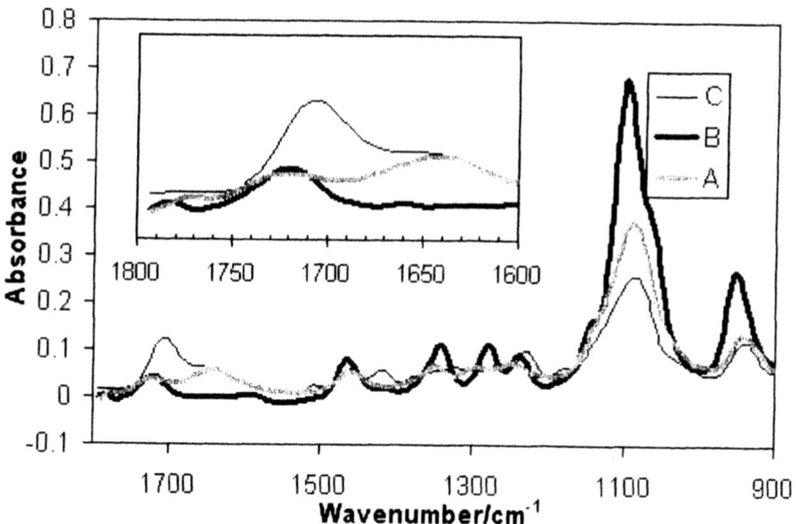

Figure 6. FTIR spectra extracted from areas indicated in Figure 5.

As a result of the drug barrier formation, water can only enter the dissolving tablet by diffusion though this barrier, causing the tablet to dissolve at a much slower rate. For the sample prepared by the mechanical method, all solid PEG was dissolved 17 minutes after the beginning of dissolution and all PEG was removed after 58 minutes. However, for the sample prepared by the melt method a small amount of solid PEG remained even after 41 minutes of dissolution, and a high concentration of dissolved PEG remained trapped by the drug barrier after 77 minutes from the start of water flow. The flow of water removed all PEG that had diffused away from the tablet through the drug barrier, causing the first steep decrease in PEG concentration. The second decrease in PEG concentration occurred when PEG transformed from the semi-crystalline to the amorphous form before being dissolved.

Nevertheless, imaging data of the dissolution of the tablet prepared by the melt method have shown that dissolved ibuprofen traveled a much farther distance away from the initial tablet edge compared to the distance traveled by particles of ibuprofen in the formulation prepared by the mechanical method. The movement of ibuprofen molecules is evident from the image captured after 3 minutes. Unfortunately, it was not possible to estimate how far the drug traveled since it moved beyond the field of view of the imaging setting employed in this study.

Conclusions

Dissolution of PEG/ibuprofen formulations prepared by mechanical mixing and melt methods have been studied by FTIR spectroscopic imaging. In this study, FTIR imaging in macro-ATR mode was used to simultaneously measure the spatial distribution of polymer and drug as a function of time upon contact of the prepared tablet with a flow of water. This approach enabled visualization of the processes occurring in tablets, which allowed comparison of dissolution mechanisms of tablets prepared by two different methods. The imaging data revealed that the formation of a drug layer of crystalline ibuprofen in the melt formulation is enhanced by the presence of water flow compared to the analogous experiment with stagnant water. Nevertheless, this layer did not prevent ibuprofen to be distributed farther away from the initial edge of the tablet compared to the distribution of ibuprofen during the dissolution of the tablet prepared by mechanical mixing. This study demonstrated new applications of the macro-ATR imaging approach. It has shown that this approach allowed imaging tablet preparations in situ in order to assess the initial distribution of a drug and its molecular state in the polymer matrix. Furthermore, macro-ATR imaging of tablets in contact with flowing water was possible, allowing to monitor the distribution of both, drug and polymer, and to compare two tablet preparation methods. A tremendous potential exists in the use of this method for studies of controlled release systems.

Acknowledgements

We thank EPSRC (grant GR/S03942/01) for funding.

References

1. Serajuddin, A.T.M. Solid dispersion of poorly water-soluble drugs: Early promises, subsequent problems, and recent breakthroughs. *J. Pharm. Sci.* **1999**, *88*, 1058-1066.
2. Craig, D.Q.M. The mechanisms of drug release from solid dispersions in water-soluble polymers. *Int. J. Pharm.* **2002**, *231*, 131-144.
3. Sethia, S.; Squillante, E. Solid Dispersions: Revival with Greater Possibilities and Applications in Oral Drug Delivery. *Crit. Rev. Therapeutic Drug Carrier Systems* **2003**, *20*, 215-247.
4. Guyot, M.; Fawaz, F.; Bildet, J.; Bonini, F.; Lagueny, A.M. Physico-chemical characterization and dissolution of norfloxacin/cyclodextrin inclusion compounds and PEG solid dispersions. *Int. J. Pharm.* **1995**, *123*, 53-63.
5. Leuner, C.; Dressman, J. Improving drug solubility for oral delivery using solid dispersions. *Eur. J. Pharm. Biopharm.* **2000**, *50*, 47-60.
6. Maggi, L.; Segale, M.L.; Torre, E.; Machiste, E.O.;Conte, U. Dissolution behavior of hydrophobic matrix tablets containing two different polyethylene oxides (PEOs) for controlled release of a water-soluble drug. Dimensionality study. *Biomaterials* **2002**, *23*, 1113-1119.
7. Kazarian, S.G.; Chan, K.L.A. "Chemical Photography" of Drug Release, *Macromolecules* **2003**, *36*, 9866-9872.
8. Chan, K.L.A.; Kazarian, S.G. FTIR spectroscopic imaging of dissolution of a solid dispersion of nifedipine in poly(ethylene glycol), *Molecular Pharmaceutics* **2004**, *1*, 331-335.
9. Fyfe, C.A.; Grondey, H.; Blazek-Welsh, A.I.; Chopra, S.K.; Fahie, B.J. NMR imaging investigation of drug delivery devices using a flow-through USP dissolution apparatus, *J. Contr. Rel.* **2000**, *68*, 73-83.
10. Hurrell, S.; Cameron, R.E. Polyglycolide: Degradation and drug release. Part I: Changes in morphology during degradation. *J. Mater. Sci. Mater. in Med.* **2001**, *12*, 811-816.

11. Riggs, P.D.; Kinchesh, P.; Braden, M.; Patel, M.P. Nuclear magnetic imaging of an osmotic water uptake and delivery process. *Biomaterials* **2001**, *22*, 419-427.
12. Chan, K.L.A.; Hammond, S.V.; Kazarian, S.G. Applications of Attenuated Total Reflection Infrared Spectroscopic Imaging to Pharmaceutical Formulations. *Anal. Chem.* **2003**, *75*, 2140-2147.
13. Van der Weerd, J.; Chan, K.L.A.; Kazarian, S.G. An innovative design of compaction cell for in situ FT-IR Imaging of tablet dissolution. *Vib. Spectrosc.* **2004**, *35*, 9-13.
14. Chan, K.L.A.; Kazarian, S.G. Visualization of the heterogeneous water sorption in a pharmaceutical formulation under controlled humidity via FT-IR imaging. *Vibrational Spectroscopy* **2004**, *35*, 45-49.
15. Coutts-Lendon, C.A.; Wright, N.A.; Mieso, E.V.; Koenig, J.L. The use of FT-IR imaging as an analytical tool for the characterization of drug delivery systems. *J. Contr. Rel.* **2003**, *93*, 223-248.
16. Chan, K.L.A.; Kazarian, S.G. New opportunities with micro and macro ATR-IR spectroscopic imaging: Spatial resolution and sampling versatility. *Appl. Spectrosc.* **2003**, *57*, 381-389.
17. Kazarian, S.G. Supercritical Fluid Impregnation of Polymers for Drug Delivery. In *Supercritical Fluids Technology in Drug Product Development.* York, P., Kompella, U.B.; Sjekunov, B., Eds.; Marcel Dekker: New York, **2004**; 343-366.
18. Kazarian, S.G.; Martirosyan, G.G. Spectroscopy of polymer/drug formulations processed with supercritical fluids: in situ ATR-IR and Raman study of impregnation of ibuprofen into PVP. *Int. J. Pharm.* **2002**, *232*, 81-90.

Chapter 14

Silver-Based Antimicrobial Coatings

Jörg C. Tiller

Freiburg Materials Research Center and Institute for Macromolecular Chemistry, University of Freiburg, Freiburg, Germany (email: joerg.tiller@fmf.uni-freiburg.de)

Three different antimicrobial coatings based on silver were prepared, characterized, and studied regarding their antimicrobial activity. A silver coating prepared by physical vapour deposition requires a silver content of at least 32 μg silver/cm^2 to completely inhibit the growth of the ubiquitous and infectious Gram-positive bacterium *Staphylococcus aureus.* Coatings based on silver nanoparticles, on the other hand, must contain only 10 μg silver/cm^2 to show the same efficiency. The preparation method, either formation of a coating with stabilized nanoparticles or loading a pre-existing coating with nanoparticles, or the film thickness did not influence the antimicrobial activity of the coatings.

Introduction

Materials that do not allow microbes to grow on their surfaces or even kill them are of great current interest due to an increasing demand in hygienic conditions. Such antimicrobial surfaces, which are commonly realized in form of coatings, are based on three major principles: microbe repelling, biocide releasing, and contact-activity (see Figure 1).

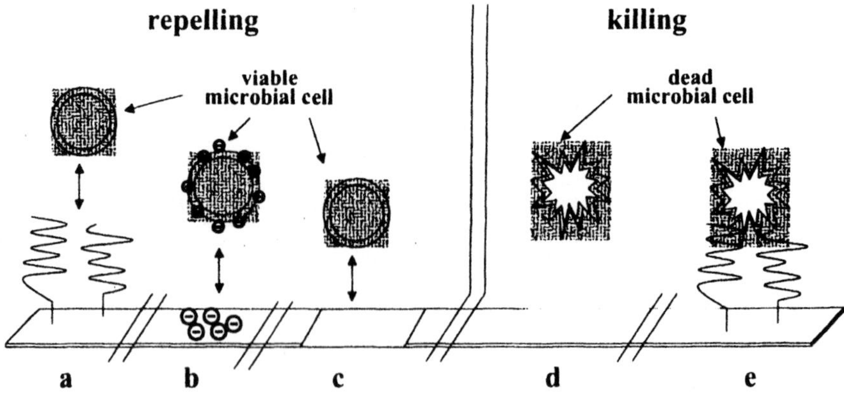

Figure 1. General principles of antimicrobial coatings: microbe-repelling by (a) antifouling polymers, (b) charge, and (c) ultrahydrophobic surfaces; microbe-killing by (d) releasing a biocide and (e) antimicrobial polymers.

Adhesion of microbes to surfaces is a complex mechanism, involving either non-specific interactions based on electrostatic, hydrophilic or hydrophobic attraction, and hydrogen bonding, or specific interactions such as protein receptors for e.g., collagen, polysaccharides or cell target proteins *(1)*. Although most bacterial and fungal cells have a net negative charge on their mostly hydrophilic surface, they also exhibit positively charged areas and hydrophobic regions. Therefore, the attachment to surfaces is a statistical phenomenon, i.e., the probability of meeting of two compatible regions.

Repelling of microbes is based on surfaces that lack binding sites to the membranes of microbial cells. One approach to create such surface materials is modifying them with antifouling polymers. These polymers are characterized by providing only hydrogen acceptor but no hydrogen donor functions. The best-known polymer with such properties is poly(ethylene glycol) (PEG) *(2)*. Surfaces modified with PEG, e.g., plastics *(3)* and collagen membranes *(4)* can lower bacterial adhesion by up to 98%. Poly(methyloxazoline), grafted onto gold surfaces, has also successfully been applied to repel bacterial cells *(5)*. Alternatively, soaking materials such as silicone or polyurethane in an albumin solution did significantly lower the microbial adhesion, indicating that the protein is capable of repelling microbes, probably due to its net negative charge *(6)*. Another approach for repelling microbes is the preparation of ultra-hydrophobic surfaces (water contact angle >150 degrees). This can be achieved by creating a surface with a high aspect ratio and hydrophobic modification, e.g., by treatment with fluorosilanes *(7)*.

Many plants keep their surfaces free of parasites by constantly renewing the surfaces. This so-called self-polishing effect could be realized for synthetic polymers, but did not prevent biofouling without release of a biocide. The release of biocides is a common concept, where a biocide is loaded to a material and then slowly released, killing all microbes in the surrounding solution. This concept is in use for most antimicrobial products, varying from medical devices over daily-life items and food packaging to coatings of ship hulls. The most common release compounds are listed in Table I.

Table I. Most common antimicrobial release coatings.

Released compound	Matrix	Application
antibiotics, chlorhexidine	silicone, polyurethanes	catheters, wound dressing, implants
triclosan	plastics	daily-life products, sutures, textiles, antifouling paints
nisin	cellulose, chitsosan	food packaging
silver ions	plastics, zeolite,	medical devices, cutlery, textiles
chlorine	N-halamine modified polymers	textiles
SO_2	plastics, cellulose	food packaging
quarternary ammonium compounds	plastics	dental materials, catheter

Since infections in medicine often occur on implants, catheters or wound dressings, a large part of the research is being done in these areas. Successful and highly effective solutions include impregnation of materials with biocides such as antibiotics, silver ion complexes, quarternary ammonium compounds, and pesticides that prevent infections by releasing the compounds effectively for up to one year *(8)*.

Many daily-life products, particularly the ones made of plastics, are protected by Triclosan, which is added to the plastics during the processing *(9)*. Recently, medical materials such as sutures, and food packaging have been loaded with this pesticide as well. However, Triclosan is a non-biodegradable, chlorinated aromatic compound, and therefore, concentrates in the environment. For this reason, an alternative biocide to this compound is inescapable. The release of silver ions, which have great antimicrobial potential, could be recently realized by the development of a coating consisting of a zeolite and a porous hydrated aluminium silicate, which is loaded with silver ions. This coating now protects numerous materials, such as cutlery and textiles, against microbial growth *(10)*. Furthermore, coatings based on polymers modified with N-halamines can be loaded and, in case of exhaustion, reloaded with chlorine or hypochloride *(11)*. Such coatings, even useful for textile protection, slowly release chlorine, which effectively kills bacteria and fungi in the proximity of the surface.

One of the most effective biocide-release systems is represented by so-called self-polishing antifouling paints that have been used for decades to prevent biofouling on materials such as ship hulls *(12)*. The most successful system is composed of poly(acrylates) esterified with tributyltin (TBT). This hydrophobic polymer coating can be penetrated by water only 10 to 100 nm in depth. The slow cleavage of the ester bonds results in the release of the strong biocide TBT. The remaining polymer, once completely metal-free, is water-soluble and washes away from the surface, creating a new and fully intact antimicrobial surface. In the beginning of 2003, however, the highly toxic TBT-based paints were banned and are now substituted by similar but less toxic compounds, which use copper, zinc, and silicon, respectively, instead of tin. These new paints are only effective when a so-called 'booster biocide', for example Irgarol 1051, diuron, dichlofluanid, zinc pyrithione, or triclosan, is added to the composition *(13)*.

Coatings containing photocatalytically active TiO_2 can be classified as not exhausting release coatings. In the presence of light and water, TiO_2 catalizes the formation of radicals that kill microbes in the proximity of the coated surface *(14)*. Addition of silver ions to the TiO_2 particles resulted in coatings that also kill microbes in the dark by releasing silver ions *(15)*.

All the release systems described above are constantly releasing the active agent. A trend in recent research is the development of triggered, "intelligent" release systems that are only active in the presence of microbes. One example of such systems is a wound dressing composed of Gentamycin that is encapsulated in poly(vinyl alcohol), crosslinked with a peptide and cleavable by the enzyme thrombin *(16)*. The antibiotic is released only in the case of a wound infection, which dramatically increases the concentration of thrombin. Grafted poly (guanidinium)-silver ion complexes were found to release silver ions only in the presence of microbes as well *(17)*. Furthermore, pH-sensitive antimicrobial release systems have been described *(18)*.

Contact-active surfaces are of particular interest in current research. Such surfaces are not releasing a biocide, which lowers the risk of creating resistant microbes, and are theoretically not exhausting. The contact-active function is achieved by immobilizing biocides via a polymeric spacer that allows transporting the toxic agent into adhering microbes, virtually killing them on contact. Successful modifications have been carried out with quarternized poly(4-vinylpyridine) *(19)* and poly(ethylene imines), (PEI), *(20)* that were grafted onto glass and plastics, resulting in efficient contact-killing surfaces. In another system it was found that poly(phosphonium salts), grafted on polypropylene, are able to kill adhering bacterial cells on contact *(21)*.

An alternative method to chemical grafting of polymers is the plasma polymerization procedure. With this method, poly(diallyldimethylammonium chloride) could be grafted onto plastic surfaces, affording antimicrobial behavior with coatings of only 2-3 nm in thickness *(22)*. Yet another contact-active surface can be obtained by treating materials with 3-(octadecyl-N,N-dimethyl-ammonium)propyltrimethoxysilane *(23)*.

Many heavy metal ions are highly toxic towards microbial cells. Particularly silver and copper ions are deadly to microorganisms but less toxic to humans. Silver salts in form of citrate, sulfadiazine, and biguanidine are often used in release systems described above, although elemental silver as well does not allow microbial growth on its surface. It is discussed that elemental silver acts as a catalytic surface that generates radicals such as toxic hydroxy radicals *(24)*, however, most authors believe that the antimicrobial potential of silver is based on the release of silver ions *(25)*. The redox potential of silver strongly depends on its environment. For example, the presence of NH_2 and SH groups lowers the redox potential of silver, resulting in easier oxidation of the metal. This way, elemental silver has the potential of being a sensitive release system, which produces silver ions specifically in presence of certain compounds such as proteins. This behaviour makes elemental silver a very interesting material for antimicrobial coatings. Unfortunately, silver is quite expensive and modification of surfaces with the metal is a rather elaborate process. In this contribution, the preparation, characterization, and antimicrobial testing of three different silver-based coatings are described. Classical CVD coatings are compared to nanoparticle-coatings regarding their antimicrobial potential in relation to their structure and silver contents.

Materials and Methods

Preparation of the Coatings

The synthesis of poly(ethylene mine)-amphiphilically modified (PEI-am) was performed by heating commercially available poly(ethylene mine) (5000 g/mol, degree of branching 60%, Hyperpolymers, Freiburg, Germany) and methylstearate. Loading PEI-am with silver nanoparticles was carried out by adding $AgNO_3$ (one silver ion per 4 PEI-nitrogen) to a 1% PEI-am solution in toluene, followed by reduction with $LiBHEt_3$ as described elsewhere *(26)*.

PEI-MA was synthesized by converting PEI with acryloyl chloride in acetone at 0 °C. After adding methanol, 2-hydroxyethyl acrylate (HEA), and the photoinitiator Irgacure 651 (Ciba, Basel, Switzerland), the solvents were removed and the remaining solution was photopolymerized on a methacrylate-modified glass slide to give a 20-μm coating. Then silver was loaded into the coating by immersing it into aqueous silver nitrate, washing with water, and transferring the coating into aqueous ascorbic acid as described in ref. *(27)*.

Silver coatings are achievable by sputtering, physical vapour deposition (CVD), and plating *(28)*. Also combined methods, e.g., of parallel plasma polymerization and silver CVD have been described *(29)*. Here elemental silver coatings were brought onto a standard microscope glass slide using CVD equipment from Edwards (UK). The process was performed at 10^{-6} mbar at room temperature, using thermal evaporation of the metals. After applying a 2-nm chromium layer, silver was added with a growth rate of ca. 1 nm/s. The thickness of the metal layer was controlled with a profilometer, Tencor P10.

Characterization of the Coatings

UV/vis measurements were carried out with the Perkin Elmer photo spectrometer Lambda 11. Scanning electron microscope (SEM) images were measured with an ESEM 2020 without previous sputtering. Atomic force microscope (AFM) images were recorded with a Nanoscope III scanning probe microscope (Digital Instruments), using Si cantilevers with a fundamental resonance frequency of around 300 kHz in Tapping-mode. Transmission electron microscope (TEM) measurements were carried out using a LEO 912 microscope from Zeiss, Germany, applying an acceleration voltage of 120 kV. The microscope images were taken with an Axioplan 2 (Zeiss). The silver content of the coatings was determined using the flame-atom absorption spectrometer (AAS) Vario 6 from Analytik-Jena-AG (Germany).

Antimicrobial Testing

The bacterial susceptibility measurements were carried out with the Gram-positive bacterium *S. aureus* (ATCC 25123). The cells were either sprayed onto the surface from their suspension in distilled water (10^6 cells per mL) or allowed to adhere from their PBS (phosphate buffer saline, 100 mM, pH 7.0) suspension (10^8 cells per mL) at 37 °C for 2 hours. After short air-drying, the samples were covered with growth agar (1.5 wt% agar in standard growth medium of Merck, Germany) and incubated at 37 °C in an humidified incubator as described in detail in refs. *(19,28)*.

Results and Discussion

Chemical Vapour Deposition (CVD) - CoatA

Layers of elemental silver with a thickness between 10 and 50 nm were created on glass slides. Measurements with a profilometer confirmed the respective layer thickness.

Nanoparticle Coating – CoatB

Silver colloids in are well known for a long time *(30)*. Usually they are prepared in water with the aid of detergents, linear polymers, and more recently, with block-copolymers and dendrimers *(31)*. Silver nanoparticles can be synthesized in organic solvents using suitably modified polymers, e.g., block-copolymers *(32)* or detergents *(31)*. In order to achieve stable organo-soluble particles that are useful for coatings, dendrimers and hyperbranched polymers are modified following the concept of core-shell particles *(33)*. The core of these particles consists of a hydrophilic polymer that binds silver ions, while the hydrophobic shell makes the particles soluble in the desired organic solvent. Following this approach, $AgNO_3$ can be dissolved within an organic medium up to the saturation of the core's binding capacity for the metal ions. Subsequently the metal ions are reduced with an appropriate reducing agent such as $LiHBEt_3$, resulting in nanoparticles of 1 to 2 nm in diameter.

A well-suited polymer for this approach is hyperbranched·PEI, which was modified with stearic acid and subsequently loaded with silver nanoparticles in toluene *(26)*. Clear brownish Ag-PEI-am solutions were cast onto commercial glass slides to give transparent films, containing between 1 and 20 µg silver/cm^2 as confirmed by AAS measurements. The thickness of the coatings was measured with a profilometer, giving a thickness of about 1.5 µm for a film containing 10 µg silver. The comparison of UV/vis spectra of an Ag-PEI-am solution and the resulting film revealed that the silver plasmon resonance band appeared at 420 nm in both cases, indicating that silver clusters are located within the PEI-core in both solution and film (Figure 2).

Nanoparticles formed within a Coating - CoatC

The formation of nanoparticles within a polymer coating is rather limited, because metals tend to form larger particles due to their high surface energy *(34)*. So far nanoparticle loading could be achieved by dissolving or dispersing $AgNO_3$ within polymer solutions or mixtures, forming the coating, and reducing the salt within the solid polymer *(35)*. Another approach involves dispersing stabilized nanoparticles within a polymer solution *(36)* or melt *(37)* prior to forming the coating. In addition, mixing or dissolving of silver salts within monomers followed by parallel polymerization and reduction in UV light has

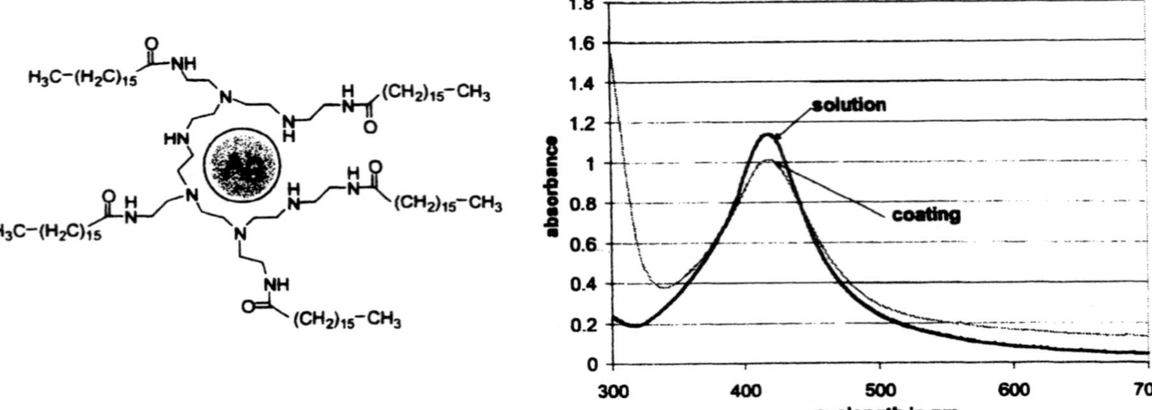

Figure 2. Structure of PEI-am and UV/VIS spectra of a solution of Ag-PEI-am in toluene and of a coating formed from this solution.

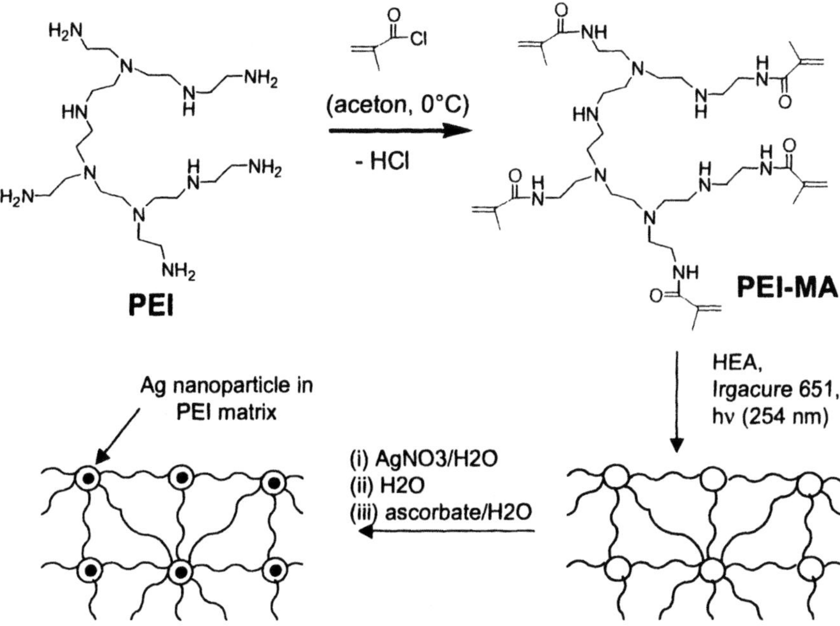

Figure 3. Principle of a polymer network as template for silver nanoparticles.

been described *(38)*. An alternative elegant approach is the formation of silver clusters by PVD in the presence of a monomer in the gas phase, followed by plasma polymerization that encapsulates the particles *(39)*. All these ways have in common that the loading must occur *before* formation of the coating, i.e., during the processing step.

Here, an approach is described that involves the loading with silver nanoparticles *after* formation of the coating. The concept is based on the previously mentioned core-shell principle used for the formation of silver nanoparticles in solution. As seen in Figure 3, the multivalent crosslinker PEI-MA is copolymerized with HEA to give a solid film that has no other binding sites for silver ions besides the dispersed PEI-molecules, i.e., the network represents a template for the metal ions.

PEI with a molecular weight of 5000 g/mol and a degree of branching of 60% was quantitatively modified with methacryloyl chloride. The derivative was not isolated, because drying resulted immediately in an insoluble substance. The phenomenon is caused by crosslinking, which unavoidably occurs because of the high density of double bonds. Mixing with HEA was successful upon solvent exchange. The resulting mixture of HEA and 1 to 10 wt% of PEI-MA was photopolymerized on a glass slide, previously modified with methacrylate groups to give a covalently attached transparent, colorless film of some 20 μm in thickness *(27)*. Up to 10 wt% of the crosslinker glycoldimethacrylate were added to the mixture of HEA/PEI-MA to enhance the mechanical strength of the coatings.

The coatings were loaded with silver ions in water, washed with water, and then treated with aqueous ascorbic acid solution. In all cases the silver reduction was completed within seconds and resulted in films that were slightly yellow to brown, depending on the PEI content. UV/vis spectra revealed a silver plasmon resonance band at 427 nm, indicating that nanoparticles were formed within the PEI compartments of the network. The absorbance signal and the silver content measured by AAS were perfectly proportional to the PEI-content *(26)*. Obviously, silver nanoparticles were formed according to the concept shown in Figure 3.

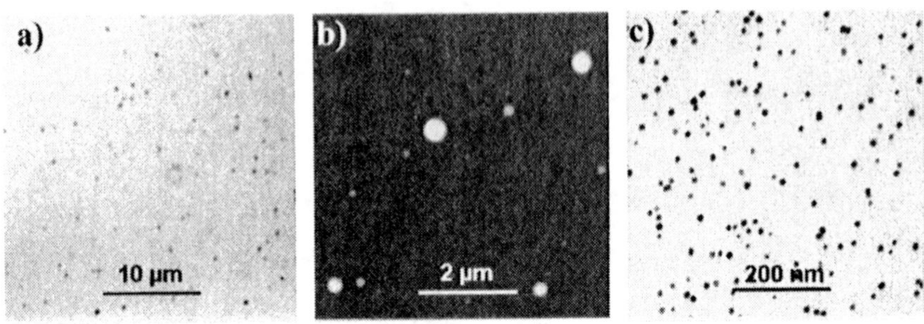

Figure 4. Microscope images of Ag-HEA/PEI-MA coatings, (a) light microscope (20 wt% PEI); (b) SEM (10 wt% PEI); and (c) TEM (5 wt% PEI).

In order to find out if truly nanoparticles were formed, the yellow Ag-HEA/PEI-MA coatings were observed by means of light microscopy, scanning electron microscopy, and transmission electron microscopy. While nothing was observable in coatings with less than 10 wt% PEI, some larger particles of 0.5 to 2 µm in diameter could be found under the light microscope observing coatings with 10 wt% PEI and even more with 20 wt% (Figure 4a). The silver-loaded polymer coatings could be observed with SEM without further treatment, i.e., they are conductive. The SEM image shown in Figure 4b reveals the existence of particles with less than 30 nm in diameter (10 wt% PEI). Finally, the TEM image of a cross-section of an Ag-HEA/PEI-MA coating with 5 wt% PEI exhibits nanoparticles within the network. The nanoparticles, some 10 nm in size, are well distributed. No larger silver clusters could be found in this coating. Even coatings with less than 5 wt% PEI did not contain particles larger than 50 nm.

Observing the surface of an Ag-HEA/PEI-MA coating (5 wt% PEI), which had been additionally crosslinked with 10 wt% glycoldimethacrylate, with AFM in phase mode shows phase separation of the two crosslinkers and HEA (Figure 5). Even the silver particles on the surface can be seen and are exclusively within the PEI-containing phases. All Ag-HEA/PEI-MA coatings with 5 wt% or less PEI contained nanoparticles on the surface and within the bulk material as shown by TEM and AFM. Coatings with more than 5 wt% PEI contained also larger silver clusters, possibly formed in microscale-separated PEI-phases.

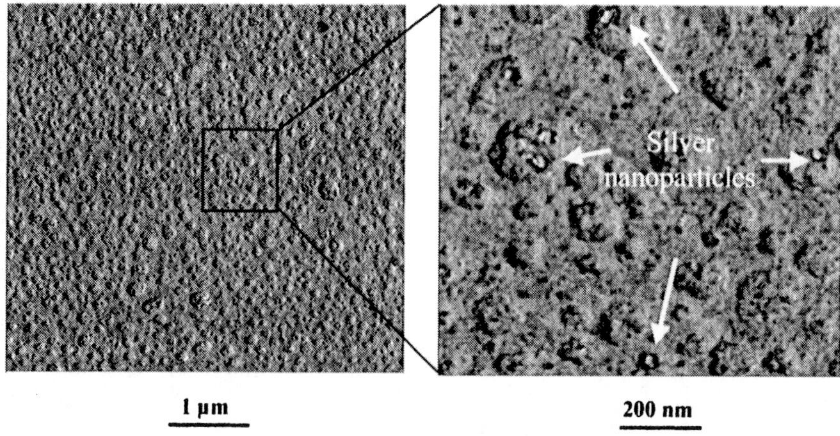

Figure 5. Phase mode AFM image of an Ag-HEA/PEI-MA coating (5 wt% PEI), additionally crosslinked with glycoldimethacrylate (10 wt%), in tapping mode.

Antimicrobial Testing

Numerous methods are in use to test antimicrobial surfaces. Due to the great variety of applications, it is literally impossible to find a standard method that meets all requirements. The oldest and easiest test is the so-called suspensions test, where the samples are given to an inoculated microbial suspension, and the number of cells is counted after different incubation times *(23)*. This test provides relatively few information on the true mechanism of the materials' mode of action.

More information can be gained when microbes are attached to surfaces, removed after a suited incubation time, and then the amount of surviving cells is counted. In this approach, cell adhesion as well as cell killing can be determined. Unfortunately, such adhesion tests take not into account that microbes significantly change their gene expression upon attaching, and grow quite differently on surfaces *(1)*. Therefore, we used a test system that allows following the microbial growth directly on the surface *(4)*. To this end, bacterial cells were either sprayed on the surface or allowed to adhere from their suspension, and then growth agar was added to the surface, fixing the bacterial cells in their position. Upon incubation at 37 °C, the formation of single colonies over the whole surface was followed over a time period of several weeks.

Silver coatings were sprayed with the ubiquitous *S. aureus* cells, incubated under growth agar as described, and the colonies were counted each day. It was observed that the bacteria were inhibited in growth or did not grow at all, depending on the amount of silver. In the case of CoatA (CVD), the growth was dependent on the film thickness, with the critical thickness of about 30 nm, i.e., 32 μg/cm² (Figure 6a). Similar results were obtained for silver coatings on poly (ethylene) *(39)*. In this case, the bacterial cells were tested using an adhesion test. The authors concluded that the limited antimicrobial potential was due to a minimal release of silver ions, which killed the surrounding microbes. Here, however, it could be shown that the silver is active directly on the surface.

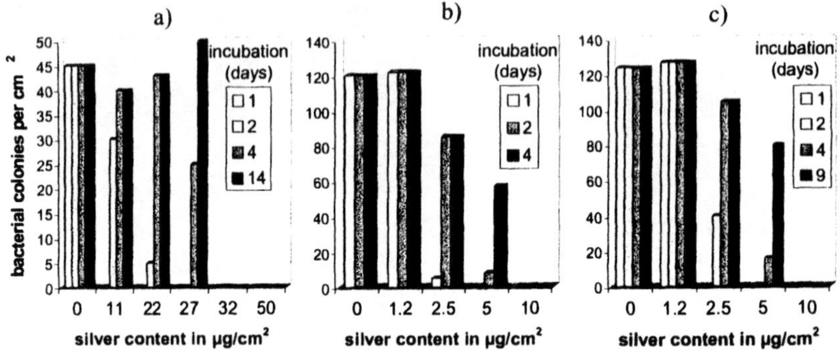

Figure 6. Growth of S. aureus colonies after being sprayed onto (a) CoatA, (b) CoatB, and (c) CoatC.

The nanoparticle coatings CoatB and CoatC were washed for 6 hours with PBS (phosphate buffered saline, 140 mM NaCl, 10 mM phosphate), pH 7.0, to remove any non-reduced traces of silver ions. The color intensity in both films did not decrease during washing, indicating that the nanoparticles were not dissolved under these conditions. After spraying *S. aureus* cells onto the coatings, bacterial cells grew on the surfaces of coatings with up to 5 μg silver/cm². With increasing silver content, however, the bacterial growth was delayed (Figures 6b and c). CoatB showed longer delay periods than CoatC. Figure 7 displays the growth of *S. aureus* colonies sprayed on CoatB (5 μg silver/cm²), which started growing after 4 days, while bacteria grew on the surrounding non-coated glass after 12 hours of incubation. Both silver nanoparticle coatings with a silver content of 10 μg/cm² completely prevented bacterial growth, which was continuously observed for up to 28 days. Apparently the effect is only dependent on the silver amount and not on the thickness of the coatings. The same

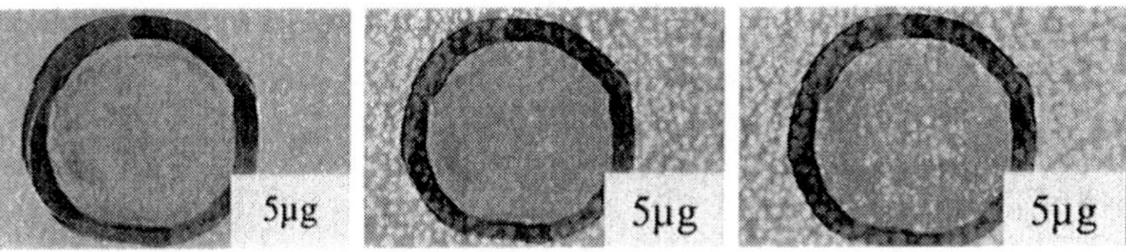

Figure 7. Growth of S. aureus colonies on CoatB (circle) loaded with 5 µg silver/cm² after 1, 4, and 9 days of incubation.

amount of nanoparticles in CoatB of 1.5 µm in thickness and CoatC of 20 µm in thickness was required to render the surfaces antimicrobial.

S. aureus cell growth was also observed in experiments where the cells adhered from their suspension onto CoatB and CoatC surfaces with up to 5 µg Ag/cm^2. As found in the spraying experiments, the microbial growth was delayed with increasing silver content, with no microbial growth on CoatB and CoatC at a silver content of 10 µg/cm^2. The *S. aureus* suspensions contained the same number of viable cells before and after immersing the silver nanoparticle coatings in it, indicating that immersing the coatings did not significantly alter the suspensions. This indicates that at least no significant amount of silver ions was released into the bulk suspension even in case of the highest silver loadings.

Conclusions

It could be shown that the preparation of silver-based coatings via CVD, nanoparticles in solution, and loading of photopolymerized coatings with nanoparticles led to comparable antimicrobial surfaces, as tested with the ubiquitous bacterium *S. aureus*. All coatings required a critical amount of silver to entirely prevent microbial growth. While massive silver coating, as prepared in CoatA, prevented microbial growth with 32 µg silver/cm^2 and higher, the silver nanoparticles-based coatings required only 10 µg/cm^2. The latter can be explained by the fact that the surface of the silver-releasing area is larger for nanoparticles than for a solid silver film. However, no toxic silver release into the surrounding solution could be observed. This fact could be interpreted as a somehow triggered release behavior, originating from the chemical nature of the microbes surfaces. Surprisingly, the density of the nanoparticle films did not significantly influence the biocidal activity of their surfaces. Altogether, coatings with silver nanoparticles either from solution or by loading them into a pre-existing coating is a true alternative to solid silver coatings, requiring about three times less silver to be antimicrobial.

Acknowledgements

This work was financially supported by the Deutsche Forschungsgemeinschaft (Emmy-Noether-Program and SFB 427) and the Fonds der Chemischen Industrie. The author thanks C.H. Ho and J. Tobis for their help with the nanoparticle films, Prof. Dr. S. Mecking, Dr. C. Aymonier, and L. Antonietti for providing the PEI nanoparticles solutions, Dr. R. Thomann for the AFM images, Dr. R. Landers for the SEM measurements, Mrs. Hirth-Walter for the AAS analysis, and Dr. Pelz for providing the *S. aureus* cells.

References

1. Fletcher, M. *Bacterial adhesion: Molecular and Ecological Diversity*, Fletcher, M., Ed.; Wiley: New York, NY, **1996**, pp 1-24.
2. Brash, J.L.; Uniyal, S. Dependence of albumin-fibrinogen simple and competitive adsorption on surface properties of biomaterials. *J. Polym. Sci., Polym. Symp.* **1979**, *66*, 377-89; for review see Kingshott, P.; Griesser, H.J. Surfaces that resist bioadhesion. *Curr. Opin. Solid State & Mater. Sci.* **1999**, *4*, 403-412.
3. Desai N.P.; Hossainy, S.F.; Hubbell, J.A. Surface-immobilized polyethylene oxide for bacterial repellence. *Biomaterials* **1992**, *13*, 417-420; Bridgett, M.J.; Davies, M.C.; Denyer, S.P.; Eldridge, P.R. In vitro assessment of bacterial adhesion to Hydromer-coated cerebrospinal fluid shunts. *Biomaterials* **1993**, *14*, 184-188; Park, K.D.; Kim, Y.S.; Han, D.K.; Kim, Y.H.; Lee, E.H.B.; Suh, H.; Choi, K.S. Bacterial adhesion on PEG modified polyurethane surfaces. *Biomaterials* **1998**, *19*, 851-859.
4. Tiller, J.C.; Bonner, G.; Pan, L.C.; Klibanov, A.M. Improving biomaterial properties of collagen films by chemical modifications. *Biotechnol. Bioeng.* **2001**, *73*, 246-252.
5. Chapman, R.G.; Ostuni, E.; Liang, M.N.; Meluleni, G.; Kim, E.; Yan, L.; Pier, G.; Warren, H.S.; Whitesides, G.M. Polymeric Thin Films That Resist the Adsorption of Proteins and the Adhesion of Bacteria. *Langmuir* **2001**, *17*, 1225-1233.
6. Pratt-Terpstra, I.C.H.; Weerkamp, A.H.; Busscher, H.J. Adhesion of oral streptococci from a flowing suspension to uncoated and albumin-coated surfaces. *J. General Microbiol.* **1987**, *133*, 3199-3206.
7. Estarlich, F.F.; Lewey, S.A. ;Nevell, T.G.; Thorpe, A.A.; Tsibouklis, J.; Upton, A.C. The surface properties of some silicone and fluorosilicone coating materials immersed in seawater. *Biofouling* **2000**, *16*, 263-275.

8. Chatzinikolaou, I.; Finkel, K.; Hanna, H.; Boktour, M.; Foringer, J.; Ho, T.; Raad, I. Antibiotic-coated hemodialysis catheters for the prevention of vascular catheter-related infections: A prospective, randomized study. *Am J. Med.* **2003**, *115*, 352-357.
9. Kalyon, B.D.; Olgun, U. Antibacterial efficacy of triclosan-incorporated polymers. *Am. J. Infect. Control* **2001**, *29*, 124-125; Cutter, C.N. The effectiveness of triclosan-incorporated plastic against bacteria on beef surfaces. *J. Food Prot.* **1999**, *62*, 474 - 479.
10. Chainer, J. Home steel home. AK steel partners with AgION to build world's first antimicrobial steel house. *AISE Steel Technol.* **2001**, *78*, 59-60
11. Worley, S.D.; Eknoian, M.; Bickert, J.; Williams, J.F. Novel antimicrobial N-halamine polymer coatings. in *Polymeric Drugs & Drug Delivery Systems*, Ottenbrite, R.M.; Kim, S.W., Eds. Technomic Publishing Co., Inc.: Lancaster, PA, **2001**, pp 231-238.
12. Lewis, J.A. TBT antifouling paints are now banned! What are the alternatives and what of the future? *Surf. Coat. Australia* **2003**, *40*, 12-15.
13. Kuo, P.L.; Chuang, T.F.; Wang, H.L. Surface-fragmenting, self-polishing, tin-free antifouling coatings. *J. Coat. Technol.* **1999**, *71*, 77-83.; Thomas, K. V.; Fileman, T.W.; Readman, J.W.; Waldock, M.J. Antifouling Paint Booster Biocides in the UK Coastal Environment and Potential Risks of Biological Effects. *Marine Pollution Bull.* **2001**, *42*, 677-688.
14. Sunada, K.; Kikuchi, Y.; Hashimoto, K.; Fujishima, A. Bactericidal and Detoxification Effects of TiO2 Thin Film Photocatalysts. *Environ. Sci. Technol.* **1998**, *32*, 726-728; Ohko, Y.; Utsumi, Y.; Niwa, C.; Tatsuma, T.; Kobayakawa, K.; Satoh, Y.; Kubota, Y.; Fujishima, A. Self-sterilizing and self-cleaning of silicone catheters coated with TiO2 photocatalyst thin films: a preclinical work. *J. Biomed. Mater. Res.* **2000**, *58*, 97-101.
15. Zhang, L.; Yu, J.C.; Yip, H.Y.; Li, Q.; Kwong, K.W.; Xu, A.-W.; Wong, P.K. Ambient Light Reduction Strategy to Synthesize Silver Nanoparticles and Silver-Coated TiO2 with Enhanced Photocatalytic and Bactericidal Activities. *Langmuir* **2003**, *19*, 10372-10380.
16. Tanihara, M.; Suzuki, Y.; Nishimura, Y.; Suzuki, K.; Kakimaru, Y.; Fukunishi, Y. A Novel Microbial Infection-Responsive Drug Release System. *J. Pharm. Sci.* **1999**, *88*, 510-514.
17. Subramanyam, S.; Yurkovetsiky, A.; Hale, D.; Sawan, S.P. A chemically intelligent antimicrobial coating for urologic devices. *J. Endourol.* **2000**, *14*, 43-48.
18. Tang, M.; Zhang, R.; Bowyer, A.; Eisenthal, R.; Hubble, J. A reversible hydrogel membrane for controlling the delivery of macromolecules. *Biotechnol. Bioeng.* **2003**, 82, 47-53.

19. Tiller, J.C., Liao, C.J., Lewis, K., Klibanov, A.M. Designing surfaces that kill bacteria on contact. *Proc. Natl. Acad. Sci. U.S.A.* **2001**, *98*, 5981-5985.
20. Lin, J.; Qiu, S.; Lewis, K.; Klibanov, A.M. Bactericidal properties of flat surfaces and nanoparticles derivatized with alkylated polyethylenimines. *Biotechnol. Progr.* **2002**, *18*, 1082-1086.
21. Kanazawa, A.; Ikeda, T.; Endo, T. Polymeric phosphonium salts as a novel class of cationic biocides. III. Immobilization of phosphonium salts by surface photografting and antibacterial activity of the surface-treated polymer films. *J. Polym. Sci., Part A: Polym. Chem.* **1993**, *31*, 1467-1472.
22. Thome, J., Hollander, A., Jaeger, W., Trick, I., Oehr, C.H. Ultrathin antibacterial polyammonium coatings on polymer surfaces. *Surf. Coat. Technol.* **2003**, *174-175*, 584-587.
23. Gettings, R.L.; White, W.C. Progress in the application of a surface-bonded antimicrobial. *Polym. Mater. Sci. Eng.* **1987**, *57,* 181-185.
24. Meduski, J.D.; Meduski, J.W.; De la Rosa, J.; Coetz, A. Chemiluminescence experiments on a model of the oligodynamic activity of silver. *Biolumin. Chemilumin.* **1981**, *2*, 639-644.
25. Sant, S.B.; Gill, K.S.; Burrell, R.E. Novel duplex antimicrobial silver films deposited by magnetron sputtering. *Philos. Mag. Lett.* **2000**, *80*, 249-256.
26. Aymonier, C.; Schlotterbeck, U.; Antonietti, L.; Zacharias, P.; Thomann, R.; Tiller, J.C.; Mecking, S. Hybrids of silver nanoparticles with amphiphilic hyperbranched macromolecules exhibiting antimicrobial properties. *Chem. Comm.* **2002**, *24*, 3018-3019.
27. Ho, C.H.; Tobis, J.; Sprich, C.; Thomann, R.; Tiller, J.C. Nanoseparated multifunctional antimicrobial networks. *Adv. Mater.* **2004**, *16*, 957-961.
28. Blair A. Silver plating. *Plating Surf. Finishes* **2000**, *87*, 62-63; Szabo, N.J.; Winefordner, J.D. Surface-Enhanced Raman Scattering from an Etched Polymer Substrate. *Anal. Chem.* **1997**, *69*, 2418-2425.
29. Biedermann, H.; Hlidek, P.; Pesicka, J.; Slavinska, D.; Stundzia, V. Deposition of composite metal/C:H films - the basic properties of Ag/C:H. *Vacuum* **1996**, *47*, 1385-1389.
30. Friedenthal, H. Absolute and relative disinfection power of elements and chemical compounds. *Biochem. Z.* **1919**, *94*, 47-68.
31. Forster, S. Amphiphilic block copolymers. *Ber. Bunsen-Gesell.* **1997**, *101*, 1671-1678.; Schmid, G.; Maihack, V.; Lantermann, F.; Peschel, S. Ligand-stabilized metal clusters and colloids: properties and applications. *J. Chem. Soc., Dalton Transactions: Inorg. Chem.* **1996**, *5*, 589-595.; General, S.; Thunemann, A.F. pH-sensitive nanoparticles of poly(amino acid) dodecanoate complexes. *Int. J. Pharm.* **2001**, *230*, 11-24.; Zhao, M.; Crooks, R.M. Intradendrimer Exchange of Metal Nanoparticles. *Chem. Mater.* **1999**, *11*, 3379-3385.

32. Saponjic, Z.V.; Csencsits, R.; Rajh, T.; Dimitrijevic, N.M. Self-Assembly of TOPO-Derivatized Silver Nanoparticles into Multilayered Film. *Chem. Mater.* **2003**, *15*, 4521-4526.
33. Groehn, F.; Bauer, B.J.; Amis, E.J. Hydrophobically Modified Dendrimers as Inverse Micelles: Formation of Cylindrical Multidendrimer Nanostructures. *Macromolecules* **2001**, *34*, 6701-6707.
34. Borel, J. P. Thermodynamical size effect and the structure of metallic clusters. *Surf. Sci.* **1981**, *106*, 1-9.
35. Fritzsche, W.; Porwol, H.; Wiegand, A.; Bornmann, S.; Kohler, J.M. In-situ formation of Ag-containing nanoparticles in thin polymer films. *Nanostruct. Mater.* **1998**, *10*, 89-97.
36. Bell, S.E.J.; Spence, S.J. Disposable, stable media for reproducible surface-enhanced Raman spectroscopy. *Analyst* **2001**, *126*, 1-3.; Chapman, R.; Mulvaney, P. Electro-optical shifts in silver nanoparticle films. *Chem. Phys. Lett.* **2001**, *349*, 358-362.
37. Yeo, S.Y.; Jeong, S.H. Preparation and characterization of polypropylene/ silver nanocomposite fibers. *Polym. Int.* **2003**, *52*, 1053-1057.
38. Zhang, Z.; Han, M. One-step preparation of size-selected and well-dispersed silver nanocrystals in polyacrylonitrile by simultaneous reduction and polymerization. *J. Mater. Chem.* **2003**, *13*, 641-643.
39. Dowling, D.P.; Donnelly, K.; McConnell, M.L.; Eloy, R.; Arnaud, M.N. Deposition of anti-bacterial silver coatings on polymeric substrates. *Thin Solid Films* **2001**, *398-399*, 602-606.

Engineered Drug Particles

Chapter 15

Engineering of Composite Particles for Drug Delivery Using Supercritical Fluid Technology

B. Y. Shekunov*, P. Chattopadhyay, and J. Seitzinger

Pharmaceutical Technologies, Ferro Corporation, 7500 East Pleasant Valley Road, Independence, OH 44131
***Corresponding author: shekunovb@ferro.com**

Challenges associated with the production of micro- and nanoparticles for drug delivery can be addressed using supercritical fluids (SCFs) or liquefied gases. These fluids offer the benefits of low processing temperatures, efficient organic solvent extraction, environmentally benign processing and cost effectiveness. In this chapter, we discuss the different technological approaches which employ SCFs, in relation to potential applications such as manufacturing of particulate systems for controlled or sustained release, coating and taste-masking, respiratory formulations and nanoparticles for increased solubilization.

Introduction

The incorporation of a drug in polymer particles for sustained or controlled release has numerous advantages for drug delivery, including increased drug efficacy, reduced dosing frequency, enhanced patient compliance and reduced costs. Such products typically employ artificial or natural polymers and/or other excipients that are biocompatible and biodegradable. From yet another perspective, the absorption of some poorly water-soluble drugs can be significantly improved by producing formulations of the drug with water-soluble or certain biodegradable polymers or lipids. The efficient size control of micron-size particles, and particle modifications such as production of drug-encapsulated biodegradable porous or hollow structures, are essential in providing controlled release respiratory formulations with enhanced aerosolization performance.

Several techniques such as emulsion solvent evaporation or extraction, co-acervation and spray-drying have been suggested for the preparation of these materials. Although predominantly used now and on a relatively small scale, these methods are not free from limitations such as use of elevated temperatures and limited size control during spray-drying, and the manufacturing complexities of emulsion-based techniques. The major disadvantage associated with all the above techniques is the removal of organic solvents, which are used to dissolve polymers and active compounds. These residual solvent impurities can lead to reduction in the efficacy of the pharmaceutical product, increased toxicity, undesirable side effects and other serious complications which can create significant regulatory hurdles in product development *(1)*.

Supercritical fluids (SCFs) have the following fundamental properties, which can be utilized for process engineering:

1. Efficient extraction of small organic molecules, which are miscible or at least partially soluble in SCFs, can provide an effective and environmentally benign process for cleaning of biodegradable polymers and pharmaceutical compounds at low temperature. This extraction process can be controlled by tuning the SCF density, pressure and temperature. Multiple impurities (e.g. residual solvents and monomers) can be removed by changing the SCF density. The most commonly used SCF in these applications is supercritical CO_2 due to its low critical temperature, non-toxic inert nature and low cost.
2. Increased diffusivity and reduced viscosity contributing to increased mass-transfer and penetration ability of SCFs are important for enhanced extraction and cleaning, especially when intensive agitation or mixing is required but difficult. Drying of porous matrixes and extraction in powder beds are examples of such applications.
3. Most pharmaceutical drug substances, reagents, and excipients have very low solubility in low-temperature SCFs, including CO_2. This is a major limitation as far as the extraction process is concerned, however, it also makes possible employing SCFs as non-solvents (antisolvents) in precipitation and crystallization processes, greatly reducing use of organic solvents, increasing product purity and providing benefits of a more controlled precipitation mechanism.
4. SCFs are very efficient plasticizers of polymers, leading to a significant reduction of the glass transition temperature, T_g, and increased molecular diffusivity in polymers. The T_g depression of most biodegradable and

pharmaceutically acceptable polymers in the presence of CO_2 is a function of pressure and temperature, and typically varies between 20-100K. This property of SCFs enables applications such as low-temperature impregnation of polymers with bioactive materials, preparation of porous polymer matrixes and particles.

5. Finally, an important property of the supercritical state is its high compressibility. Solvents, both miscible and partially miscible, when expanded together with SCFs, can be rapidly atomized into small droplets. In this role, SCF can be more efficient than compressed gases for nebulization. In certain cases, it can also be used for nebulization followed by freezing of solutions due to the Joule Thomson effect. The frozen droplets formed can be subsequently lyophilized in order to yield the dry product with interesting morphology.

Together with these advantages, there are also several challenges associated with large-scale applications of SCFs in the pharmaceutical industry. The low solubility of most pharmaceutical ingredients in supercritical CO_2 means that more complex processes than spraying by expansion have to be developed to produce particles. Another technical challenge is the immiscibility between water and CO_2, preventing efficient drying or precipitation of many biological and water-soluble substances using pure SCFs. The most important barrier however is the very nature of pharmaceutical product development, which is cautious to any new and unproven technology. A new technology must clearly show some formulation advantages compared to more traditional techniques, and/or lead to significant improvement of the product quality and process economics.

The following chapter introduces different technological approaches with SCFs, with emphasis on those which can be utilized for production of polymer-drug composite materials. We also introduce a new technique of supercritical fluid extraction from emulsions (PSFEE), which can be used to produce fine particles for various drug delivery applications.

Results and Discussion

SCF Methods

Supercritical Antisolvent (SAS) Precipitation

Antisolvent methods constitute the majority of the current SCF particle formation techniques *(1-3)*. They are based on a simple principle, whereby a drug and excipient (e.g. polymer or lipid) are co-precipitated together, using the antisolvent (non-solvent) properties of supercritical carbon dioxide, since most polymers and drugs are not soluble in CO_2. This method is likely to succeed when polymer and drug molecules are compatible with each other, forming ordered solid solutions, molecular dispersions or amorphous mixtures by strong hydrogen bonding or complex formation *(3,4)*. In the event of solid phase separation or re-crystallization, a non-homogeneous drug dispersion or partial coating could occur, leading to a characteristic "burst" drug release in dissolution studies. An example of successful co-precipitation is microencapsulation of budesonide into poly(L-lactic acid) (PLLA) in an amorphous form *(5)*. Another example includes production of a solid solution of theophylline in ethyl cellulose at drug loadings up to 35% *(6)*. In certain cases some of the drugs can be precipitated with polymers using a hydrophobic ion pairing (HIP) approach *(7)*, by which ionic species can be directly solubilized in non-aqueous solutions using pairing of charged molecules with oppositely charged surfactants. Such complexes may prevent drug re-crystallization and promote uniform release profiles.

A significant problem with the antisolvent process is that SCFs, including most commonly used CO_2, are excellent plasticizers of many polymers and lipids, in particular those with a significant amorphous phase and high molecular weight, causing swelling and agglomeration in particles. Amorphous high-molecular weight polymers are often preferable for drug formulation, allowing for more uniform release and more efficient drug loading than polymers of semi-crystalline structure or low-molecular weight. Thus a compromise has to be found by optimizing processing conditions to prevent agglomeration in CO_2 whilst retaining sufficient drug-release properties of such polymers *(8)*.

The antisolvent techniques are difficult to use for water-soluble compounds because they have to be dissolved in the same solvent as the polymer. However, an example in Figure 1 illustrates that suspensions of such compounds in a polymer solution can be used for particle coating, with potential applications for taste-masking or sustained release formulations. Coated particles of an extremely bitter and highly water-soluble drug were prepared in our laboratory for use in fast disintegrating tablets. Crystals of the drug, with particle sizes ranging between 50 and 150 μm in size, where coated with ethyl cellulose with drug loading between 40-60% to obtain a defined release profile, allowing for 90% drug dissolved within 40 minutes (Figure 1). The bitter taste of the original water-soluble drug was significantly reduced by this formulation. The particles also exhibited sufficient mechanical strength for the tablet compression.

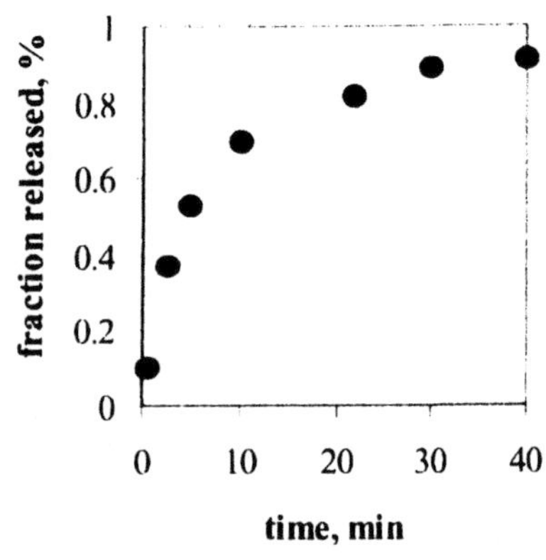

Figure 1. (a) Particles of a bitter drug coated with ethyl cellulose using antisolvent precipitation with CO_2 and (b) correspondent release profile.

Rapid Expansion of Supercritical Solutions (RESS)

The RESS technique involves precipitation of particulate material by expansion of a solution in SCFs *(9)*. Its applications are limited to compounds which have an appreciable solubility in the SCFs. As a rule, when a compound is soluble in supercritical CO_2 at concentrations above 10^{-4} mole fraction, RESS is preferred as it provides a simple, direct, solvent-free and continuous process to particle production. It can be optimized to achieve a relatively narrow particle size distribution, and can also be used for the coating of micron particles with CO_2-soluble polymers such as low-molecular weight (M_w <6000) poly(lactic-glycolic acid) (PLGA) and poly(L-lactic acid) (PLLA) *(10)*, lipids *(11)* or waxes *(12)*. Addition of an organic co-solvent, miscible with CO_2, such as acetone *(9)* can improve the solubility of the drug or formulation excipient, however, efficient removal of these solvents after expansion may prove to be a challenge if more than a few percents are used. Solvent can be, for example, removed by using an additional stream of warm nitrogen or air. If this process is applied to relatively high co-solvent mole fractions, the RESS process becomes rather an accelerated spray-drying process, where nebulization with SCFs may result in a reduced drying temperature. Such a technique (CO_2 - assisted nebulization and bubble drying or CAN-BD) has been applied for production of composite particles, using suspensions of model water-soluble compounds in PLGA solutions *(12)*.

Plasticization of Polymers in Gas-Saturated Solutions (PGSS)

High solubility of SCFs in polymers, lipids or some other materials (typically between 5 and 40% mole fraction) can result in lower viscosity and enhanced segmental and chain mobility of the polymer, leading to depressed melting and glass-transition points. This property of SCFs is beneficial for enhanced processability of composite materials during mixing, homogenization or impregnation. The expansion of melts under reduced pressure in the PGSS process can produce particles, often with micro-porous structures *(13,14)*. The challenge here is typically particle size control due to the high melt viscosities, which may suppress the formation of fine particles. This technology has been utilized for solvent-free production of polymer particles and matrixes with drug loading, for controlled release or taste-masking applications *(15)*. An example is shown in Figure 2.

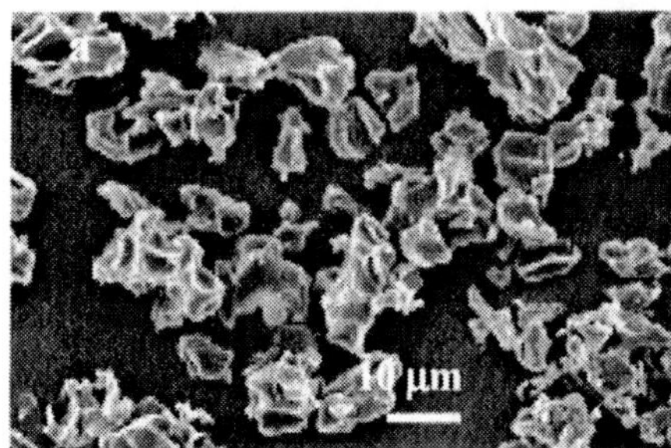

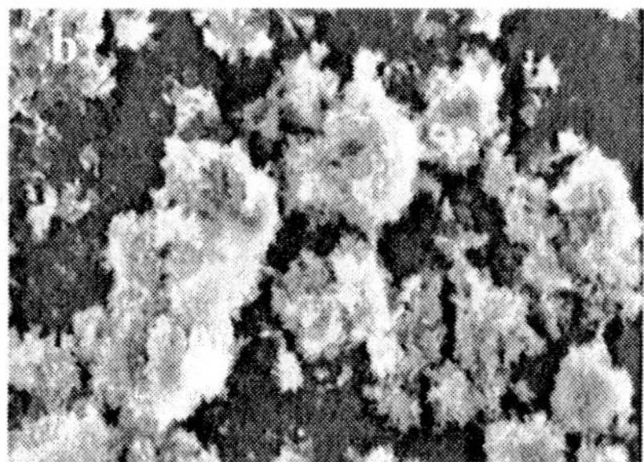

Figure 2. Examples of composite particles produced by plasticization-expansion of excipient/drug mixture: (a) polymer (Eudraget RS100) with 20% encapsulated acetaminophen and (b) solid lipid particles (tripalmitin) with 10% ketoprofen.

Particle sizes of such composites may vary from submicron (as in the case of the lipid in Figure 2b) to hundreds of microns, or even fibers for viscous polymers, depending on the polymer structure and expansion conditions. Recent advances of this technology also enable obtaining drug formulations in both a crystalline form and in the form of a molecular dispersion. Importantly, impregnation of the polymers can be achieved with significant loading, even if the drug is not soluble in CO_2. High affinity of the drugs to certain polymers results in the preferential partitioning of the drug into the polymer phase as compared to the SCF phase *(16)*. This plasticization-impregnation process is also applicable for formulation of many therapeutic proteins, because CO_2 is typically inert to biomolecules and the processing temperature can be sufficiently low (between about 30 and 50°C), thereby preserving the biological activity of the protein used *(1,14)*.

Preparation of Liposomes

Liposomes are lipid-based vesicles used in drug delivery by injection. The traditional methods of preparing liposome formulations involve usage of organic solvents in a solvent evaporation-hydration process. Characteristically, such a method suffers from poor encapsulation efficiency, low scalability, low consistency and difficulties in removal of residual organic solvents on the industrial scale. Thus supercritical solvents represent a viable economical option for such preparations. Several processing schemes have been suggested, including RESS and SAS processes, in which the removal of organic solvent was shown to be complete *(17)*. A modified RESS method involves expansion of supercritical solution of phospholipid and cholesterol in CO_2 into an aqueous solution, containing a water-soluble lipophilic marker (fluorescein isothiocyanate) for encapsulation (18). This method reduced the solvent use by an order of magnitude and led to improved encapsulation efficiency. Similarly, a multi-fold increase of the encapsulation efficiency of liposomes was observed for other water-soluble compounds and hydrophobic drugs *(18)*. All these data therefore suggest that SCFs can indeed be used as an efficient dispersing and/or extracting agents to prepare uniform liposomal systems.

Production of Micro- and Nano-Particles Using Supercritical Fluid Extraction of Emulsions (PSFEE)

PSFEE Technique

The SCF techniques discussed above can be successfully used for specific applications to produce composite particles for drug delivery. However, the principal shortcoming of all these methods lies in the particle size control. It is possible to show, both experimentally and theoretically *(19)*, that for all precipitation techniques, SAS and RESS included, limitations exist on the particle size as well as on the widths of the particle size distribution. This phenomenon is related to the competitive nucleation-growth-aggregation mechanism in solutions. In addition, strong particle aggregation for most pharmaceutical excipients used in controlled release makes the production of discrete particles a very difficult task. Typically, only a small fraction of such particles are produced in the primary particle size range, as confirmed by different particle size analysis combined with scanning electron microscopy (SEM).

The PSFEE process addresses these challenges and is able to produce a wide variety of pure and composite (e.g. polymeric or lipid) particles in the required size range. The process is based on extraction of oil-in-water (o/w), oil-in-oil (o/o) or different multiple emulsions using SCFs. In particular, water and carbon dioxide, two of the most inexpensive and benign solvents, can be used in conjunction with each other to provide the emulsification and extraction media as shown in Figure 3.

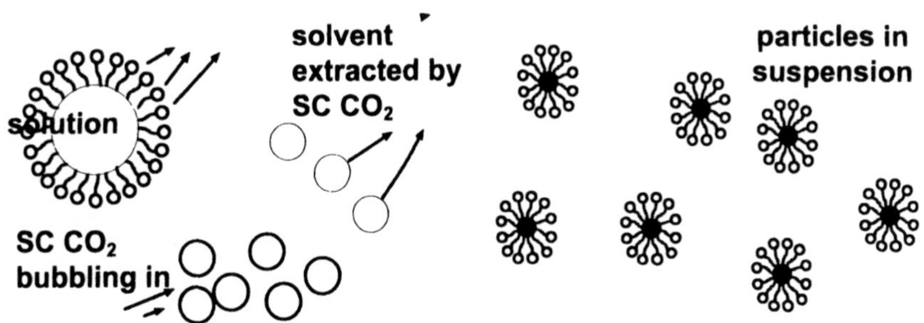

Figure 3. Mechanism of the PSFEE process.

In this system, organic solvents immiscible or partially miscible with water are used, thus the majority of organic solvents are suitable for dissolving both the excipients and drug substances. According to Figure 3, the process consists of the following stages:

1. Preparation of an emulsion by dispersing a solution of a drug and polymer in a suitable organic solvent in an aqueous solution of surfactants or stabilizing agents.
2. The emulsion prepared above is then injected into an extraction column, in which supercritical carbon dioxide (SC CO_2) typically flows countercurrent to the emulsion flow. The SC CO_2 removes the organic solvent as well as other impurities, such as monomers, soluble in the supercritical mixture of CO_2 and organic solvent. A suspension of solid particles stabilized in water is obtained.
3. The aqueous suspension obtained is continuously removed from the column and, if required, pure dry powders are produced using processes of in-line high-pressure filtration, centrifugation and/or lyophilization. A novel spray-freezing technique with supercritical CO_2 *(20)* can be conveniently applied at the column exit in order to remove water with greater efficiency than a standard lyophilization process.

The PSFEE method combines the flexibility of particle engineering using different emulsion systems with the efficiency of large-scale, continuous extraction using SCF. This combination also overcomes the limitations of the other techniques with respect to the control of particle size distribution, shape and structure. As each emulsion droplet serves as a microreactor, it enables formation of uniform particles of defined size and morphology, for example, spherical, prismatic or platelet. Coating or encapsulation is more efficient in such a process because precipitation is confined within the droplet volume, and the particles do not agglomerate. This precipitation also occurs relatively slow, providing the opportunity to obtain thermodynamically stable solid structures of a drug or drug-excipient mixtures. Further dissolution of CO_2 in polymers followed by gas expansion in the suspension may lead to porous or hollow structures useful for some formulations. Although PSFEE is a truly continuous process, this comes at the expense of obtaining products in the form of solvent-free aqueous suspensions with solid content between 0.5 and 10%. Surfactant, which is typically a pharmaceutical excipient, is retained in the aqueous phase. More highly concentrated suspensions as well as pure dry powders can be obtained by the consecutive processing. Some of the specific applications of the PSFEE for different drug delivery systems are discussed below.

Micro- and Nano-Particles for Controlled Release

SCF extraction of emulsions using the PSFEE process provides consistent manufacturing of micro- and nanoparticles, including polymer-drug particles down to about 50 nm VMD. The important advantage is that this process can be made continuous and also implemented on a large scale.

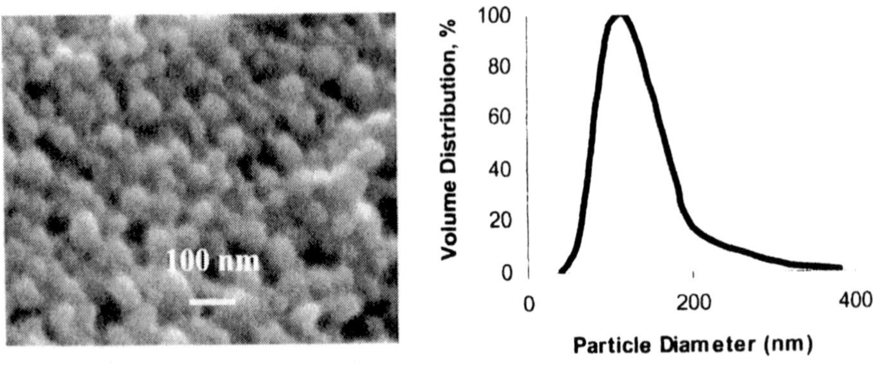

Figure 4. PLGA nanoparticles produced using PSFEE process.

Figure 4 shows an example of PLGA nanospheres, which can be used for injectable formulations, with VMD = 170 nm. The residual solvent content for such particulate systems are consistently below 50 ppm, independent of the organic solvent used, and well below the regulatory-accepted levels. For particles shown in Figure 4, a drug (such as anti-fungal or anti-cancer agents) is incorporated within the polymer matrix. The drug loading efficiency for water-insoluble molecules ranges between 90-100%.

Similarly, microspheres with VMD in the range between 1-1000 μm can be produced by adjusting the size of the emulsion droplets. Plasticization and extraction processes with SCF provide favorable conditions to obtain the most stable solid form for a given substance, such as a stable solid dispersion in amorphous or semi-crystalline polymer matrix, or nanocrystals incorporated into the polymer, dependent on the polymer molecular weight and molecular structure. Both water-soluble and water-insoluble drug molecules can be encapsulated into polymeric microcapsules using PSFEE with either multiple (e.g. w/o/w) emulsions, water-free (e.g. o/o) emulsions or suspensions, where solid drug particles are dispersed in the internal phase. Several different particle structures can be obtained in this fashion.

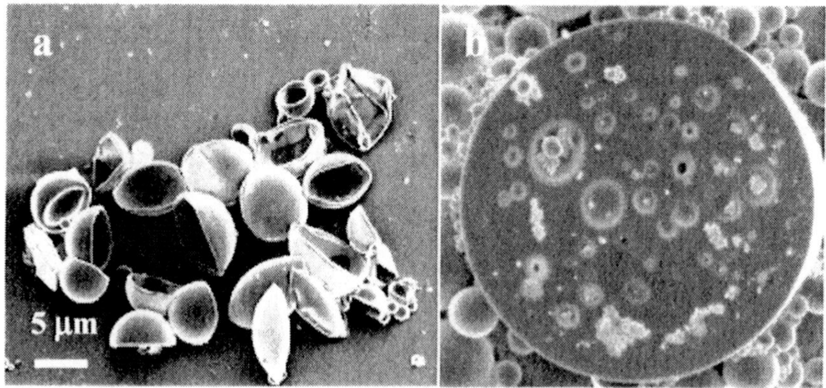

Figure 5. (a) Hollow PLGA particles produced using PSFEE with w/o/w emulsion and (b) encapsulation of bovine serum albumin into PLGA using a similar emulsion system.

The example in Figure 5a shows cross-sections of hollow PLGA spheres, in which the active molecules can be encapsulated in the polymeric shell and/or precipitated in the core of the particle. Encapsulation of a protein (BSA) into a polymer matrix in a form of a micro-dispersion is shown in Figure 5b. The efficiency of encapsulation in this case was sufficiently high at about 93%, despite the fact that this compound is soluble in the external aqueous phase. Similarly, porous structures can be produced by means of creating multiple cavities within a polymer structure.

Particles for Respiratory Delivery

Efficient respiratory delivery requires non-aggregated, easily dispersible and uniform particles with mass-median aerodynamic diameter (MMAD) between 1 and 5 μm. It is also well-known that porous or hollow particles are advantageous for aerosol drug delivery because of their relatively large volume mean diameter (VMD) and small MMAD. In combination, these parameters lead to enhanced particle dispersion and deep-lung deposition *(20)*. Although many inhalation drugs are for immediate release, there are several reasons to utilize sustained release formulations. For example, pulmonary absorption of poorly water-soluble drugs can be improved by using a co-formulation with muco-adhesive and water-soluble polymers. Composite particles, for example, biodegradable polymer microcapsules, can also provide selective controlled or prolonged release, and increased drug permeability, combined with protection against enzymatic hydrolysis. Although controlled drug release by pulmonary (or nasal) administration is a new field, there is a growing interest in particle technologies that can produce such formulations. A detailed analysis of the applicability of SCF technologies, such as SAS, for respiratory formulations is given elsewhere *(21)*. PSFEE is another method which can provide uniform, crystalline water-insoluble drug particles, microspheres or microcapsules within the 1–2 μm size range for dry-powder and metered-dose inhalers *(20)*. In addition, nanosuspensions of water-insoluble drugs formed using this process can be used for liquid respiratory formulations such as nebulizers and soft mist inhalers (unpublished results).

Lipid-Based Formulations

Lipid-based systems employed in drug delivery are oil-in-water emulsions, liposomes and solid-lipid particulate systems. Among these, solid-lipid nanoparticles (SLN) have been considered as a viable alternative to polymeric nanoparticles *(22)*. The technology typically employed to produce such formulations involve high-pressure homogenization of molten lipid droplets in which the drug is dissolved. This is an organic solvent-free process with, however, several important limitations. Firstly, many drugs cannot be dissolved in lipids and thus a practical drug-loading capacity cannot be obtained. Secondly, the melting temperature of some lipids may be sufficiently high to produce a damaging effect on certain biological molecules such as proteins, DNA and peptides. Finally, the control of the particle size is limited by the properties of molten lipids and surfactants used with them. Typically, particles in the upper nano-size range between 200-1000 nm are produced *(22)*.

Supercritical CO_2 allows a significant depression of the lipid's melting point, in the range between 10-30°C, and can readily extract any organic solvents from a lipid-drug mixture. Thus the PSFEE method enables the production of solid-lipid nanoparticles in form of a concentrated, stable aqueous suspension, which can be, in turn, transformed into redispersible solid formulations if required. Particle size control for these systems is excellent. Figure 6 shows an example of cumulative particle size distribution with VMD = 23 nm of the water-insoluble drug indomethacin, encapsulated into a lipid. Drug loadings up to 30% can be produced in this system.

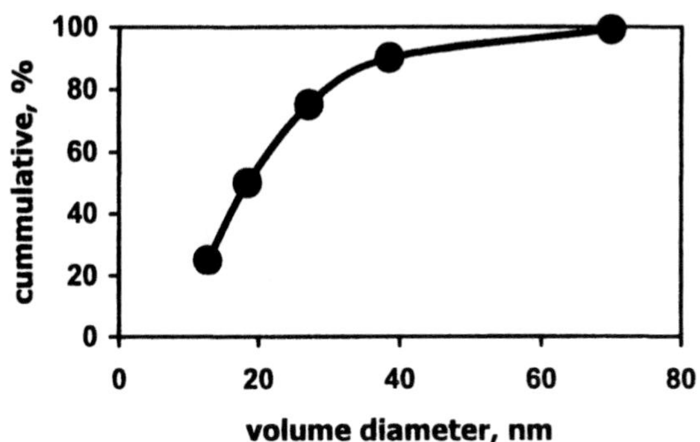

Figure 6. Particle size distribution of composite Gelucire 50/13-indomethacin nanoparticles (20% loading), obtained using PSFEE.

The rheological properties of this suspension such as viscosity, filtration and dispersion are almost equivalent to those of a solution phase. This behavior opens up new opportunities for use of such formulations for injectable and respiratory drug delivery designed for solution phases.

Drug Nanoparticles

Nanoparticulate drugs can be required for many formulations to increase consistency and uniformity of such dosage forms as oral suspensions, tablets or capsules or as drug delivery formulations via the injection or pulmonary routes.

The most obvious application of nanoparticles is administration of poorly water-soluble drugs, which presents a significant challenge due to their high toxicity, irregular absorption and low bioavailability. The dissolution rate of these drugs can be enhanced by comminuting them into nanoparticles with VMD between 100-800 nm and increased surface area. Here the PSFEE process can offer a continuous, high-output, controlled production of drug nanoparticles in the form of concentrated aqueous suspensions. Dry powders can be obtained by high-pressure filtration and/or spray-freezing techniques *(20)*.

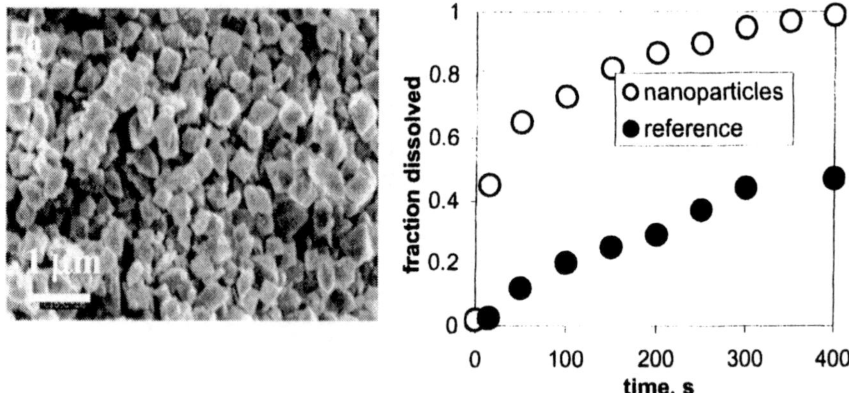

Figure 7. (a) Microphotograph of crystalline griseofulvin nanoparticles with VMD = 760 nm and number-weighted mean diameter, NMD = 224 nm as determined by dynamic light scattering technique, and (b) dissolution profile of these particles compared to reference amorphous material with VMD = 2 μm.

Figure 7 shows an example of griseofulvin, a water-insoluble anti-fungal agent chosen as a model drug because of its persistent acicular morphology and difficulties in size reduction using both antisolvent precipitation or milling techniques. These powders show consistent reconstitution into suspensions after storage. The drug nanoparticles are usually crystalline (Figure 7a), and therefore, stable in both powdered and suspension forms. This property compares favorably, in terms of the physical stability, with amorphous materials often obtained by milling, solubilization or rapid precipitation techniques, with no significant changes detected over the period of several months. A particle size change would be expected due to the Ostwald ripening effect or particle aggregation. The uniform particle size and high drug crystallinity, combined with the effect of surfactant, provides the required stability in suspension. The products obtained using PSFEE typically exhibit increase in the dissolution rate between 3 to 20 times when compared to jet-milled and amorphous reference materials, as shown for griseofulvin in Figure 7b. It is also noticeable that more regular, prismatic particles are formed, compared to acicular shapes produced by other precipitation techniques. Thus the microenvironment during the SCF extraction process allows controlling both the particle morphology and particle size.

Conclusions

The present work provides a brief overview of the different technologies utilizing supercritical fluids, and introduces a new PSFEE method based on SCF extraction in dispersed systems such as oil-water-CO_2. The proposed technique facilitates production of particles with well-defined size and structure, including coated, hollow or porous materials. With growing demand for new and more advanced drug formulations of water-insoluble molecules, biopharmaceuticals and composite materials for controlled release, there will also be a need for more consistent and economical particle engineering techniques. Unlikely though that any "universal" method will ever be created. The particle formation techniques are complementary to each other and have to be optimized for a specific drug delivery system. When tailored for such specific applications, SCF methods can offer significant processing advantages and provide a viable economical alternative to the existing manufacturing techniques.

References

1. Bandi, N.; Roberts, C.B.; Gupta, R.B.; Kompella, U.B. Formulation of controlled-released drug delivery systems. In: Supercritical Fluid Technology for Drug Product Development. Eds: York, P.; Kompella, U.B. and Shekunov B.Y. Marcel Dekker Series: Drugs and the Pharmaceutical Sciences **2004**, *138*, 367-409.
2. Bodmeier, R.; Wang, H.; Dixon, D.J.; Mawson, S.; Johnston, K.P. Polymeric microspheres prepared by spraying into compressed carbon dioxide. *Pharm. Res.* **1995**, *12*, 1211-1217.
3. Kompella, U.B.; Koushik, K. Preparation of drug delivery systems using supercritical fluid technology. *Crit. Rev. Ther. Drug Carrier Syst.* **2001**, *18*, 173-199.
4. Kazarian, S.G.; Martirosyan, G.G. Spectroscopy of polymer/drug formulations processed with supercritical fluids: in situ ATR-IR and Raman study of impregnation of ibuprofen into PVP. *Int. J. Pharm.* **2002**, *232*, 81-90.
5. Martin, T.M; Bandi, N.; Shultz, R.; Roberts, C.B.; Kompella, U.B. Supercritical fluid technology-derived budesonide and budesonide-pla microparticles for respiratory delivery. *AAPS Pharm. Sci.* **2001**, *3(3)*, abstract.
6. Wilkins, S.; Shekunov B.Y.; York, P. Theophylline:ethylcellulose co-precipitates formed by solution enhanced dispersion by supercritical fluids (SEDS) and solvent co-evaporation – structural analysis by synchrotron powder X-ray diffraction. *AAPS Pharm. Sci.* **2001**, *3*, 1136.

7. Zhou, H.; Lengsfeld, C.; Claffey, D.J.; Ruth, J.A.; Hybertson, B.; Randolph, T.W.; NG, K-Y; Manning, M.C. Hydrophobic ion pairing of isoniazid using a prodrug approach. *J. Pharm. Sci.* **2002**, *91*, 1502-1511.
8. Shekunov B.Y.; Edwards, A.D. Crystallization and plasticization of poly(L-lactide) (PLLA) with supercritical CO_2. *Proceedings of the 6th International Symposium on Supercritical Fluids,* Versailles, France, **2003**, *3*, 1801-1806.
9. Debenedetti P.G.; Tom J.W.; Yeo S.-D.; Lim, G.-B. Application of supercritical fluids for the production of sustained delivery devices. *J. Controlled Release* **1993**, *24*, 27-44.
10. Ribeiro Dos Santos I.; Richard J.; Thies, C.; Pech B.; Benoit J.-P. A supercritical fluid-based coating technology. 3: Preparation and characterization of bovine serum albumin particles coated with lipids. *J. Microencapsulation* **2003**, *20*, 110-128.
11. Wang, T.-G.; Tsutsumi, A.; Hasegawa, H.; Mineo T. mechanism of particle coating granulation with RESS process in a fluidized bed. *Powder Technology* **2001**, *118*, 229-235.
12. Sievers, R.E.; Villa, J.A.; Cape, S.P.; Alargov, D.K.; Rinner, L.; Huang, E.T.S.; Quinn, B.P. Synthesis and coating respirable particles with a low-volume mixing cross or tee. *Proceedings of the Conference on Respiratory Drug Delivery*, Palm Springs, CA IX, **2004**, *2*, 765-768.
13. Mandel, F.S.; Wang, J.D. Pharmaceutical material production via supercritical fluids employing the technique of particles from gas-saturated solutions. *Proceedings of the 7th Meeting on Supercritical Fluids*, Antibes, France, **2000**, *1*, 35-45.
14. Howdle, S.M.; Watson, M.S.; Whitaker, M.J.; Popov, V.K.; Davies, M.C.; Mandel, F.S.; Wang, J.D; Shakesheff, K.M. Supercritical fluid mixing: preparation of thermally sensitive polymer composite containing bioactive materials. *Chem. Commun.* **2001**, 109-110.
15. Shekunov, B.Y.; Chattopadhyay, P.; Seitzinger, J.S. Preparation of composite particles for taste masking and controlled release using liquefaction with supercritical CO2. *AAPS Pharm. Sci.* **2003**, *5*, W5101.
16. Kazarian S.G. Supercritical fluid impregnation of polymers for drug delivery. In: Supercritical Fluid Technology for Drug Product Development. Eds: York, P.; Kompella, U.B. and Shekunov B.Y. Marcel Dekker Series: Drugs and the Pharmaceutical Sciences **2004**, *138*, 343-365.
17. Badens, E.; Magnan, C.; Charbit, G. Microparticles of soy lecithin formed by supercritical process. *Biotechnol. Bioeng.* **2001**, *72*, 194-204.
18. Frederiksen, L.; Anton, K.; van Hoogevest, P.; Keller, H. R.; Leuenberger, H. Preparation of liposomes encapsulating water-soluble compounds using supercritical carbon dioxide. *J. Pharm. Sci.* **1997**, *86*, 921-928.

19. Baldyga, J.; Henczka, M.; Shekunov, B. Y. Fluid dynamics, mass-transfer and particle formation in supercritical fluids. In: Supercritical Fluid Technology for Drug Product Development. Eds: York, P, Kompella, UB and Shekunov BY. Marcel Dekker Series: Drugs and the Pharmaceutical Sciences **2004**, *138*, 91-157.
20. Shekunov, B. Y.; Chattopadhyay, B.; Seitzinger, J. Production of respirable particles using Spray-Freeze-Drying with Compressed CO_2. *Proceedings of the Conference on Respiratory Drug Delivery*, Palm Springs, CA IX, **2004**, *1*, 489-492.
21. Shekunov, B. Y. Production of powders for respiratory drug delivery. In: Supercritical Fluid Technology for Drug Product Development. Eds: York, P, Kompella, UB and Shekunov BY. Marcel Dekker Series: Drugs and the Pharmaceutical Sciences **2004**, *138*, 247-282.
22. Müller, R. H.; Mader, K.; Gohla, S. Solid lipid nanoparticles (SLN) for controlled drug delivery – a review of the state of the art. *Eur. J. Pharm. Biopharm.* **2000**, *50*, 161-177.

Chapter 16

Using a Supercritical Fluid-Based Process: Application to Injectable Sustained-Release Formulations of Biomolecules

J. Richard[1–3], F. Deschamps[2], A. M. De Conti[1], and O. Thomas[2]

[1]Ethypharm, 194 Bureaux de la Colline, Saint-Cloud 92213, France
[2]Mainelab, 8 rue André Boquel, Angers 49100, France
[3]Permanent address: Serono International, Via di Valle Caia 22, I–00040 Ardea, Rome, Italy

The market development of fragile biomolecules such as recombinant proteins is impeded by the lack of an appropriate delivery method due to the proteins' poor oral bioavailability and short plasma half-life. Therefore, a strong need for a new microparticulate parenteral delivery system exists, allowing sustained release over weeks and preventing loss of protein activity during formulation and administration. Classical microencapsulation processes display major drawbacks related to the use of organic solvents and heat that can denaturate or cause aggregation of the proteins. Conversely, supercritical fluid (SCF)-based processes exhibit major advantages for designing and producing solvent-free microparticles with outstanding applications in the field of drug delivery. Special emphasis is given on the SCF-based process that we have recently developed to coat lyophilized protein particles with lipids, and obtain injectable sustained-release microparticles. Results concerning the physicochemical features of the microparticles produced, as well as the first *in vivo* pharmacokinetic effects in animals are presented and discussed.

Introduction

Therapeutic proteins are fragile biomolecules, which display a short plasma half-life after injection and cannot be administered by the oral route due to their poor stability in the gastrointestinal (GI) tract and their poor oral bioavailability. These proteins need an appropriate injectable delivery system to maintain therapeutic levels in the blood stream over long periods of time and hence avoid repeated injections, typically several times a week. Moreover, a sustained-release system is also expected to restrict the peak plasma concentration responsible for major clinical issues that are still unsolved. Microparticulate systems with specific physicochemical and size properties are currently developed and used for drug delivery. These microparticles consist of biocompatible polymers that contain drugs and can be administered to humans safely. Biodegradable polymers are commonly used to prepare injectable sustained-release delivery microparticles because they allow controlling the drug release profile by properly choosing their physicochemical features. These systems have been extensively investigated for their capability of achieving the release of therapeutically active proteins in a controlled way. Current methods for the preparation of protein-containing polymer microparticles are mainly based on emulsion-solvent extraction, phase separation or spray-drying. The main drawback of these techniques is the extensive use of organic solvents to either dissolve the polymer or induce phase separation and particle hardening. These solvents are suspected to be partly responsible for biological inactivation of the proteins incorporated in microspheres. Successful incorporation of proteins in poly(lactic-*co*-glycolic acid) (PLGA) copolymer microparticles has been reported with respect to loading and encapsulation efficiency, as well as microparticle size and morphology. However, protein instability in the polymer microparticles has been recognized as a major problem, resulting in incomplete release of native protein, which is denaturized or aggregated. Upon denaturation or aggregation, protein species become therapeutically inactive and may also induce side effects such as immunogenicity or toxicity, even if there is only a low amount of degraded protein. Thus, development of new protein-loaded microparticle production processes should be focused on full preservation of the structure and biological activity of the native protein during preparation, storage and release.

The use of supercritical fluids (SCF) for engineering of drug-loaded microparticles with or without limited use of organic solvents is a recent development. A SCF is a fluid whose pressure and temperature are higher than those at the critical point, that is, the end-point of the liquid-gas phase transition line, as it appears in a Pressure-Temperature phase diagram. Several features of SCF make them versatile and appropriate for production of drug-loaded polymer microparticles. SCF display some liquid-like properties such as a high density, and gas-like ones such as low viscosity and high diffusivity. The most important property of a SCF is its large compressibility near the critical point. This specific property leads to a fluid whose solvent power can be continuously tuned from that of a liquid to that of a gas applying small variations of pressure. CO_2 is the most widely used SCF in pharmaceutical development and processing because of its low critical temperature (T_c = 31.1°C), its environmentally benign nature and other advantages. CO_2 allows working at moderate temperatures and leaves no toxic residues since it turns back to its gas phase at ambient conditions. Due to its unique properties, CO_2 is used routinely in large-scale operations for extraction of food ingredients. Many different approaches have been reported in the literature for the production of pure drug or drug-loaded microparticles, using an SCF either as a solvent for the drug and the polymer (RESS process), or as an anti-solvent (SAS, GAS, ASES, SEDS processes), or even as a swelling and plasticizing agent for the polymer (PGSS process) *(1,2)*. In addition, coating processes of preformed solid drug microparticles have been recently developed in our laboratory to produce sustained-release microparticles of fragile biomolecules. These processes use SC or liquid CO_2 either as a solvent or a non-solvent for the coating material, which can be a solid lipid compound or a polymer *(3,4)*.

Materials and Methods

Materials

Interferon α-2b (IFNα-2b) particles to be coated were produced by milling and sieving (<25 μm) a commercially available freeze-dried powder of human recombinant interferon α-2b (INF®, Gautier Cassara, Argentina). The mean particle diameter (D4,3) of milled and sieved IFNα-2b particles was ~15 μm. The IFNα-2b titer of these particles was measured using an ELISA method, and found to be 2.866 x 10^6 pg per mg of particles.

A blend of glycerides, provided as a fine, white, free-flowing powder of glyceryl palmitostearate, was used as the coating material. The melting point of this material was found to be 56.2°C. The phase behavior of lipids in CO_2 was also studied using trymyristin (Dynasan® 114, Sasol, Germany) as model compound. The trymyristin content of Dynasan® 114 is higher than 96%, and its melting point was measured at 58.4°C.

Phase Behavior Study

The phase behavior of coating material/CO_2 systems was studied using a high-pressure view cell (SITEC, Maur/Zurich, Switzerland), equipped with a magnetic stirrer, whose volume can be varied in a controlled way using a pressure balance piston. A known amount of coating material was added to the view cell chamber. The chamber was then sealed with the piston and CO_2 was pumped into the chamber until a clear, single-phase solution was obtained. The cloud point was determined visually at a given temperature (T) by slowly lowering the pressure (P) upon moving the pressure balance piston until the solution becomes cloudy. The phase behavior of coating material/CO_2 systems under the conditions of the coating process was studied by loading the view cell with a known amount of coating material and CO_2 until defined T/P conditions are reached (typically 45°C/20 MPa), then modifying the T/P conditions in order to study the T/P pathway used in the coating process. The objective was to identify the appropriate T/P pathway to get a coating material whose morphology, crystallinity and thermal properties are consistent with the features of a sustained-release formulation.

Coating Process

Lyophilized protein particles were coated using an apparatus described elsewhere *(5,6)*. Known amounts of freeze-dried protein particles and coating material were introduced in a high-pressure stirred 1-L vessel. The vessel was closed, and CO_2 was pumped into it until the T/P conditions for solubilization of coating material were reached. After 1 hour agitation, an isochoric cooling was carried out, typically until a temperature of 20°C was reached. CO_2 was then slowly vented. Microparticles were harvested, sieved with a 150-µm sieving plate and kept under a nitrogen atmosphere.

In Vivo Study

In vivo evaluation of coated IFNα-2b microparticles was carried out using a series of 70 male Swiss mice, 8 to 10 week old. The series of animals was divided into 7 sets of 10 mice (labelled A to G). Prior to subcutaneous injection, the protein titre in the microparticles was assayed using a commercially available immunoenzymatic (ELISA) kit (R&D Systems, France). Then the microparticles were redispersed in an aqueous vehicle for injection, containing carboxymethylcellulose. The particle size distribution and physical stability of the dispersions were characterized using a laser beam-scattering size analyzer (Mastersizer-S, Malvern, France), equipped with a 120-mL Qspec small volume dispersion unit (Malvern, France). Stirring speed was set at 1400 rpm. The particle size distribution of uncoated IFNα-2b particles was measured from a dispersion of particles in heptane containing 0.05% v/v of Montane®85 (Seppic, France). For coated protein particles, the particles were dispersed in a PBS pH 7.4 buffer containing 0.01% v/v Montanox® 20PHA (Seppic, France). Dose uniformity was checked by assaying protein content in 8 samplings of the dispersion. Each animal received a dose of 500,000 IU IFNα-2b that is

~3,414,000 pg protein, administered as a subcutaneous injection of a 0.9 mL suspension of microspheres. A standard 0.9-mL solution of non-coated IFNα-2b, containing the same dose, was administered to the reference group for comparison. A 400-mL blood sample was collected from the retro-orbital venous sinus of the animals, alternatively in set A to G of the series. The collected blood was centrifuged immediately to harvest the serum, which was then frozen and stored at –20°C until analysis. Circulating IFNα-2b concentration was measured using the ELISA method, and pharmacokinetic study was carried out over a period of 6 days. The limit of quantification of the ELISA method was determined and found to be 12.5 IU/mL.

Results and Discussion

The encapsulation process involves the following three successive steps:

1. Dissolution of the coating material in SC CO_2 under defined T/P conditions and dispersion of insoluble freeze-dried protein particles in this SC solution under stirring.
2. Induction of a controlled phase separation and precipitation of the coating agent onto the dispersed particles by gradual and controlled variation of T/P conditions in the reactor. Deposition of the coating onto the microparticles occurs progressively as the solubility of the coating material in CO_2 decreases and the coating material desolvates.
3. Harvesting the composite microparticles after depressurization of the coating vessel.

The process can be described as a coating/granulation process. First, the coating of individual particles (e.g. lyophilized protein powders) results in the production of primary core-shell microcapsules. Then granulation of these primary coated particles results in the formation of microspheres, where protein particles are embedded in the lipid matrix. The final structure (microcapsule or microsphere), size distribution and release properties of coated protein particles were found to be dependent on many parameters, as quoted in Table I.

For any given lyophilized protein particles, coating material and loading ratio, the process has to be optimized regarding the structure, size and release properties of the protein-loaded microspheres. The main pre-requisite for an effective process optimization is getting an in-depth knowledge of the phase behavior of the respective coating material/CO_2 systems.

Table I. Parameters influencing the microparticle structure and properties.

Process operating conditions	- starting T/P conditions - weight fraction of coating material in CO_2 - T/P pathway - kinetics of T/P variation - stirring conditions - protein/coating material weight ratio (loading ratio of lipid/protein particles)
Coating material	- composition - melting point - physical state - phase behavior in CO_2
Lyophilized protein particles	- particle size distribution - aspect ratio - density - composition

The phase behavior of such complex systems was studied using the high-pressure view cell described above. The main objectives of this investigation were:

- Quantifying the solubility of various lipid coating materials in CO_2;
- Controlling when and how the coating material desolvates and is able to form a coating at the surface of particles dispersed in the medium.

The solubility domain of coating materials in CO_2 was determined, and the cloud point pressure of coating material/CO_2 systems was recorded at different temperatures and for different weight fraction values of the coating material in CO_2. Coating materials are typically commercial blends of glycerides. Thus, their solubility in SC CO_2 cannot be derived from the literature, where solubility studies are mainly performed with pure lipids. Moreover, the solubility of such lipid materials can be dramatically altered by the presence of minor lipid components. Hence, in order to get comprehensive control of the process, it is mandatory to quantitatively determine the solubility of each batch of coating material. The cloud point curve of a model lipid system, Dynasan® 114/CO_2, is shown in Figure 1.

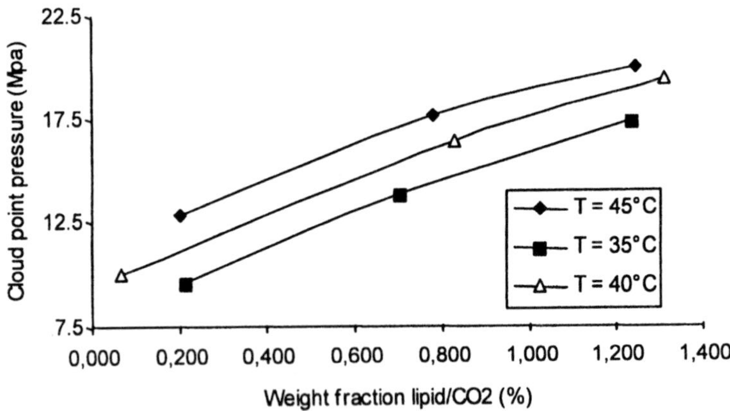

Figure 1. Cloud point curves for the model system Dynasan® 114/CO_2.

For this tryglyceride, an isothermal increase of pressure led to an increase of solubility. Conversely, at constant pressure, an increase of temperature led to a decrease of solubility, as expected with regards to the decrease of CO_2 density. This behavior was observed for a wide range of glycerides or blends of glycerides, and particularly for the blend of glycerides which was further used for coating experiments.

Solubility data were used to select the starting T/P conditions of the coating process, so as to ensure that all the coating material was dissolved. The presence of a minor part of not dissolved coating material before performing the coating operation by tuning the T/P parameters of the system could cause the final product to be inhomogeneous, and the process to be poorly reproducible.

The melting point of the model material, Dynasan® 114, was measured at 58.4°C at atmospheric pressure. However, during cloud point experiments at a temperature of 45 or 40°C (Figure 1), Dynasan® 114 was found to be in the liquid state after phase separation. Under these experimental conditions, the phase separation of the Dynasan® 114/CO_2 system performed by decreasing the pressure led to liquid droplets of Dynasan® 114 in CO_2. At 35°C, the phase separation led to solid particles of trymyristin in CO_2. The decrease of the melting point upon contact with SC CO_2 was previously reported for various lipids *(7)* and is related to the high solubility of the SC solvent in the heavy liquid phase.

The high-pressure view cell was also used to mimic the whole process without protein particles in order to determine when and how the coating material would deposit onto the host particles. The weight fraction of coating material in the coating material/CO_2 system and the starting T/P conditions were found to be key parameters for the coating process. These two parameters govern the T/P conditions under which the phase separation of the coating material/CO_2 system occurs. They also govern the physical state of the coating material, which deposits onto the host particles. Depending on the starting T/P conditions and weight fraction of coating material in the system, the phase separation of the coating material from the SC solution can lead to fine solid particles or to fine liquid droplets of coating material dispersed in SC CO_2. An efficient coating, leading to sustained-release behavior, was obtained by selecting an appropriate weight fraction of coating material in CO_2 and starting T/P conditions to obtain liquid droplets of coating material upon phase separation. The melted coating material further solidifies during the subsequent isochoric cooling step. The experimental phase behavior for an optimized process pathway is shown in Figure 2. Starting from a SC solution of coating material (0.37% weight fraction of coating material) at 45°C and 20 MPa, the isochoric cooling resulted in the formation of fine liquid droplets of coating material at 42.4°C, then to a solid coating at 37.5°C. It is worth reminding that the melting point of the coating material is 56.2°C at atmospheric pressure. Because high melting point lipids can melt at relatively low temperature in the presence of CO_2 under a high pressure, the SC coating process makes it possible to coat particles with a melted coating material at a temperature far below its melting point. An efficient coating can thus be performed under mild temperature conditions, avoiding the denaturation of fragile compounds such as therapeutic proteins.

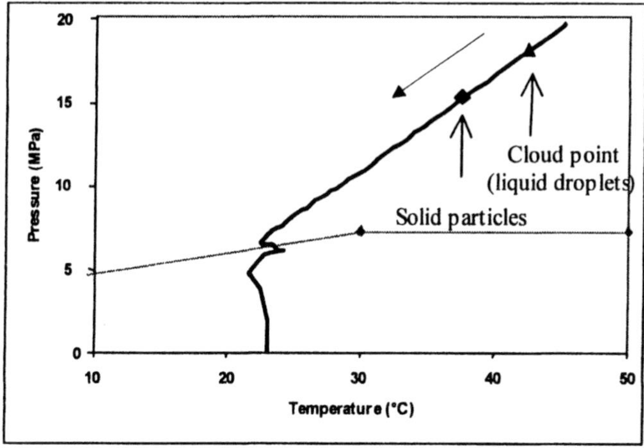

Figure 2. Phase behavior of coating material/CO_2 system during an optimized thermodynamic pathway of the coating process (starting T/P conditions: 45°C, 20 MPa; weight fraction of coating material: 0.37%; the system is cooled down to ~22°C, then CO_2 is vented).

These optimized operating conditions were then used to produce microspheres loaded with IFNα-2b particles. 3 g of coating material and 157.45 mg of IFNα-2b particles were added to a high-pressure 1-L vessel. Initial conditions were 45°C and 20 MPa, and the weight fraction of coating material was 0.37%. The temperature was decreased to 17°C during 31 minutes. CO_2 was then vented during 45 minutes. During the venting operation, the temperature was kept close to ~20°C. Composite microparticles were harvested and sieved using a 150-μm sieve plate. The sieving yield was 91 wt%.

The IFNα-2b content of coated microparticles, as determined by ELISA, was 146.10^3 pg of IFNα-2b per mg of microparticles. Thus, composite microparticles contained 4.1 wt% protein lyophilizate. The matrix structure of coated microparticles was evidenced using optical microscopy. Figure 3 shows the microparticles at ambient temperature in the left panel. The *in situ* melting behavior of the microparticles' coating material was observed using an optical microscope, equipped with a heating plate set at 70°C. After melting the coating material, it became apparent that each lipid microparticle contained several individual lyophilized protein particles, embedded in the coating material matrix (Figure 3 right panel).

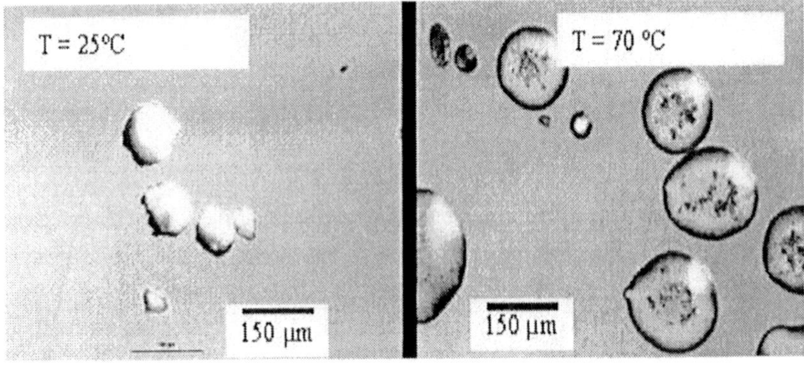

Figure 3. Optical microscopy of coated INF® particles. Left: observation at ambient temperature. Right: observation at ~ 70°C (heating plate); protein particles appear in the melted coating material matrix.

The particle size distribution of protein-loaded microspheres is shown in Figure 4. A narrow and unimodal particle size distribution is noticed, with a mean particle diameter (D4,3) of 101.5 μm.

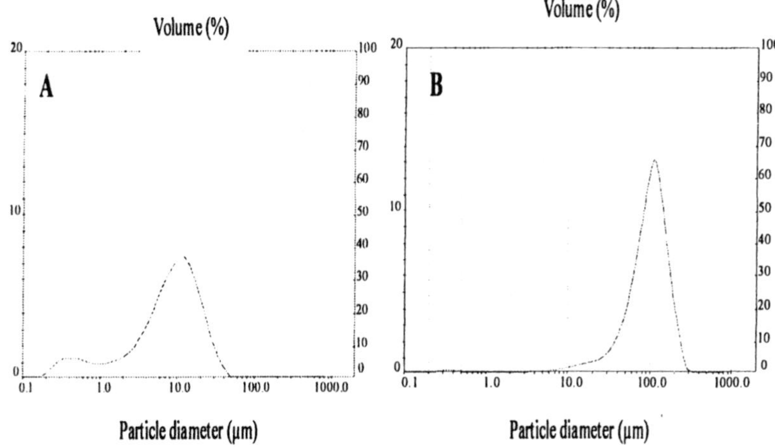

Figure 4. Particle size distribution of (A) lyophilized protein particles before coating, and (B) microspheres loaded with lyophilized protein particles.

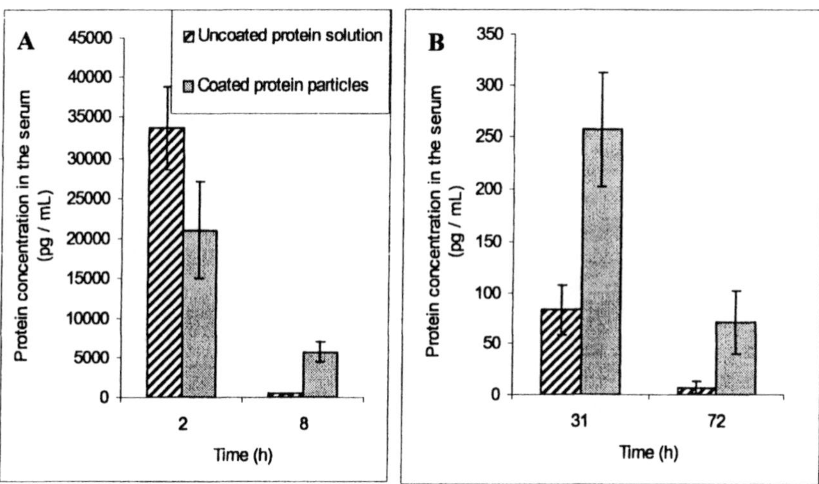

Figure 5. Concentration of circulating IFNα-2b in the blood of the mice versus time for animals receiving either IFNα-2b-containing microspheres or reference solution of uncoated IFNα-2b. (A) Sampling at 2 and 8 hours, and (B) sampling at 31 and 72 hours. The dose was 314,414 pg (500,000 IU)/mouse.

The concentration of circulating IFNα-2b in the blood *versus* time for mice receiving either IFNα-2b-containing microspheres or the reference solution of uncoated IFNα-2b is shown in Figure 5. The data clearly show that the use of microspheres strongly reduces the plasma peak concentration ("burst effect") and provides sustained release of the protein in the blood. The concentration of circulating protein remained significant and close to ~70 pg/mL for 72 hours after injection of the IFNα-2b-containing microspheres, whereas it reached the detection limit in animals receiving the uncoated protein at this time point. This experiment further revealed that the time needed to release all protein from these glyceride-based microspheres into the animal blood was ranging from 4 to 6 days.

Conclusions

We have developed a SC CO_2-based coating process that makes it possible to completely avoid the use of organic solvents and hence is well-adapted for the formulation of fragile biomolecules. The process has been optimized by properly selecting the appropriate weight fraction of coating material in CO_2 and T/P starting conditions to obtain liquid droplets of coating material upon phase separation, that solidify during a subsequent isochoric cooling step. The optimization step had been carried out through an in-depth study of the phase behavior of coating material/CO_2 systems, performed using a high-pressure view cell of variable volume. IFNα-2b-containing microparticles have been prepared using this optimized process and shown to display a matrix structure, the lyophilized protein being dispersed in a solid lipid matrix, enabling sustained release. *In vivo* experiments in mice confirmed the prolonged release of the therapeutic protein over 4 to 6 days after subcutaneous injection of the microspheres, together with strong reduction of the burst effect and plasma peak concentration. These results are highly promising with regard to the clinical development of new sustained-release formulations of therapeutic proteins that will be able to fulfill still unmet medical needs related to high peak concentration and high injection frequency.

References

1. Richard, J., Deschamps, F. Supercritical fluid processes for polymer particle engineering – Applications in the therapeutic area, in Colloid Biomolecules, Biomaterials, and Biomedical Applications, Elaissari, A. (Ed.), Marcel Dekker Inc., New York, 2003, pp 429-475.

2. Deschamps, F., Richard, J. Processing of pharmaceutical products using supercritical fluids: industrial considerations and scaling-up, in Supercritical Fluids and Materials, Bonnaudin, N., Cansell, F., Fouassier, O. (Eds.), INPL, Nancy, 2003, pp 331-367.
3. Benoit, J.P., Rolland, H., Thies, C., Vandevelde, V. Method of coating particles. Eur. Patent EP 0 784 506 B1, 1999; US Patent US 6,087,003, 2000.
4. Benoit, J.P., Richard, J., Thies, C. Process for preparing microcapsules of active material coated with a polymer and microcapsules obtained using the process. Eur. Patent 0930 936 B2, 2003; US Patent US 6,183, 783, 2001.
5. Ribeiro Dos Santos, M.I., Richard, J., Pech, B., Thies, C., Benoit, J.P. Microencapsulation of protein particles within lipids using a novel supercritical fluid process. *Int. J. Pharm.* **2002,** *242*, 69 -78.
6. Thies, C., Ribeiro Dos Santos, M.I., Richard, J., Vandevelde, V., Rolland, H., Benoit, J.P. A supercritical fluid-based particle coating technology. *J. Microencapsulation*, **2003**, *20*, 87-96.
7. Hammam, H., Sivik, B. Phase behavior of some pure lipids in supercritical carbon dioxide. *J. Supercrit. Fluids*, **1993**, *6*, 223-227.

Chapter 17

Paclitaxel Nanoparticles: Production Using Compressed CO_2 as Antisolvent: Characterization and Animal Model Studies

Fenghui Niu[1], Katherine F. Roby[2], Roger A. Rajewski[3], Charles Decedue[3], and Bala Subramaniam[4,*]

[1]CritiTech, Inc., 1321 Wakarusa Drive, Lawrence, KS 66049
[2]Department of Anatomy and Cell Biology, University of Kansas Medical Center, Kansas City, KS 66160
[3]Higuchi Biosciences Center, University of Kansas, Lawrence, KS 66045
[4]Department of Chemical and Petroleum Engineering, University of Kansas, Lawrence, KS 66045
[*]Corresponding author: bsubramaniam@ku.edu

Nanoparticles of paclitaxel (termed NanoTax) were successfully produced from acetone solution sprayed into compressed CO_2, an antisolvent for the drug. The mean particle sizes of the NanoTax range from 0.60-0.64 μm and 0.77-0.84 μm based on the number average and volume average distributions, respectively. NanoTax suspensions in physiological saline were found to be effective in the treatment of mice bearing ovarian cancer. Intravenous delivery of NanoTax suspension exhibited the same therapeutic effect in mice as Taxol®, but without the adverse effects of Cremophor®. In contrast, cancer-bearing mice treated intraperitoneally with NanoTax suspension had the longest survival time as compared to similar treatments with either macroparticulate paclitaxel or Taxol®. Overall, the progression of ovarian cancer in mice treated with NanoTax was dramatically reduced as compared to all other dosing modes, resulting in increased survival duration. Additionally, treatment-related toxicity in this mouse model of ovarian cancer was reduced with NanoTax compared to Taxol®.

Introduction

Paclitaxel is a known inhibitor of cell division or mitosis and is widely used in the treatment of ovarian, breast, lung, esophageal, bladder, and, head and neck cancers. Paclitaxel is a natural product originally purified from the bark of yew trees, but now obtained by semisynthesis from 10-desacetylbaccatin, a precursor purified from yew leaves. Paclitaxel, however, is poorly water soluble and is conventionally solubilized in Cremophor® EL, a formulation comprising 50% ethyl alcohol and 50% polyethoxylated castor oil. Cremophor® EL is believed to result in histamine release in certain individuals and patients receiving paclitaxel in that delivery method must normally be protected with a histamine H_1-receptor antagonist, an H_2-receptor antagonist and a corticosteroid to prevent severe hypersensitivity reactions. Many other anti-cancer compounds cannot be effectively administered because they are not soluble in solvents that are Generally Recognized As Safe (GRAS). As a result, these anti-cancer agents are unavailable for use in cancer prevention or treatment using conventional methods of administration.

While anti-cancer compounds are commonly administered by intravenous injection to patients in need of treatment, it is also known to inject cisplatin and carboplatin into the peritoneal cavity as a means of direct delivery to peritoneal cancers. A comparative study of intravenous versus intraperitoneal administration of cisplatin has been published by Alberts *et al.*[1] Dedrick *et al.*, have published a pharmacokinetic rationale for the advantage of intraperitoneal versus intravenous administration of cisplatin.[2] Similarly, intraperitoneal delivery of cisplatin as an infusion is discussed elsewhere.[3] To date, however, there do not appear to be any published reports of intraperitoneal delivery of suspensions of poorly water-soluble anti-cancer compounds.

The pharmaceutical industry is increasingly interested in developing technologies for the production of nano- and microparticles for drug delivery applications. Such applications require controlled particle size distribution and product quality (crystallinity, purity, morphology). As reviewed elsewhere,[4~6] Precipitation with Compressed Antisolvents (PCA) has been receiving increased attention as a technique to produce particles with controlled properties. An increasing number of drugs processed using PCA technology can be found in many publications.[7~12] In this study, a continuous process for producing and harvesting paclitaxel nanoparticles is described. An ultrasonic converging-diverging nozzle[13] that employs *sc*CO_2 as the energizing medium, or an ultrasonic probe is used to form droplets of drug solution. The *sc*CO_2 also selectively extracts the solvent from the droplets, precipitating the drug. The paclitaxel particles are separated from the solvent-laden *sc*CO_2 in a membrane vessel external to the crystallizer.[14] The nanoparticles not only improve dissolution rates but also may be injected as a suspension.

In this study, suspensions of paclitaxel nanoparticles (NanoTax) were injected intravenously or intraperitoneally into mice in which cancer had been induced by the intraperitoneal injection of tumor cells derived from a mouse ovarian tumor. The experimental results showed that the intraperitoneal administration of the NanoTax suspension gave substantial increases in animal survival time with a single treatment.

Materials and methods

Industrial grade carbon dioxide was supplied by Airgas Ltd. Paclitaxel was obtained from NaPro BioTherapeutics, Inc. Acetone and methylene chloride were purchased from Fisher Scientific. The particle size distributions were measured with an Aerosizer® Particle Size Analyzer (API, Amherst), while the particle morphology and structural information were obtained by SEM (LEO 1550) photomicrographs.

In Vivo Mouse Model of Ovarian Cancer

Tumorigenic mouse ovarian surface epithelial cells were developed following a spontaneous transformation event *in vitro*, and a clonal cell line (ID8) was used for these studies[15]. Injection of ID8 cells into immune competent female mice results in the development of multiple tumor implants throughout the peritoneal cavity and the production of ascites fluids, similar to late stage disease in women. Female C57BL6 mice 8 weeks of age were injected intraperitoneally (IP) with 6 $x10^6$ ID8 cells. Forty-five days after tumor cell injection, when macroscopic tumor implants are visible in the peritoneal cavity, drug treatment was initiated. For IP administration, mice were administered NanoTax (18, 36, 48 mg/kg); Taxol® (12, 18, 36 mg/kg, Bristol-Myers Squibb), macroparticulate paclitaxel (36 mg/kg), Cremophor® (at the final percent and volume equal to the 36 mg/kg dose of Taxol®), or saline (at the volume equal to the 48 mg/kg dose of NanoTax) once every 2 days for a total of 3 doses. For intravenous (iv) administration, mice were anesthetized with subcutaneous pentobarbital (50 mg/kg). NanoTax (12 mg/kg); Taxol® (12 mg/kg, Bristol-Myers Squibb), Cremophor® (at the final percent and volume equal to 12 mg/kg of Taxol®), or saline (at the volume equal to 12 mg/kg NanoTax) were injected via the tail vein once every 2 days for a total of 3 doses. Mice were observed daily for signs of toxicity and were sacrificed when 'end-stage' disease was reached, when ascites accumulation caused peritoneal swelling and the coat became rough. Mice were sacrificed by cervical dislocation. The trunk blood and ascites fluids were collected, and multiple organs and tissues were fixed in Bouins fixative for subsequent histological analysis. All animal studies were approved by the Institutional Animal Care and Use Committee at the University of Kansas Medical Center and adhered to the "Principles of Laboratory Animal Care" (NIH publication #85-23, revised 1985).

Formation of nanoparticles

Experimental setup

A schematic of the experimental unit for producing nanoparticles is shown in Figure 1. The unit is composed of a pressure vessel (crystallizer), a membrane for harvesting particles, a surge tank, a Haskel (Model# AGD-7) booster pump (for CO_2) and an ISCO (model# 260D) pump (for the solution). The pressure vessel, surge tank and membrane are housed in a water bath to maintain steady temperature. The pressure vessel has two view windows for observing the spray pattern and particle formation. An ultrasonic nozzle was fixed in the top cap of the vessel. The supercritical CO_2 is used as antisolvent medium for the drug but is completely miscible with the solvent. CO_2 from a cylinder is pressured/pumped by a booster pump through a surge tank, for dampening the flow fluctuations, then through the ultrasonic nozzle into the crystallizer. The drug solution is pumped through a coiled 1/16" tube for preheating, then through the ultrasonic nozzle/capillary nozzle to form small droplets. Small drug particles can be formed when the droplets are super-saturated upon mixing with supercritical CO_2. The particles and the solvent-laden CO_2 are led from the crystallizer into a Graver® (model# 1-625A-1P) membrane filter that separates particles from the solvent-laden CO_2 stream. The particles are retained in the membrane while the solvent-laden CO_2 stream passes through a back pressure regulator into a condenser, in which the solvent is separated from the CO_2 stream. The CO_2 stream flows through a flow meter before being vented. A LabVIEW program is used for monitoring temperatures and pressures as well as controlling the heaters.

Experimental conditions

Paclitaxel recrystallization was conducted using an ultrasonic nozzle at supercritical conditions (75 – 83 bar, 34 – 38^{0}C) with respect to the solvent/CO_2 binary. Paclitaxel concentration in organic solvent (acetone or methylene chloride) varied from 5 to 20 mg/mL. The solution spray rate was 2 mL/min while the CO_2 flow rate was 75 sL/min (0.161 kg/min). About 14 – 45 mL solution was sprayed into the system.

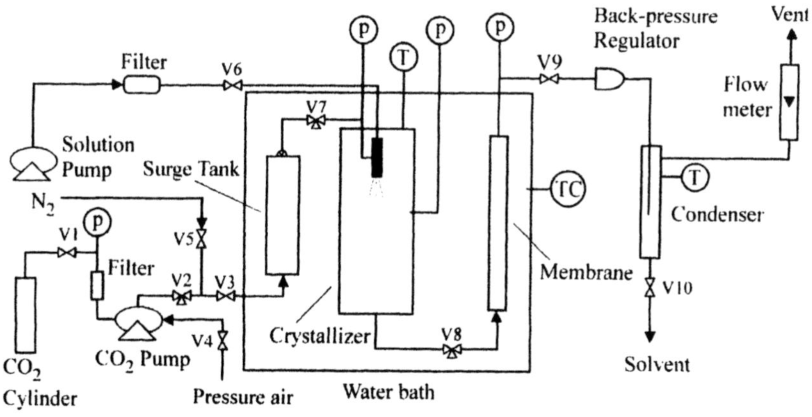

Figure 1. Schematic of experimental setup

Table 1. Summary of particle sizes of NanoTax obtained under different processing parameters .

Experimental parameters		Particle size (μm) by Aerosizer analysis					
		Number distribution			Volume distribution		
		Mean	10%	95%	Mean	10%	95%
Pressure (bar)	83	0.62	0.45	0.98	0.78	0.54	1.26
	75	0.64	0.44	1.05	0.84	0.57	1.41
	Acetone as solvent, 38 ^{0}C, 20 mg/mL						
Solvent #1-Acetone #2-CH_2Cl_2	#1	0.62	0.45	0.98	0.78	0.54	1.26
	#2	0.60	0.43	0.97	0.77	0.53	1.24
	82 bar, 38^{0}C, 20 mg/mL						
Drug concentration (mg/mL)	20	0.62	0.45	0.98	0.78	0.54	1.26
	5	0.63	0.45	0.99	0.78	0.55	1.19
	Acetone as solvent, 83 bar, 38 ^{0}C						
Temperature (^{0}C)	38	0.63	0.45	0.99	0.78	0.55	1.19
	34	0.73	0.50	1.19	0.93	0.65	1.44
	Acetone as solvent, 83 bar, 5 mg/mL						

Results

The effects of pressure, temperature, solvent and drug concentration on PCA-generated particle characteristics are summarized in Table 1. In the range of parameters investigated, nano-sized particles with a unimodal, narrow particle size distribution were typically obtained. The samples contained few particles either below 0.43 μm (10% or less) or above 1.05 μm (5% or less) based on the number average particle distribution.

From the dry powder Aerosizer® analysis, the unprocessed compound has a mean particle size of 12.7 μm based on volume average distribution. For the processed NanoTax, the mean particle size was typically reduced to a mean particle size of around 0.8 μm based on volume average distribution.

From Table 1, it is inferred that parameter variation within the range of investigated operating parameters resulted in relatively minor changes in the particle size. The mean particle sizes of the sample processed at 83 bar are slightly smaller than at the lower pressure. The particle size distribution at lower pressure is slightly broader than at higher pressure. Reducing the drug concentration from 20 to 5 mg/mL had no effect on the particle size. Mean particle sizes at the lower temperature are larger than at the higher temperature. These effects are qualitatively consistent with complementary experimental and modeling studies with other systems.[16]

Figure 2 shows the SEM images of unprocessed (left figure with a lower magnification) and NanoTax (right with a higher magnification) produced in this study. The unprocessed paclitaxel has rod-shaped particles (tens of microns) and a rather broad particle size distribution. In contrast, the processed paclitaxel has a narrow particle size distribution centered between 0.4 – 1.0 μm. The size inferences from SEM micrographs are consistent with the Aerosizer® measurements.

Animal studies and results

Three sets of experiments were carried out to assess the efficacy of NanoTax and to directly compare the efficacy of

Figure 2. SEM images of unprocessed (left) and NanoTax (right)

NanoTax to Taxol® in the mouse model of ovarian cancer. In all experiments mice were injected intraperitoneally with mouse ovarian epithelial cancer cells (ID8). Drug treatments were initiated after macroscopic tumors were established within the peritoneal cavity. In Study #1, intravenous delivery of NanoTax and Taxol® were compared. In Study #2, intraperitoneal delivery of NanoTax and Taxol® were compared and finally in Study #3 intraperitoneal delivery of NanoTax, macroparticulate Paclitaxel, and Taxol were compared.

Study#1

Four groups of mice with ovarian cancer were each treated differently with one of the following: 1) phosphate buffered saline alone, used as a control; 2) Cremophor EL solution alone, used as a control; 3) Taxol®; and 4) NanoTax suspended in phosphate buffered saline. All treatments were administered intravenously.

The results of this IV injection study are shown in Figure 3. Survival of mice treated with NanoTax or Taxol® IV tended to be longer when compared to control saline or Cremophor® treated mice (Figure 3). Although the effects were not significant when analyzed by Kaplan-Meier and subsequent Rank Tests, repeat experiments in both the NanoTax and Taxol® groups revealed a trend toward increased survival in each experiment.

It is noteworthy that the mice survived the direct IV injection of the suspension of NanoTax in phosphate buffered saline without the need to add anti-clotting agents such as heparin or agents such as surfactants or emulsifiers to prevent aggregation of the particles. While the preferred formulations would include these additional ingredients to further reduce the opportunity for clotting, the NanoTax formulation did not appear to cause blockage or infarction of fine capillaries. Rather surprisingly, it was determined that IV injection of the NanoTax suspension was as effective as the solution of Taxol® in Cremophor EL® in lengthening the survival time for mice inoculated with cancer cells. Consequently, it may be possible to deliver a suspension of NanoTax intravenously with the same therapeutic effect as Taxol®, but without the adverse effects of Cremophor®.

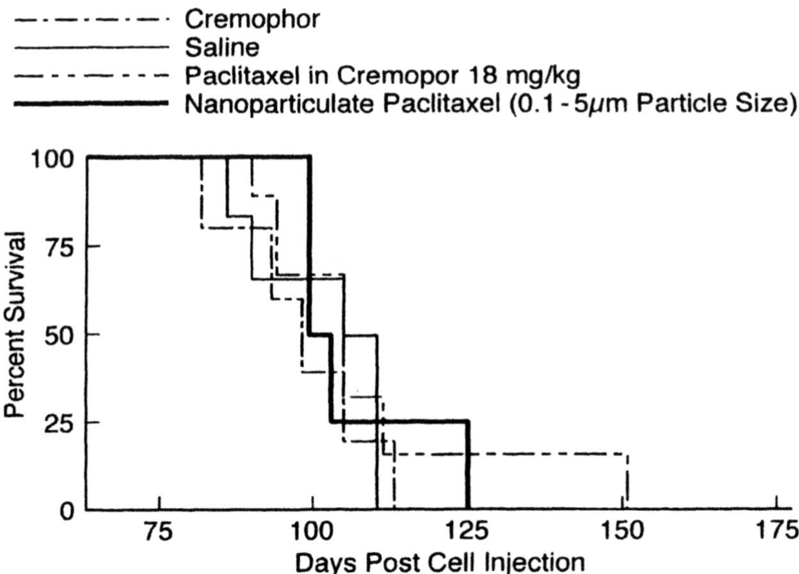

Figure 3 Results from four group of mice treated with NanoTax administered intravenously compared with Taxol® and saline and Cremophor® controls solution.

Study#2

Eight groups of mice with ovarian cancer were each treated differently with one of the following: 1) phosphate buffered saline alone, used as a control; 2) Cremophor EL® solution alone, used as a control; 3) Taxol®, 12 mg/kg; 4) Taxol®, 18 mg/kg; 5) Taxol®, 36 mg/kg; 6) NanoTax, 18 mg/kg; 7) NanoTax e, 36 mg/kg; or 8) NanoTax 48 mg/kg. All treatments were administered intraperitoneally.

The results of this group injection study are shown in Figure 4. NanoTax administered mice survived significantly longer than control saline or Cremophor® treated mice and significantly longer than Taxol® treated mice (Figure 4). At the end of the experiment, the survival duration was directly related to the dose of NanoTax administered. In addition to the increased efficacy, the NanoTax also demonstrated reduced toxicity compared to Taxol®.

Administration of 36 mg/kg Taxol® once every 2 days for three doses proved to be the ED_{50}. Fifty percent of the animals treated with 36 mg/kg Taxol® died due to toxicity within 7 days (this dose is excluded from Figure 4). Using the same dosing schedule, mice treated with the highest dose of NanoTax (48 mg/kg) exhibited no signs of toxicity. Together, these results indicate NanoTax is more effective than Taxol® in inhibiting the progression of ovarian cancer and increasing the survival duration. In addition, NanoTax exhibited reduced toxicity compared to Taxol®.

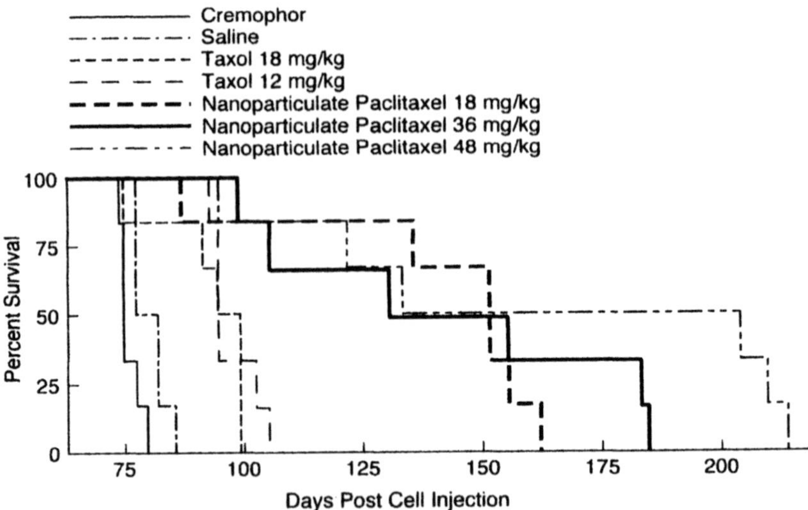

Figure 4. Results from a group of mice treated with NanoTax administered intraperitoneally compared with controls and paclitaxel in Cremophor® solution.

Comparison of the results from Studies #1 and #2 indicate that IP administration provides a significant increase in survival rates compared to IV administration.

It was determined that intraperitoneal injection of the NanoTax suspension significantly lengthened the survival time of the mice in comparison to the intraperitoneal injection of solubilized paclitaxel in Cremophor EL®. Note also that the survival time of mice treated IP with NanoTax greatly exceeded the survival time of both groups treated IV.

The progression of the ovarian cancer in NanoTax treated mice was compared to control treated mice. Mice were treated intraperitoneally with 48 mg/kg NanoTax or saline control, as described in Study #2, were sacrificed when the control treated animals had progressed to end-stage disease. As can be seen from Figure 5 (left panel), control mice at end-stage disease exhibit multiple tumor implants on the surface of the peritoneal wall. In contrast, mice treated with NanoTax (Figure 5, right panel), exhibit only a few, very small tumors.

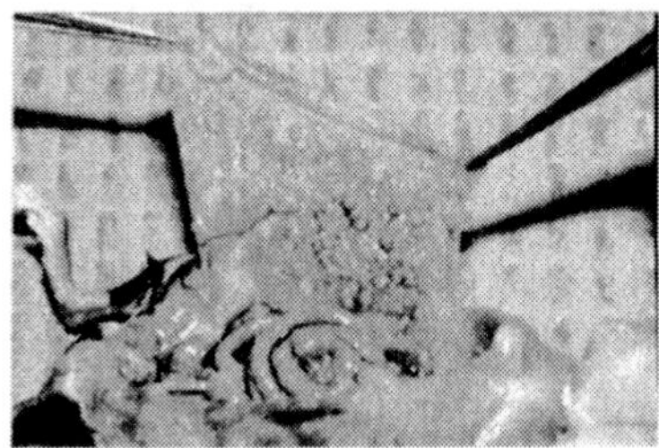

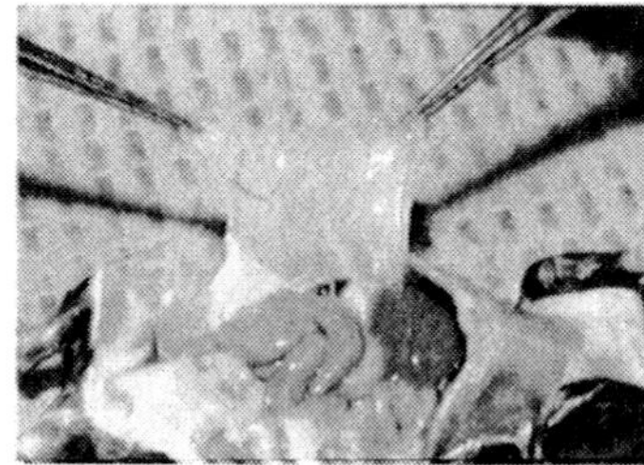

Figure 5. Comparison of the body wall of a mouse treated with 48 mg/kg of NanoTax in suspension administered intraperitoneally (right) with the body wall of a mouse treated with the saline control (left).

In this mouse model of ovarian cancer, tumors develop on multiple tissues and organs within the peritoneal cavity, similar to that observed in women with ovarian cancer. Tumor development in control mice and the effects observed with NanoTax treatment, as shown in Figure 5, were observed throughout the peritoneal cavity. Another example is Figure 6, where numerous cancerous tumors on the diaphragm of mice treated with saline as a control is illustrated (left panel). In stark contrast is the mouse illustrated in the right panel where a significant reduction in tumor burden on the diaphragm was observed following NanoTax treatment.

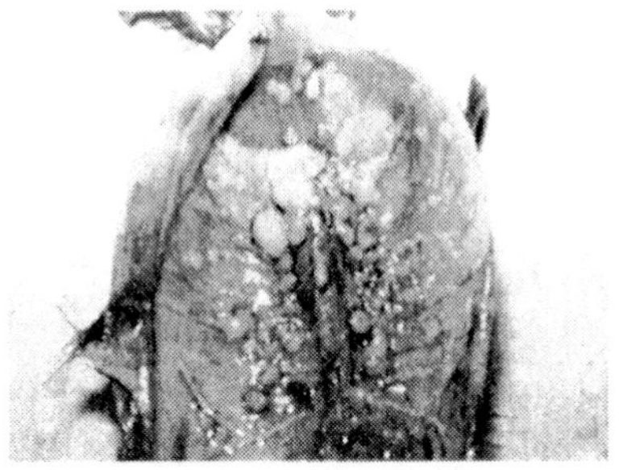

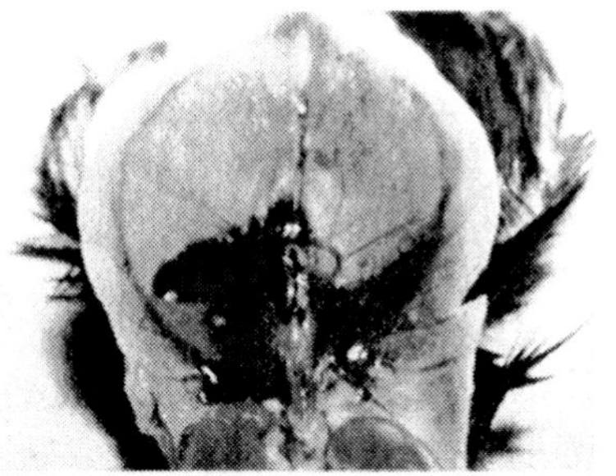

Figure 6. Comparison of the diaphragm of a mouse) treated with 48 mg/kg of NanoTax (right) with the diaphragm of a mouse treated with the saline control (left).

Similarly, as seen in Figure 7, the peritoneal organs of mice treated with NanoTax exhibit significantly fewer and smaller tumors compared to control treated mice.

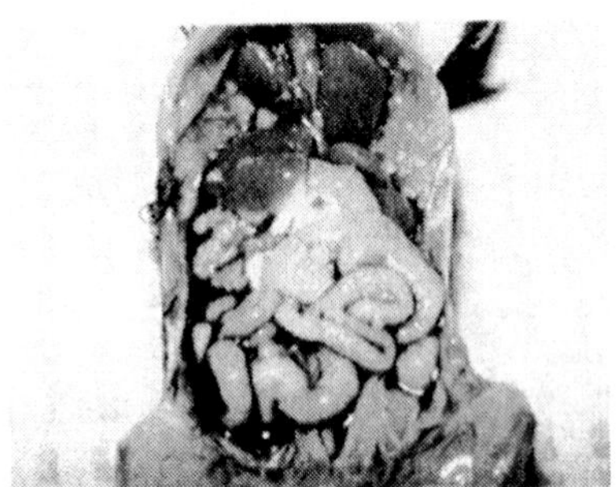

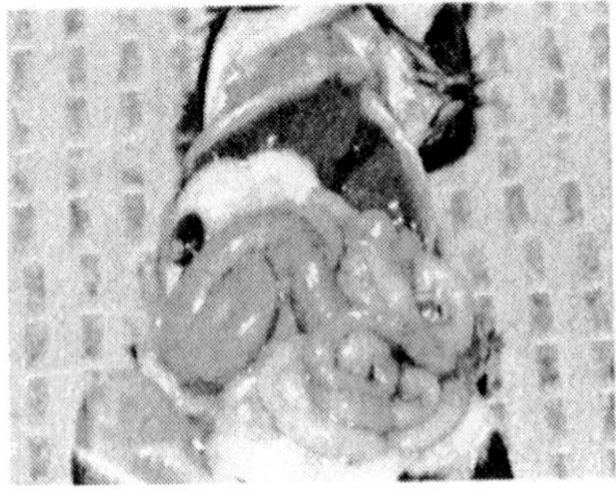

Figure 7. Comparison of the cancerous growths on the peritoneal organs of the mice treated with the saline control (left) and with NanoTax (right).

Study#3

Nine groups of mice with cancer were each treated differently with one of the following: 1) phosphate buffered saline alone, used as a control; 2) Cremophor EL solution alone, used as a control; 3) Taxol®, 18 mg/kg; 4) macroparticulate paclitaxel, 18 mg/kg; 5) macroparticulate paclitaxel, 36 mg/kg 6) macroparticulate paclitaxel, 48 mg/kg; 7) NanoTax, 18 mg/kg; 8)

NanoTax, 36 mg/kg; or 9) NanoTax, 48 mg/kg. All treatments were administered intraperitoneally.

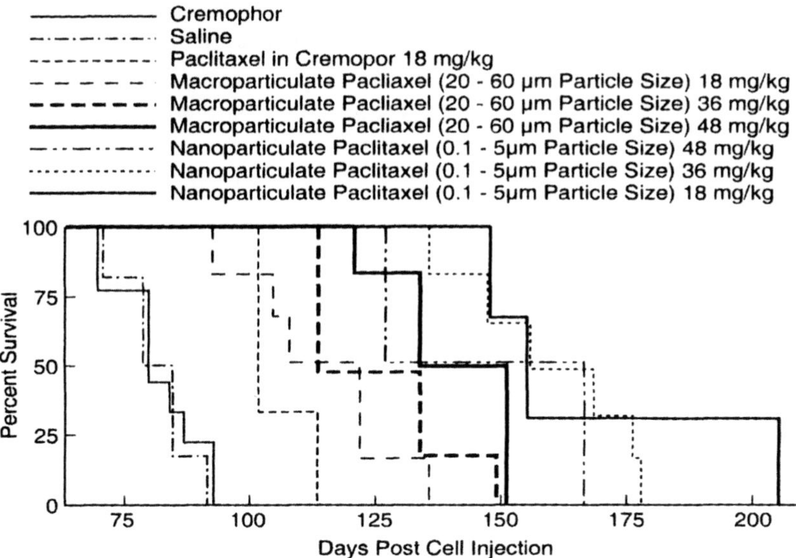

Figure 8. Results from a study of mice treated with NanoTax administered intraperitoneally compared with controls, paclitaxel in Cremophor® solution and macroparticulate paclitaxel, 20 to 60 microns in size

The results of Study #3 are shown in Figure 8. As anticipated, NanoTax treated mice survived significantly longer than both Taxol® and control (saline or Cremophor®) treated mice. When comparing equivalent doses of NanoTax and macroparticulate paclitaxel (20-60 µm particle size), the NanoTax treatment resulted in significantly greater survival duration. Interestingly, the survival of macroparticulate treated mice was greater than that of Taxol® but not as great as the survival duration in response to NanoTax treatment.

Conclusions

The results presented here demonstrate the feasibility of producing NanoTax using organic solvent and compressed CO_2 as antisolvent. The mean particle size of produced samples is 0.60-0.64 μm based on the number distribution and 0.77-0.84 μm based on the volume distribution. Suspensions of these NanoTax particles in physiological saline were found effective in the treatment of groups of mice bearing ovarian cancer. For the animal studies, intravenous delivery of a suspension of NanoTax exhibited the same therapeutic effect as Taxol®, but without the adverse effects of Cremophor®. The mice treated intraperitoneally with NanoTax suspension had the longest survival time as compared to macroparticulate paclitaxel, Taxol® and the controls. Overall, the progression of ovarian cancer in mice treated with NanoTax was dramatically reduced as compared to all other dosing modes, resulting in increased survival. Importantly, treatment-related toxicity in this mouse model of ovarian cancer was reduced with NanoTax compared to Taxol®.

References

1. Alberts, D.; Liu, P.; Hannigan, E.; Toole, R.O. Intraperitoneal Cisplatin plus Intravenous Cyclophosphamide versus Intravenous Cisplatin plus Intravenous Cyclophosphamide for Stage III Ovarian Cancer. *New England Journal of Medicin,*. **1996**, *335*, 1950-1955
2. Dedrick, R.; Myers, C.; Bungay, P.; DeVita, V., Pharmacokinetic Rationale for Peritoneal Drug Administration in the Treatment of Ovarian Cancer. *Cancer Treatment Report,* **1978**, *62*, 1-11
3. Atkinson, A.; et al, Principles of Clinical Pharmacology. *Academic Press*, **2001**
4. Subramaniam, B.; Rajewski, R.; Snavely, K, Pharmaceutical Processing with Supercritical Carbon Dioxide. *J. Pharm. Sci.*, **1997**, *86*, 885-890
5. Perrut, M., Supercritical Fluids Applications in the Pharmaceutical Industry. *STP Pharma Sciences*, **2003**, *13(2)*, 83-91
6. Foster, N.; Mammucari, R; et al, Processing Pharmaceutical Compounds Using Dense Gas Technology. *Industrial & Engineering and Industrial Chemistry*, **2003**, *42(25)*, 6476-6493
7. Muhrer, G.; Mazzotti, M., Precipitation of Lysozyme Nanoparticles from Dimethyl Sulfoxide using Carbon Dioxide as Antisolvent. *Biotechnology Progress*, **2003**, *19(2)*, 549-556

8. Snavely, W.; Subramaniam, B.; Rajewski, R., Micronization of Insulin from Halogenated Alcohol Solution using Supercritical Carbon Dioxide as an Antisolvent. *J. Pharm. Sci.*, **2002**, *91(9)*, 2026-2039
9. Rajewski, R.; Subramaniam, B.; Snavely, W.; Niu, F., Precipitation of Proteins from Organic Solutions to form Micron-sized Protein Particles. *U.S.Patent 6,562,952*, Issued May 13, **2003.**
10. Reverchon, E.; De Marco, I.; Caputo, G.; Della Porta, G., Pilot Scale Micronization of Amoxicillin by Supercritical Antisolvent Precipitation. *Journal of Supercritical Fluids*, **2003**, *26(1)*, 1-7
11. Fusaro, F.; Mazzotti, M.; Muhrer, G., Gas Antisolvent Recrystallization of Paracetamol from Acetone using Compressed Carbon Dioxide as Antisolvent. *Crystal Growth & Design*, in press, **2004**
12. Rehman, M.; Shekunov, B.; York, P.; et. al., Optimisation of Powders for Pulmonary Delivery Using Supercritical Fluid Technology. *Euro. J. Pharm. Sci.*, **2004**, *22(1)*, 1-17
13. Subramaniam, B.; Saim, S.; Rajewski, R.; Stella, V. J., Methods and Apparatus for Particle Precipitation and Coating Using Near-Critical and Supercritical Antisolvents. *U. S. Patent 5,833,891*, Issued Nov. 10, **1998**
14. Subramaniam, B.; Bochniak, D. J.; Rajewski, R., Methods for Continuous Particle Precipitation and Harvesting, *U. S. Patent 6,113,795*, Issued Sept. 5, **2000**
15. Roby, K.F., Taylor, C.C., Sweetwood, J.P., Cheng, Y., Pace, J.L., Tawfik, O., Persons, D.L., Smith, P.G., Terranova, P.F., Development of a syngeneic mouse model for events related to ovarian cancer. *Carcinogenesis*, **2000,** *21*, 585-591
16. Fusaro, F.; Hänchen, M.; Mazzotti, M.; Muhrer, G.; Subramaniam, B., Dense Gas Antisolvent Precipitation: A Comparative Investigation of the GAS and PCA Techniques. *Industrial & Engineering Chemistry Research*, submitted, **2004**.

Chapter 18

Nanoprecipitation of Pharmaceuticals Using Mixing and Block Copolymer Stabilization

Brian K. Johnson, Walid Saad, and Robert K. Prud'homme*

Department of Chemical Engineering, Princeton University, Princeton, NJ 08544
***Corresponding author: prudhomm@princeton.edu**

A new technology to form nanoparticles of hydrophobic organic actives at high concentration and yield, and methods to characterize the process are presented. In the *Flash Nano Precipitation* process, an organic active and an amphiphilic diblock copolymer are dissolved in an organic phase and mixed rapidly with a miscible anti-solvent to trigger precipitation of the active with a narrow particle size distribution and controlled mean particle size (50-500 nm). The enabling components are a novel "analytical" (quantified mixing time) Confined Impinging Jets (CIJ) mixer for millisecond stream homogenization and amphiphilic diblock copolymers, which alter the organic nucleation and growth, provide steric stabilization for the particles, and offer a functional surface for the composite. Methods to quantify fundamental time scales of the process and their relation to component thermodynamics are provided. The technology is useful for applications in enhanced pharmaceutical delivery, dye preparation, and pesticide formulation.

Introduction

The production of submicron particles of hydrophobic, water-insoluble organic compounds at high solids loading is important in applications involving pharmaceuticals, dyes, and pesticides. For pharmaceuticals, the rate of dissolution of hydrophobic compounds can be controlled by the particle surface area, and therefore, particle size. For drugs used in cancer therapy, the immature vasculature in the cancer tumor allows passive targeting of the drug if particle sizes are in the range of 100-200 nm. In addition, all of the particles must be below 220 nm to allow sterile filtration. The color intensity for pigments and insoluble dyes increases with decreasing particle size, as does the resolution for ink jet printing applications. While it is relatively easy to produce small inorganic particles owing to their higher surface energy and charge stabilization, organics are significantly more difficult *(1)*. In nanoparticale formation via direct precipitation, the short length scales and small mass of nanoparticles result in rapid formation (ms) and dissolution kinetics (ms to s) for individual particles. Creation of a tailored nanoparticle size distribution requires rapid processing for tight control of nucleation, growth, and equilibration (ripening).

Our goal is to develop a process and the understanding required for producing nanoparticles of organic compounds at high solids concentration, high productivity, and low colloidal stabilizer content. We also wish to tailor the surface properties of the nanoparticles through the formation of unique composite organic and block copolymer nanoparticles. For example, we investigate particles with a surface presented to the human body of poly(acrylic acid), a mucoadherent for enhanced drug retention, or poly(ethylene glycol), known to extend the lifetime of particles in the blood stream. Here, we present and rationalize an avenue for the production of highly tailored nanoparticles to meet the above applications.

Our approach comprises the *Flash NanoPrecipitation* process as described in Figure 1 *(2,3)*. The process contains several key components, the first of which is a rapid mixing time smaller than the formation time, τ_{Flash}, for a nanoparticle. This corresponds to a Damköhler number for precipitation (Da_p), which is the ratio of these two mixing times, and which should be <1. Therefore, the mixing time must be less than the induction time for formation of a block copolymer nanoparticle, τ_{agg}, and the induction time for nucleation and growth, τ_{n+g}, of organic particles smaller than one micron. The ratio of the induction times can be tuned effectively by changing the active concentration or the molecular architecture of the diblock copolymer. When these two precipitation times are properly tuned to match one another, the insoluble portion of the protective colloid (block copolymer) is deposited on the surface of the growing organic particle to freeze the size distribution at that desired for a particular formulation. In addition, the proper selection of block copolymer can act as a nucleation promoter and further control the production of nanoparticles. Since the block copolymers are amphiphilic, a functional surface is created for the purposes stated above and to provide steric stabilization. Figures 2 and 3 provides one such unique composite where the same organic active and block copolymer was used to form a narrow particle size distribution at a variety of sizes.

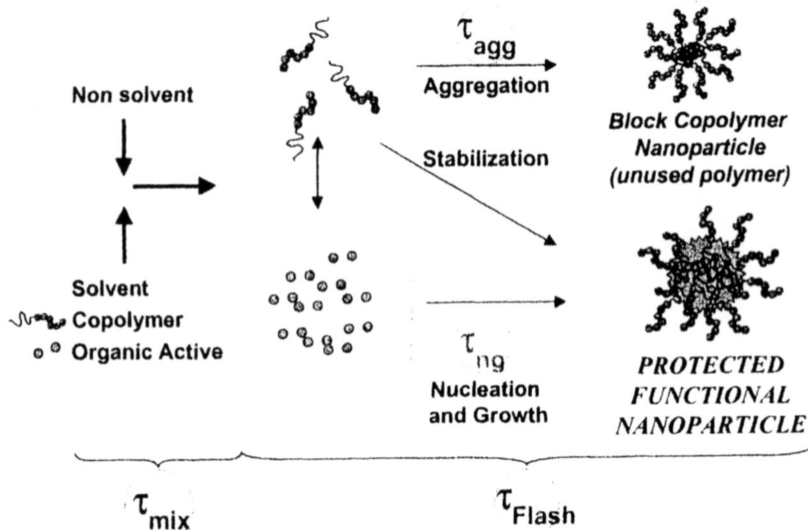

Figure 1. Flash NanoPrecipitation. An organic active and an amphiphilic diblock copolymer are molecularly dissolved in an organic phase and mixed rapidly with a miscible anti-solvent for the active and one block of the copolymer. This affords precipitation of the active into nanoparticles. When water is the anti-solvent, a hydrophobic active and the hydrophobic portion of the copolymer precipitate simultaneously to form nanoparticles. They are protected from aggregation and sterically stabilized by the hydrophilic portion of the block copolymer, which would be soluble in the solvent after mixing if a homopolymer. The mixing time, τ_{mix}, must be small compared to the composite process, τ_{Flash}, to obtain "homogeneous" kinetics free of mixing effects and a narrow particle size distribution. The precipitation process can be further subdivided into the characteristic induction times for copolymer aggregation, τ_{agg}, and active organic nucleation and growth, τ_{ng}. The "reactions" compete and when times are matched, the block copolymer can interact with the growing active particle to alter nucleation and growth and offer colloidal stabilization.

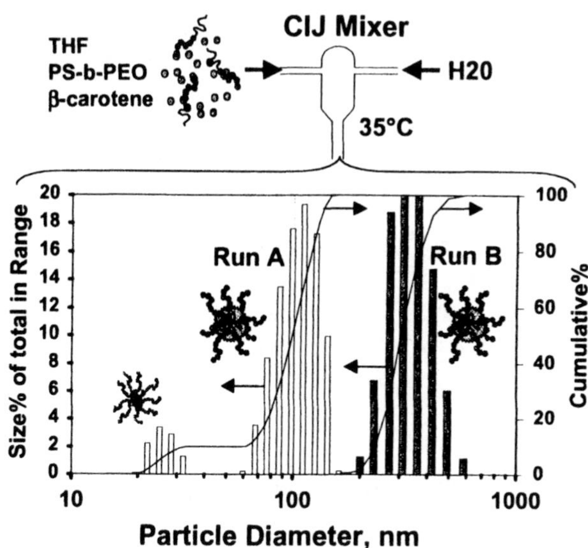

Figure 2. Flash NanoPrecipitation of β-carotene and poly(styrene)$_{10}$-b-poly (ethylene oxide)$_{68}$. Run A corresponds to 0.52 wt% of both components in tetrahydrofuran (THF) fed at 2.8 m/s. Run B corresponds to 2.6 wt% and 4.3 m/s, respectively. For each run, the momentum of the water jet and the organic jet is matched and $Da_p = \tau_{mix}/\tau_{Flash} \leq 1$. Since the aggregation time of this diblock copolymer was determined to be a weak function of concentration (not shown here), the size at which the particles are stabilized can be controlled by the concentration of the active. A lower concentration corresponds to a slower time for nucleation and growth and stabilization occurs at a smaller size. For Run A, the nature of the competitive "reactions" in Figure 1 is signified by a distinct population of nanoparticles at the size range for diblock copolymer aggregated without any organic active (~25 nm). The smaller particles were not plotted for Run B.

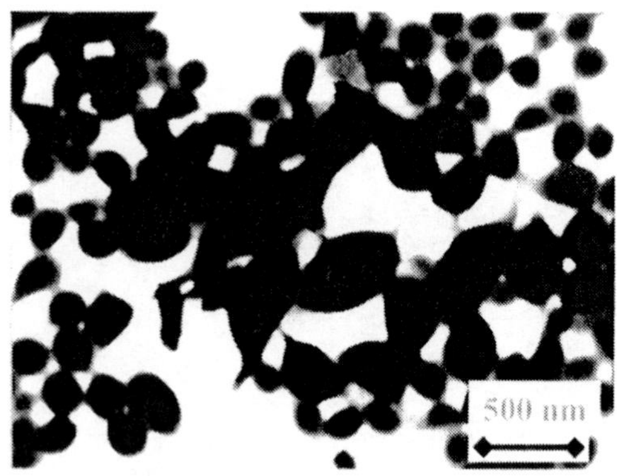

Figure 3. Transmission Electron Micrograph (TEM) of Run B of Figure 2. For this run, ~100% of the particles are less than 1000 nm in diameter, and the nanoparticle yield approaches 100% based on the solubility of each component at the conditions after mixing. This TEM was taken by H. Auweter of BASF after OsO_4 staining.

Results and Discussion

Design Considerations for NanoPrecipitation of Organics

Producing organic nanoparticles via rapid precipitation is difficult without enabling additives. Even if nanoparticles can be produced initially, their life-times are fleeting since in the absence of stabilizers subsequent processes lead to microparticulate systems via aggregation or ripening.

Table I displays the mean particle growth rate based on the equivalent spherical diameter for (a) stepwise molecular growth, where the rate is found based on the assumption that the surface integration of molecules is immediate and the growth is only limited by diffusion of solute to the particle surface; (b) aggregation of particles, where the rate is found using Smoluchowski kinetics with no barrier to aggregation or perfect sticking; and (c) Ostwald ripening, where there is no surface integration step and mass is transferred by diffusion. Details of these calculations are provided elsewhere *(2)*, but in summary, the calculations reflect the *maximum* rates possible, corresponding to diffusion limited processes for mass transport. These values, which are representative of organics, clearly demonstrate the rapid rate of molecular processes, which must be controlled to afford nanoparticles.

Table I. Maximum particle growth rate via diffusion limited processes.

Diffusion limited process (for β-carotene)		*G* *nm/ms*	$G \propto d^n$ *n =*
G_G =	Stepwise growth at supersaturation, S = 8	150	-1
G_A =	Aggregation (Smoluchowski kinetics)	2.1	-2
G_{OR} =	Ostwald ripening in 48 wt% THF	0.14	-2
G_{OR} =	Ostwald ripening in 10 wt% THF	5×10^{-4}	-2

Estimates of the maximum growth rate for the equivalent particle diameter starting with a diameter of 50 nm following the work of Johnson *(2)*. The following values were used for β-carotene: surface energy ≅ 2 mJ/m^2, solubility ≅ 0.5 mM, $D_{AB} \cong 1 \times 10^{-9}$ m^2/s, particles shape factor = 45, molecular volume = 8.9 x 10^{-28} m^3, particle diffusion coefficient $D_l \cong k_B T / 3\pi\eta d_0$, crystal density = 1000 kg/m^3, solution density = 950 kg/m^3, M_w = 0.537 kg/mole, T = 308K, and 48 wt% tetrahydrofuran (THF) in water.

Avoiding Particle Number Reduction

Table I demonstrates that during and after particle formation the rate of aggregation and the rate of Ostwald ripening can be significant (nm/s maximum particle growth rates). These effects are particularly important for nanoparticles due to their small size, since the rates are inversely proportional to the square of the particle diameter. The net result of these processes is a larger and broader particle size distribution with time, in other words fewer particles. Thus, it is clear that these processes must be considered in any process designed to maintain a nanoparticulate size.

Aggregation appears to be the more serious problem with a maximum growth rate, which is an order of magnitude faster than that for Ostwald ripening. Aggregation leads to significant particle growth, unless the chemical structure of the organic active leads to a surface charge (zeta potential) ample to provide particle stabilization. Alternatively, a foreign colloidal stabilizer is required to maintain disperse nanoparticles.

Ostwald ripening can be limited by reducing the solubility of the product in solution. For example, changing from 48 wt% tetrahydrofuran (THF) to 10% THF with anti-solvent (water) reduces the solubility of β-carotene by a factor near 500 and slows the equilibration rate by a factor near 250. Thus, for down-stream processing of a nanoparticle stream, reducing the solvent-to-antisolvent content is an effective way to slow Ostwald ripening sufficiently for particle harvesting. As with aggregation, Ostwald ripening can also be reduced by the addition of a well designed inhibitor, such as an amphiphilic copolymer, to coat the particle surface and limit the transport of active molecules.

Promoting Particle Nucleation Over Growth

In the case of stepwise molecular growth, the rate estimated for β-carotene, 150 nm/ms, is close to that measured by Mahajan and Kirwan *(4)* for lovastatin, ~50 nm/ms, under the same supersaturation (S = concentration/solubility). This demonstrates that to form nanoparticles by direct precipitation, significant and rapid nucleation instead growth is required to consume the mass fraction of solute above the solubility at the final process conditions.

One could envision adding *solid* heterogeneous particles to promote nucleation, but there are a number of caveats. Each heterogeneous particle typically produces only one new particle. Thus, a prohibitively high number of particles are required to compete with the rate of homogeneous nucleation at high supersaturation. In practice, that number of solid particles would be difficult to disperse. Thus, one is encouraged to use a molecularly *soluble* compound to promote nucleation over growth.

Another strategy to alter nucleation and growth is to design a growth inhibitor that will have a special affinity to a nascent particle and will precipitate with the solute or promote solute precipitation itself, thus coating or changing the character of the particle surface such that additional molecules of solute will no longer be incorporated quickly. This will encourage nucleation over growth and result in a larger number of particles. Table I demonstrates that the growth rate of particles at high supersaturation can be on the order of 100 nm/ms if growth is diffusion-limited and there is no barrier to integration. Thus, the surface inhibitor must precipitate at the same time as the solute to limit growth. The requirements are less stringent for eliminating growth by aggregation (~2 nm/ms maximum rate). Knowledge of the characteristic time of aggregation for the inhibitor in the absence of the solute is important information and methods to determine this time are valuable. Likewise, the relative solubility of the additive *versus* the active will be important to tuning the process in order to achieve precipitation at the same time. Both topics are discussed further below.

Flash NanoPrecipitation as described in this chapter is run at sufficient supersaturation after mixing to encourage rapid nucleation over growth. The block copolymers chosen alters the particle nucleation and growth kinetics and provides the particle stabilization required for downstream processing of the particles before aggregation or equilibration to a microparticulate size.

Technology Advancements and Engineering

Quantification of Kinetics via Characteristic Times

Mixing Time

Mixing in a period of milliseconds, faster than the induction time for precipitation ($Da_p < 1$), by a custom-designed Confined Impinging Jets (CIJ) mixer has enabled the process to be run under a "homogeneous" starting condition, where the effect of mixing is not convoluted with the role of the precipitation times. This set-up permits the design of a narrow particle size distribution, with 100% of the particles less than one micron even at low colloidal stabilizer content. We have demonstrated 1:1 to 1:7 copolymer:active by weight (only 1:1 shown in Figure 2). In a CIJ mixer, the segregation length scale of the incoming fluids streams is rapidly reduced by impinging two collinear jet streams. The chamber is large enough to allow for an impingement plane, but confined to avoid significant bypassing. The unit is operated with jet velocities characteristic of a turbulent-like flow pattern within the cylindrical mixer, which further reduces the mixing time. A feature of the CIJ mixer is that probe tests are accomplished with <20 mL/run and the scale-up rules to m^3/day have been determined.

The absolute mixing time, τ_{mix}, has been characterized to yield a novel "analytical" CIJ mixer *(5)*. Using the techniques described below, we have demonstrated that balancing the aggregation time for the copolymers and the nucleation and growth time for the organic active allows control of the particles size distribution. This result is intuitive since if either the block copolymer or the organic active precipitated well before the other, the two processes would be independent.

Induction Time for Aggregation of Diblock Copolymers or for Precipitation

Any of the precipitation times can be determined by realizing the mixing time and the aggregation time are equivalent at the "breakpoint", as shown in Figure 4. This is the point where the rate of mixing no longer effects the final particle size distribution, a precipitation Damköhler number, $Da_p = \tau_{mix}/\tau_{agg}$, of unity. Thus, the kinetics of precipitation for the individual processes can be characterized in relation to each other. This technique is especially useful for the rapid time scales of nanoparticle formation in this technology, where there are no simple methods of characterizing kinetics.

We have applied this technique to both the production of block copolymer nanoparticles (τ_{agg}) and the *Flash NanoPrecipitation* process (τ_{Flash}) and have shown the anti-solvent and solvent streams must be mixed in less than 100 ms for optimum behavior. At higher mixing times, the particle size distribution becomes wide and a large number of particles can exceed one micron, since they are not properly stabilized by the block copolymer during the kinetic process.

We have also verified the mechanism of rapid self-assembly for the diblock copolymers alone, which is a "fusion only" process consisting of diffusion limited aggregation (or growth) and a local rearrangement (or nucleation time) *(6)*. For the large solvent quality changes studied here, kinetically frozen nanoparticles are formed since the rate of unimer chain exchange is low at the solvent conditions after mixing. This is in contrast to conditions near the solubility limit, C_{cmc}, where the chains are in facile dynamic equilibrium.

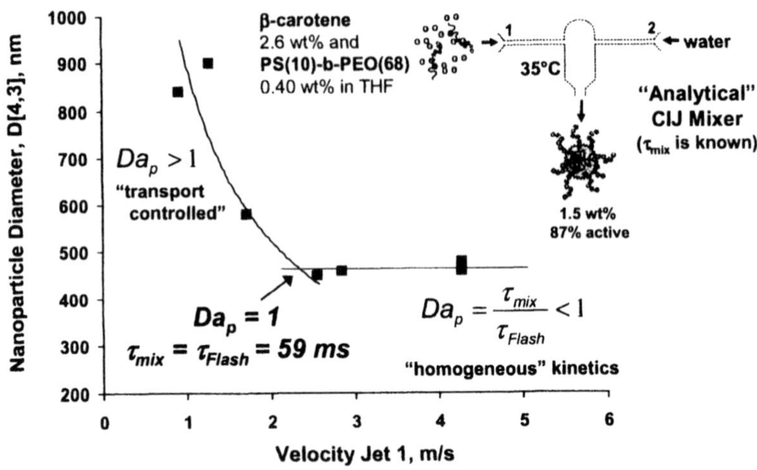

Figure 4. Measurement of τ_{Flash} for precipitation of an organic active and amphiphilic diblock copolymer as nanoparticles. Since the mixing time in the "analytical" CIJ mixer has been characterized as a function of the jet velocity and stream physical properties, the breakpoint where the $Da_p = 1$ allows definition of the characteristic time of Flash NanoPrecipitation, τ_{Flash}. The procedure also demonstrates the mixing time to reach a homogeneous starting condition, where $Da_p < 1$.

Thermodynamics of Diblock Copolymer Micellization in Mixed Solvents

We have investigated the thermodynamics of micellization of amphiphilic diblock copolymers in some detail since little data exist for mixed aqueous-organic solvents. In the case of poly(butylacrylate)-*b*-poly(acrylic acid), (PBA-b-PAA), in methanol and water, the closed association model and measurements of the critical micelle concentration, C_{cmc}, were used to generate Figure 4. Here, the entropy-enthalpy compensation yields a decreasing Gibbs free energy as the water content increases. It is interesting that the entropy will contribute to micellization at higher water levels, but it is uncertain if the enthalpy will turn positive. Overall, the thermodynamic quantities are most sensitive to the hydrophobic block length and water content. The soluble hydrophilic block has little influence *(2)*.

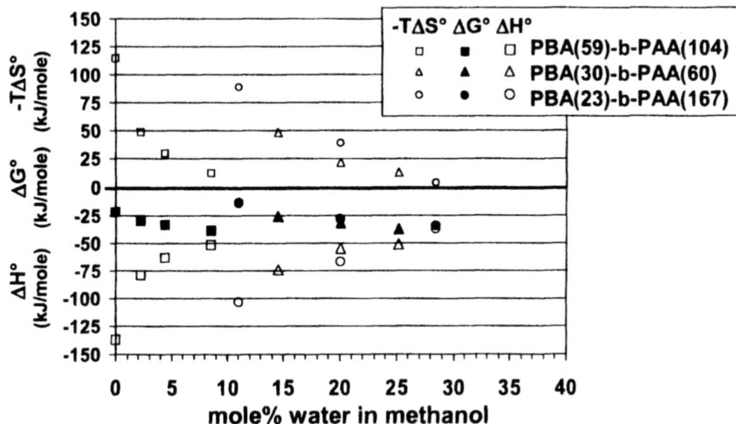

Figure 5. Thermodynamics of diblock copolymer micellization for poly(butyl-acrylate)-b-poly(acrylic acid), (PBA-b-PAA), of various block lengths (in monomer units). As the copolymer concentration is increased, the step change in static light scattering intensity upon micellization identifies the C_{cmc}. The C_{cmc} was measured at 10-35 °C for each copolymer and water content. The standard (°=298K) properties are calculated following the closed association model: $\Delta G = RTln(C_{cmc})$ and $\Delta H = \partial(Rln(C_{cmc}))/\partial(1/T)$.

Mixing Operating Lines

For PS-b-PEO as in Figure 6, the solubility (C_{cmc}) is a stronger function of water content than that for the organic active. With a mixing operating line starting at ~0.5 wt% for both compounds in THF (Run A of Figure 2), the solubility of each component is matched and precipitation begins simultaneously as the mixing process proceeds and the metastable regions are exceeded. Thus, the sensitivity to the mixing rate is reduced. It is less desirable to have the solubility of each component mismatched, since one component could precipitate prior to the other (as is often the case for precipitation by slow dialysis), but if the mixing rate is sufficiently fast, homogenous competitive kinetics still prevails.

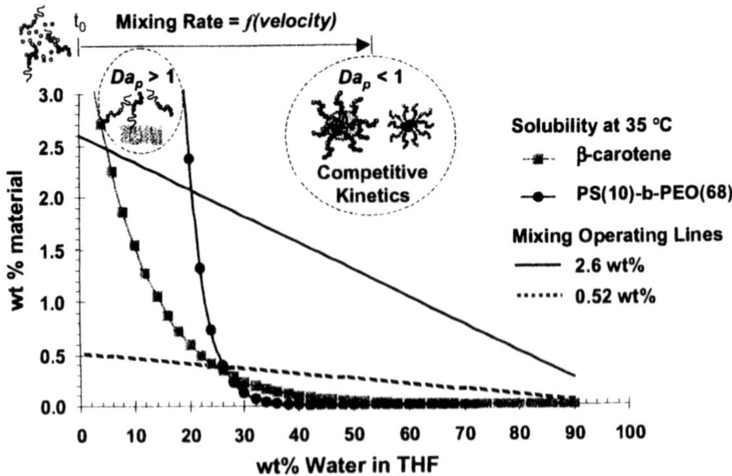

Figure 6. Mixing operating lines and solubility fits for process design. β-carotene and PS(10)-b-PEO(68) are analyzed at 35 °C. The straight lines represent mixing operating lines for dilution of a component (from an initial wt% in THF at t_0) as the mixing with water proceeds. The rate of change in the average solvent content is based on τ_{mix}. Starting at 100% THF and under a slow mixing condition and on a mixing operating line of 2.6 wt% for β-carotene and for PS(10)-b-PEO(68), the β-carotene is expected to precipitate prior to the block copolymer. By solubility measured independently, over 50% of the β-carotene could precipitate before the C_{cmc} for the block copolymer is reached. In turn, a significant amount of "unprotected" particle growth can occur. Under the case of perfect mixing or "homogeneous" kinetics, the final solvent content is rapidly established and the competitive rate at which the β-carotene nucleates and grows versus the rate at which the amphiphilic colloidal stabilizer precipitates dictates the final particle size distribution.

Downstream Processing for Pharmaceutical Applications

In most pharmaceutical systems used for precipitation, it is necessary to remove the organic solvent from the drug compound. A key issue is whether the particles will agglomerate or Ostwald ripen during this processing. As described below, we have been able to use distillation or freeze drying to maintain nanoparticles from *Flash NanoPrecipitation* at the same size as initially produced.

The ability to remove solvent by distillation has been demonstrated in our laboratory. The CIJ mixer effluent was collected in a tank containing deionized water, resulting in a final ratio of 10:1 DI water:solvent. The batch was then repeatedly vacuum-concentrated with a 50-70°C jacket temperature, followed by addition of water until the organic solvent was removed from the system. The resulting nanoparticle slurry size is provided as Run B of Figure 2. The particle size was the same before and after distillation processing and was found stable for at least a period of months.

The ability to freeze-dry and redisperse a stream of β-carotene and poly (styrene)-*b*-poly(ethylene oxide) nanoparticles has also been demonstrated. The approach led to an increase in mean particle diameter of 3 to 60% with the addition of sucrose in ratios of 50:1 to 10:1 sucrose:nanoparticles by weight. Similar results have been seen for the lyoprotection of peptide-DNA condensates using sucrose *(7)*, whereas the use of trehelose at a ratio of 2:1 is effective to lyoprotect nanoparticles with hydrophobic core and poly(ethylene oxide) surface *(8)*.

Paclitaxel Nanoparticle Size Control by Coprecipitation with Homopolymer

Poly(ε-caprolactone)-*b*-poly(ethylene glycol), (PCL-b-PEG), copolymer and paclitaxel is an appropriate matrix for the treatment of cancer indications, where a particle size between 100 and 200 nm is desired. Using a tangential flow mixing cell, commonly known as a vortex mixer, and a copolymer of average molecular weight 2,900-*b*-5,000 g/mole, or PCL(25)-*b*-PEG(114) in monomer units, we have created paclitaxel nanoparticles while coprecipitating poly(ε-caprolactone) homopolymer. Table II displays the potential to tailor the particle size to that desired by the addition of homopolymer, without changing the feed concentration of active compound.

A vortex mixer was employed for this application since streams of unequal flow rates can be employed. This is a requirement of some systems to get the desired nanoprecipitation, especially when conducting probe studies at low organic active concentration on compounds with relatively high solubility in the product stream if mixed at 1:1 flow ratio. As with the CIJ mixer, we have previously demonstrated methods to characterize the mixing time of this confined mixer, also built in our laboratory. When operating under similar turbulent-like conditions, we have shown tangential mixers to be significantly slower *(9)*, but the mixing time is sufficient for the low feed concentration of active organic in this example.

Table II. Size of paclitaxel nanoparticles formed in the presence of hydrophobic homopolymer in a vortex mixer.

Poly (ε-caprolactone) Mw (g/mole)	*Mean Diameter (nm)*
1, 250	80
2, 000	95
10, 000	148

A stream of paclitaxel in tetrahydrofuran was co-fed with Poly(ε-caprolactone) homopolymer and PCL(25)-*b*-PEG(114) copolymer with each component dissolved at 0.3 wt%. The organic stream was rapidly mixed with water entering a vortex mixer at 10.2 m/s resulting in a ratio of 20:1 deionized water:THF. The results demonstrate that particle size can be controlled independently of drug concentration by the addition of appropriate additives, such as a hydrophobic homopolymer.

Conclusions

This chapter presents a new technology to form functional nanoparticles of hydrophobic organic actives, and methods to characterize the process. The enabling components are an amphiphilic diblock copolymer as a colloidal stabilizer and a rapid confined mixer for mixing and precipitation in tens of milliseconds. The process is run at relatively high concentration and low stabilizer content, and it easily fits into conventional systems available in the pharmaceutical or specialty chemical industries. More importantly, the significant technological barriers to understanding the precipitation of an amphiphilic diblock copolymer (or other stabilizers) in tandem with active organic compounds have been broken. The novel analytical CIJ mixer provides a technique for measuring an absolute precipitation time, and it provides the mixing requirement to achieve homogeneous competitive kinetics *(6)*.

Several challenges still remain to extend *Flash NanoPrecipitation* technology. The optimization of block copolymer molecular weights is currently under study. Methods of coating nanoparticles to control Ostwald ripening and dissolution kinetics are also under investigation. Finally, we are actively pursuing controlled release of active compounds from nanoparticles using novel release mechanisms.

References

1. Horn, D.; Rieger, J. Organic nanoparticles in the aqueous phase-Theory, experiment and use, *Angew. Chem. Int. Ed.* **2001**, *40*, 4331-4361.

2. Johnson, B.K. Ph.D. Thesis – Volume I, Princeton University. NJ, **2003**.
3. Johnson, B.K.; Prud'homme, R.K. Flash nanoprecipitation of organic actives and block copolymers using a confined impinging jets mixer, *Aust. J. Chem.* **2003**, *56*, 1021-1024.
4. Mahajan, A.J.; Kirwan, D.J. Nucleation and growth kinetics of biochemicals measured at high supersaturations, *J. Crystal Growth* **1994**, *144*, 281-290.
5. Johnson, B.K.; Prud'homme, R.K. Chemical processing and micromixing in confined impinging jets, *AIChE Journal*, **2003**, *49(9)*, 2264-2282.
6. Johnson, B.K.; Prud'homme, R.K. Mechanism for rapid self-assembly of block copolymer nanoparticles, *Phy. Rev. Lett.*, **2003**, *91*, 118302.
7. Kwok, K.Y.; Adami, R.C.; Hester, K.C.; Park, Y.; Thomas, S.; Rice K.G. Strategies for maintaining the particle size of peptide DNA condensates following freeze-drying, *Inter. J. Pharm.* **2000**, 203, 81-88.
8. De Jaeghere, F.; Allemann, E.; Feijen, J.; Kissel, T.; Doelker, E.; Gurny, R. Freeze-drying and lyopreservation of diblock and triblock poly(lactic acid)poly(ethylene oxide) (PLA-PEO) copolymer nanoparticles, *Pharm. Dev. and Technol.*, **2000**, *5*, 473-483.
9. Johnson, B.K. Ph.D. Thesis - Volume II, Princeton University. NJ **2003**.

Chapter 19

Particle Engineering of Poorly Water Soluble Drugs by Controlled Precipitation

E. J. Elder[1], J. E. Hitt[1], T. L. Rogers[1], C. J. Tucker[2], S. Saghir[3], G. B. Kupperblatt[1,4], S. Svenson[2,5], and J. C. Evans[1]

[1]Dowpharma, [2]Chemical Sciences, and [3]Toxicology and Environmental Research and Consulting, The Dow Chemical Company, Midland, MI 48674
[4]Current address: Mylan Technologies, 110 Lake Street, St. Albans, VT 05478
[5]Current address: Dendritic Nanotechnologies, 2625 Denison Drive, Mt. Pleasant, MI 48858

Controlled precipitation is a particle engineering technology under development for enhancing the aqueous dissolution rate and bioavailability of poorly water soluble drugs. Experiments with danazol and other model drugs showed that *in vitro* solubilization rate and *in vivo* absorption of the drug were improved by this technology.

Introduction

Many active pharmaceutical ingredients (API) suffer from low bioavailability due to poor solubility and/or poor dissolution rates in water (*1,2*). The preferred way to improve the solubility is to increase the surface area of the API particles, which is accomplished by reducing the particle size. A commercial technique for producing crystalline drug nanoparticles involves wetmilling the drug particles for extended times in the presence of stabilizers (*3*). However, this approach is limited to water-soluble stabilizers with low viscosity, and there is a risk of contamination from milling media. In addition, milling may reduce the degree of crystallinity of the drug particles (*4*).

Various precipitation techniques have been employed in pharmaceutical applications (*5-14*). Controlled precipitation is a particle engineering technology under development for enhancing the aqueous dissolution rate and bioavailability of poorly water soluble drugs (*15*). This process involves precipitating the drug into an aqueous solution in the presence of crystal growth inhibitors to form drug nanoparticles. Advantages and features include:

- Crystalline drug particles from molecular solution enable a polish filtration.
- Process is fast and scalable with conventional process equipment.
- Levels of residual solvents are low.
- Excipients used are generally pharmaceutically acceptable.

This paper describes preparation of three model drugs (danazol, naproxen, and ketoconazole) by controlled precipitation and presents the resulting *in vitro* dissolution and *in vivo* bioavailability in dogs.

Materials and Methods

Materials

Table I lists the characteristics of the three model drugs. Crystal growth inhibitors included poloxamer 407, polyvinyl pyrrolidone, and polyvinyl alcohol, all pharmaceutically acceptable excipients.

Table I. Properties of Model Drugs Danazol, Naproxen, and Ketoconazole

Property	*Danazol*	*Naproxen*	*Ketoconazole*
Activity	androgenic steroid	NSAID/analgesic	antifungal
Melt point, °C	226	152	150
Particle size, μm	~5	~15	~25
Aqueous solubility, mg/mL (25°C)	0.001	0.023	0.017

Assessment of commercial ketoconazole tablets (Nizoral, Janssen-Ortho, Lot 93P0241E) was also conducted in this study.

Process Description

Figure 1 illustrates the controlled precipitation process. Table II gives the experimental parameters for the three model drugs. The model drug and the crystal growth inhibitor (CGI) were dissolved in methanol, and in the case of ketoconazole, crystal growth inhibitor was also dissolved in the water phase. The organic phase was precipitated into chilled water. The controlled precipitation process was conducted in a continuous manner allowing for slurry concentration and solvent removal. The evaporated solvent was recovered. The concentrated, solvent-stripped slurry was then freeze-dried (lyophylized) in the case of laboratory-scale drying or spray-dried in the case of pilot-scale drying to remove the aqueous phase and isolate the nanostructured crystalline drug substance (modified model drug).

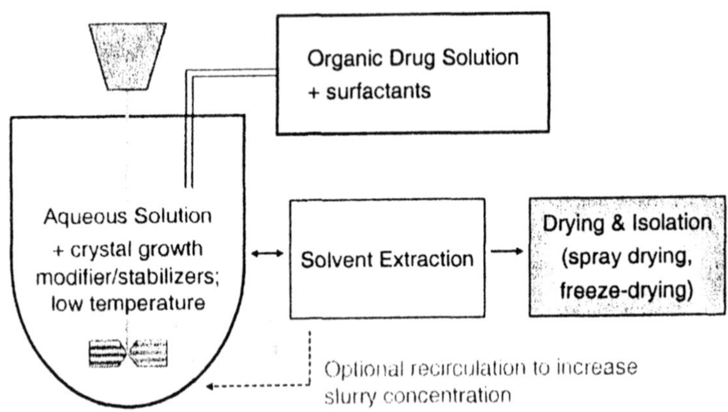

Figure 1. Process diagram for controlled precipitation. (Adapted with permission from Reference 15. Copyright 2004 Springer Science and Business Media, Inc.)

Characterization

The modified model drugs were characterized by scanning electron microscopy (SEM) and x-ray diffraction (XRD). Particle size was determined by dynamic light scattering. Dissolution rates were determined using USP apparatus II (paddles) with an aqueous media containing 0.75% sodium lauryl sulfate and 1.21% TRIS at pH 9.0. Bioavailability was determined in beagle dogs as described for two of the drugs in their respective sections.

Table II. Experimental Parameters for Controlled Precipitation

Parameter	*Danazol*	*Naproxen*	*Ketoconazole*
Organic phase CGI	poloxamer 407	polyvinyl pyrrolidone K30	polyvinyl pyrrolidone K30
Ratio drug:CGI	2:1	2:3	1:0.5
Water phase CGI and level	—	—	polyvinyl alcohol, 0.61%
Ratio organic:water during precipitation	1:5	1:5	1:5

CGI = crystal growth inhibitor

Results and Discussion

When the dissolved drug is precipitated in the presence of a crystal growth inhibitor, the inhibitor adsorbs on the crystal surface shortly after nucleation, hindering solute mass transport with the following results:

- Nucleation dominates growth in the relief of supersaturation.
- Mean crystal size is depressed.
- Slurry can be recycled with little growth on existing crystals.

Danazol

Figure 2 shows SEMs of bulk danazol compared to the modified danazol. Particle size was reduced from 5 μm to 0.46 μm as determined following redispersion of the agglomerated nanostructured particles prepared by controlled precipitation.

X-ray diffraction analysis indicated that the modified danazol was crystalline (Figure 3). Residual solvent level in the modified danazol powder was less than 250 ppm, which is well below ICH guidelines for methanol (3000 ppm).

In Vitro Dissolution

Dissolution of both danazol USP (micronized) powder and a physical blend of danazol USP and excipients required 30 minutes to reach 90% (Figure 4). In

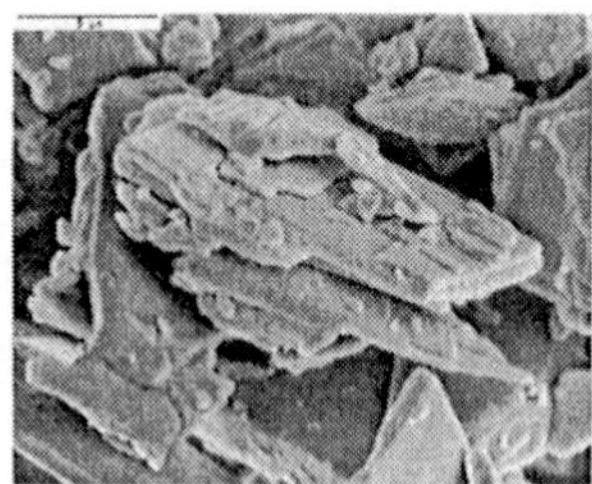

As-received bulk danazol

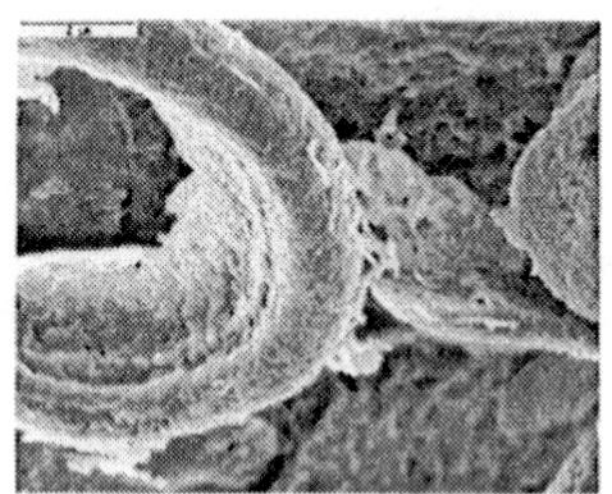

Danazol prepared by controlled precipitation

Figure 2. Scanning electron micrographs of as-received bulk danazol (5 μm scale) and agglomerated danazol (2 μm scale) prepared by controlled precipitation, which rapidly redisperse in water to 0.46 μm primary particles. (Reproduced with permission from Reference 15 and 16. Copyright 2004 Springer Science and Business Media, Inc. and Drug Delivery Technology LLC.)

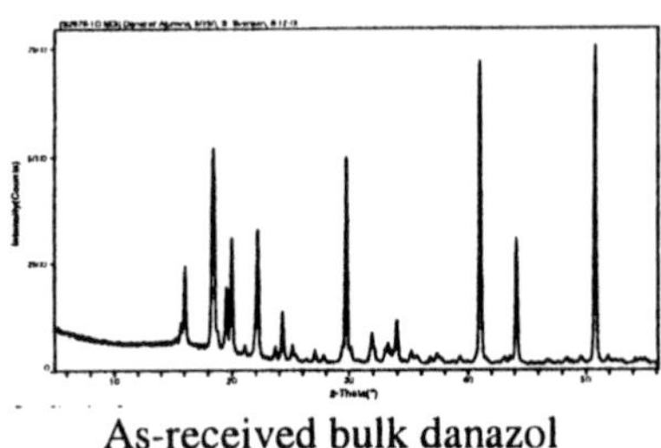

As-received bulk danazol

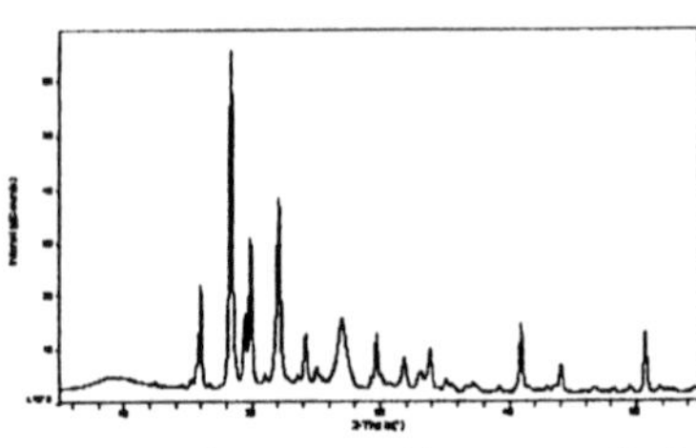

Danazol prepared by controlled precipitation

Figure 3. X-ray diffraction analysis of as-received bulk danazol and danazol prepared by controlled precipitation.

comparison, the modified danazol powder prepared by controlled precipitation was 100% dissolved within 5 minutes.

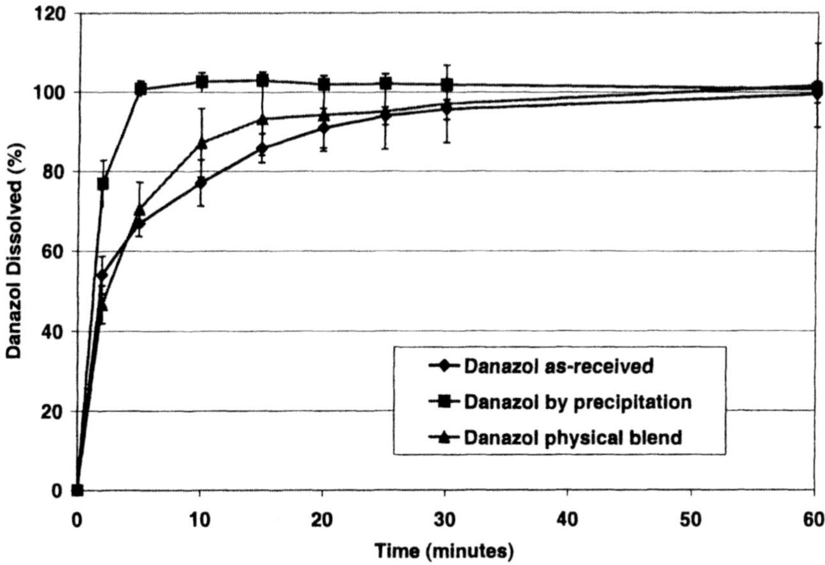

Figure 4. Dissolution of danazol (n=6). (Reproduced with permission from Reference 16. Copyright 2004 Drug Delivery Technology LLC.)

Bioavailability

The danazol powder prepared by controlled precipitation showed substantially improved bioavailability (Figure 5) compared to the drug as-received (micronized danazol USP). Tablets prepared on a Carver press from modified danazol (equivalent to 200 mg danazol) formulated with microcrystalline cellulose and carboxymethylcellulose (47.5:47.5:5) showed further enhancement in bioavailability. The increased bioavailability observed with the control is due to an excipient effect that enhances wettability of the powder.

Naproxen

Figure 6 shows SEMs of bulk naproxen compared to the modified naproxen. Particle size was reduced from 17 μm to 4 μm. X-ray diffraction analysis indicated that the modified naproxen was crystalline (Figure 7). Residual solvent level in the modified naproxen powder was approximately 150 ppm, which is well below ICH guidelines for methanol (3000 ppm).

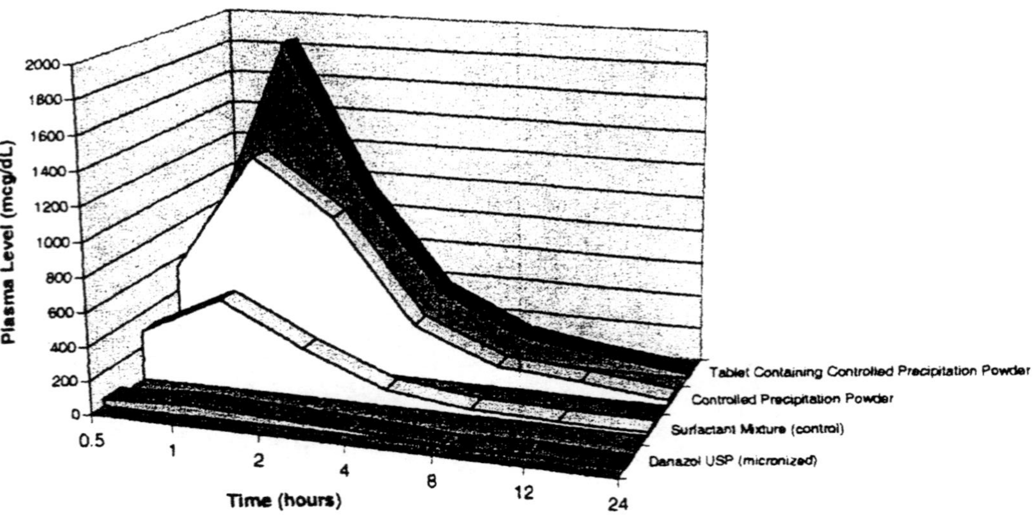

Figure 5. Mean plasma levels of danazol in dogs. (Reproduced with permission from Reference 16. Copyright 2004 Drug Delivery Technology LLC.)

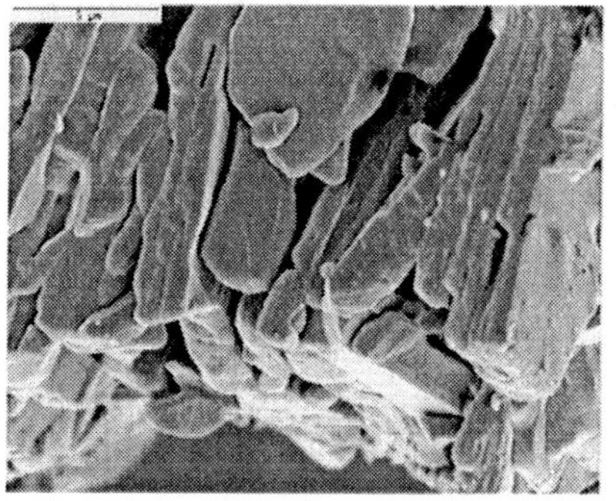

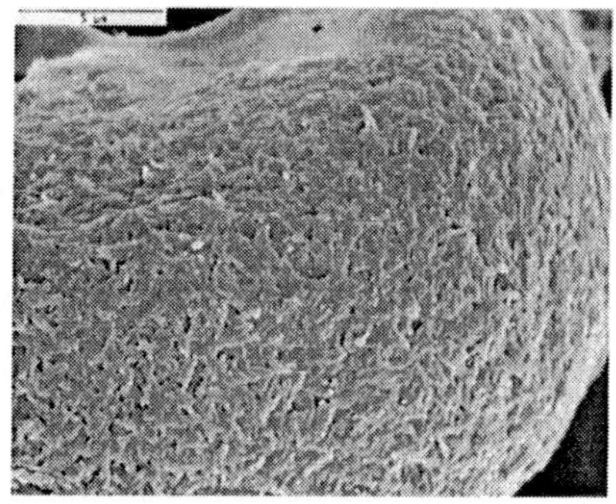

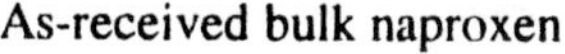

As-received bulk naproxen

Naproxen prepared by controlled precipitation

Figure 6. Scanning electron micrographs (5 μm scale) of as-received bulk naproxen and naproxen prepared by controlled precipitation. (Adapted with permission from Reference 15. Copyright 2004 Springer Science and Business Media, Inc.)

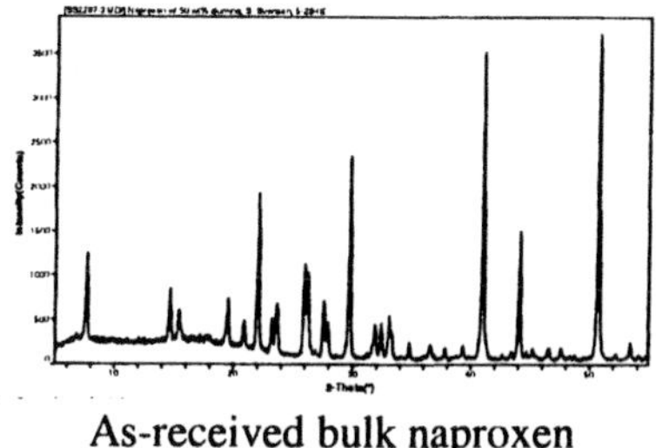

As-received bulk naproxen

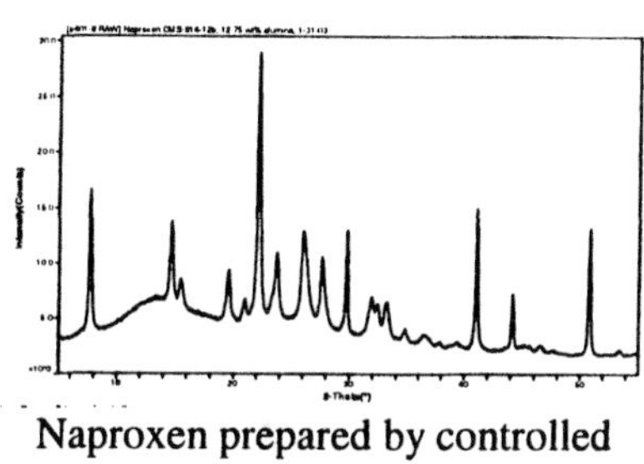

Naproxen prepared by controlled precipitation

Figure 7. X-ray diffraction analysis of as-received bulk naproxen and naproxen prepared by controlled precipitation.

In Vitro Dissolution

Dissolution of as-received naproxen was less than 60% at 60 minutes. In comparison, the modified naproxen powder prepared by controlled precipitation and either freeze-drying (lab-scale) or spray-drying (pilot-scale) was 95 to 100% dissolved within 20 minutes. Bioavailability of naproxen was not measured.

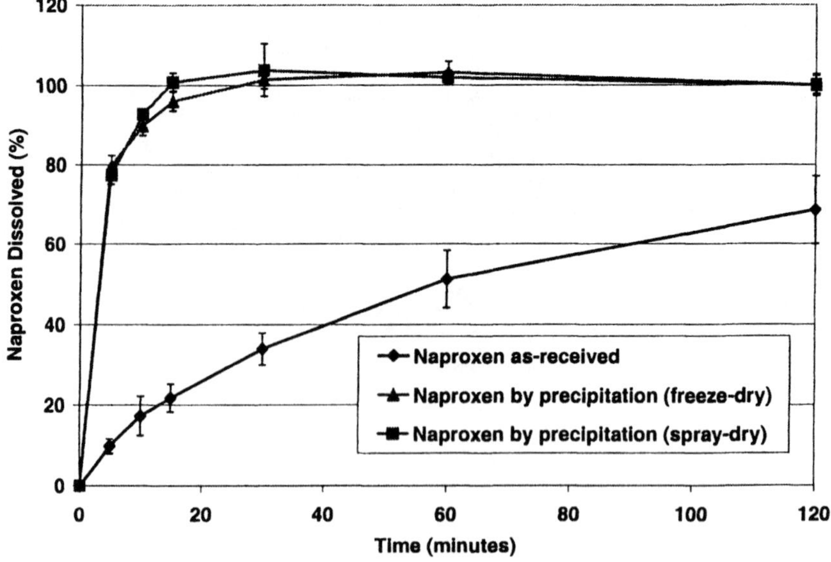

Figure 2. Dissolution of naproxen (n=6).

Ketoconazole

Figure 8 shows SEMs of bulk ketoconazole compared to the modified ketoconazole. Particle size was reduced from 15 μm to 7 μm for freeze-dried (lab-scale) ketoconazole by precipitation and 2 μm for spray-dried (pilot-scale) ketoconazole by precipitation.

X-ray diffraction analysis indicated that the modified ketoconazole was crystalline (Figure 9). Residual solvent level in the modified ketoconazole powder was well below ICH guidelines.

In Vitro Dissolution

Dissolution of ketoconazole as-received powder was less than 60% at 60 minutes (Figure 10). Ketoconazole by precipitation was greater than 80% dissolved within 10 minutes and 100% dissolved within 60 minutes. In comparison, dissolution of the commercial tablet was approximately 70% at 60 minutes.

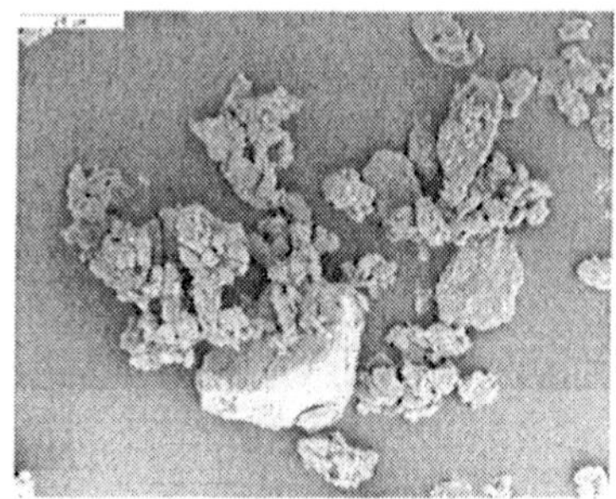

As-received bulk ketoconazole

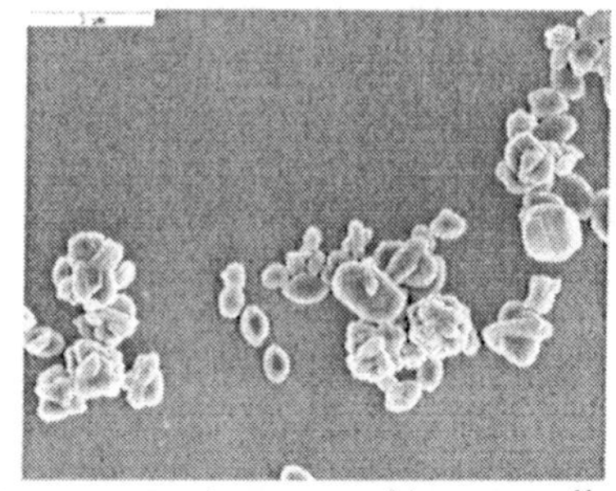

Ketoconazole prepared by controlled precipitation

Figure 8. Scanning electron micrographs of as-received bulk ketoconazole (20 μm scale) and ketoconazole prepared by controlled precipitation (5 μm scale).

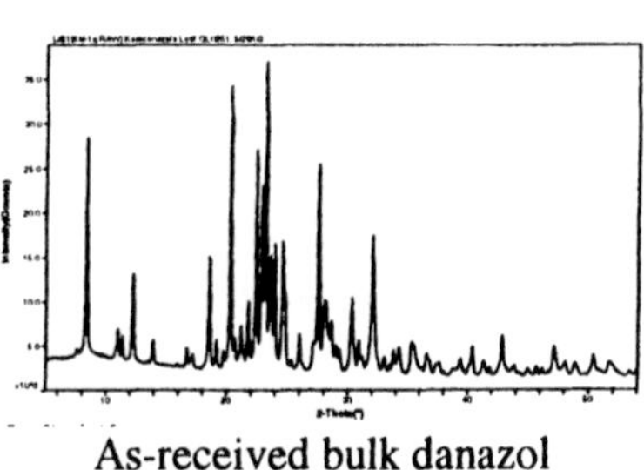

As-received bulk danazol

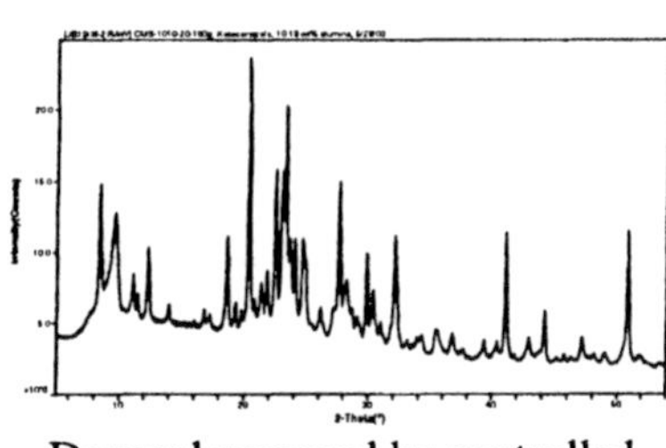

Danazol prepared by controlled precipitation

Figure 9. X-ray diffraction analysis of as-received bulk ketoconazole and ketoconazole prepared by controlled precipitation.

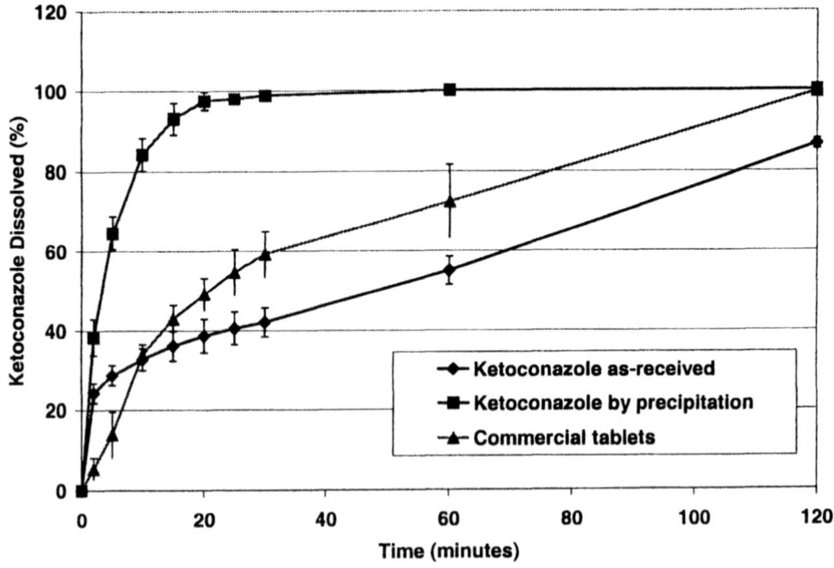

Figure 10. Dissolution of ketoconazole (n=6).

Bioavailability

Ketoconazole prepared by precipitation showed improved bioavailability over both ketoconazole as-received and a blend of ketoconazole as-received with excipients in capsules (Figure 11). Further improvement was seen when the modified ketoconazole was formulated in capsules or tablets. The capsules were prepared by blending the modified ketoconazole (equivalent to 200 mg ketoconazole) with microcrystalline cellulose and carboxymethylcellulose (47.5:47.5:5) and hand-filling into size 0 gelatin capsules. The tablets were formulated in the same manner and compressed manually on a Carver press.

Conclusions

Controlled precipitation is a viable particle engineering technology for enhancing the bioavailability of poorly water soluble drugs. Demonstration with danazol and other model drugs resulted in substantial improvements to *in vitro* dissolution and *in vivo* bioavailability.

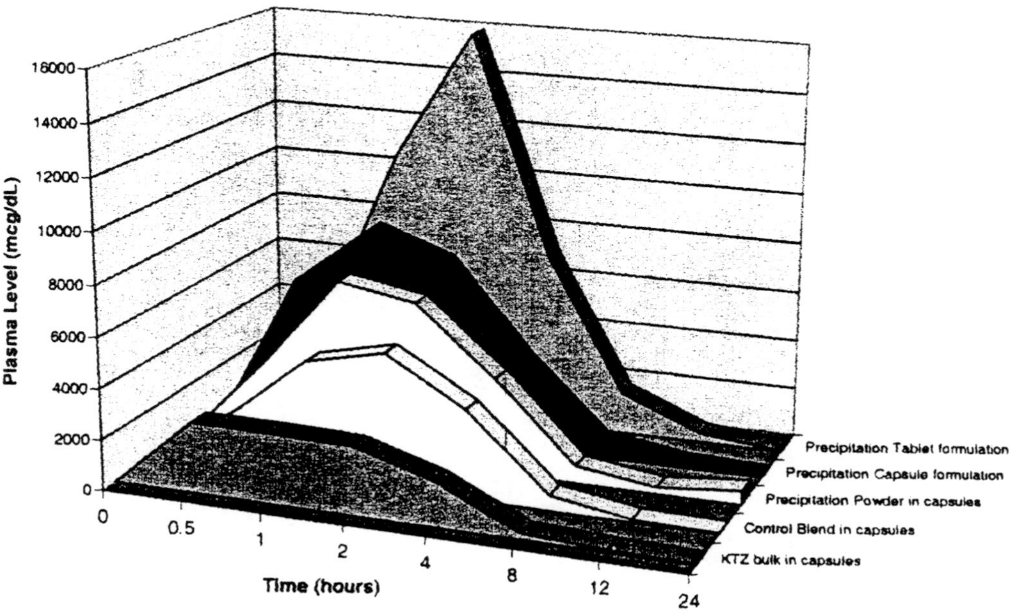

Figure 11. Mean plasma levels of ketoconazole in dogs.

Acknowledgements

The authors gratefully acknowledge the laboratory support contributions provided by Analytical Sciences and BioAqueous^{SM} Solubility Services at Dow.

BioAqueous is a service mark of The Dow Chemical Company.

References

1. Lipinski, C. *Am. Pharm. Rev.* **2002**, 5, 82–85.
2. Radtke, M. *New Drugs.* **2001**, 3, 62–68.
3. Liversidge, G.G.; Cundy, K.C. *Int. J. Pharm.* **1995**, 125, 91–97.
4. Suryanarayanan, R.; Mitchell, A.G. *Int. J. Pharm.* **1985**, 24, 1-17.
5. Goia, C.; Matijević, E. *J. Colloid Interface Sci.* **1998**, 206, 583-591.
6. Auweter, H.; Andre, V.; Horn, D.; Luddecke, E. *J. Disp. Sci. Tech.* **1998**, 19, 163-184.
7. Ruch, F.; Matijević, E. *J. Colloid Interface Sci.* **2000**, 229, 207-211.
8. Violante, M. R.; Fischer, H. W. U.S. Patent 4,997,454, 1991.
9. Bagchi, P.; Karpinski, P. H.; McIntire, G. L. U.S. Patent 5,560,932, 1996.
10. Bagchi, P.; Scaringe, R. P.; Bosch, H. W. . U.S. Patent 5,665,331 1997.
11. Bagchi, P.; Stewart, R. C.; McIntire, G. L.; Minter J. R. U.S. Patent 5,662,883, 1997.
12. Frank, S.; Lofroth, J.-E.; Bostanian, L. U.S. Patent 5,780,062, 1998.
13. Chen, X.; Young, T. J.; Sarkari, M.; Williams, R. O.; Johnston, K. P. *Int. J. Pharm.* **2002**, 242, 3-14.
14. Sarkari, M.; Brown, J. N.; Chen, X.; Swinnea, S.; Williams, R. O.; Johnston, K. P. *Int. J. Pharm.* **2002**, 243, 17-31.
15. Rogers, T.L.; Gillespie, I.B.; Hitt, J.E.; Fransen, K.L.; Crowl, C.A.; Tucker, C.J.; Kupperblatt, G.B.; Becker, J.N.; Wilson, D.L.; Todd, C.; Broomhall, C.F.; Evans, J.E.; Elder; E.J. *Pharm. Res.* **2004**, 21, 2048-2057.
16. Connors, R. C; Elder, E. J. *Drug Delivery Technology* **2004**, 4, 78-83.

Chapter 20

Improvement of Dissolution Rate of Poorly Water Soluble Drugs Using a New Particle Engineering Process: Spray Freezing into Liquid

Kirk A. Overhoff[1], Keith P. Johnston[2], and Robert O. Williams III[1,*]

[1]College of Pharmacy and [2]Department of Chemical Engineering, University of Texas at Austin, Austin, TX 78712
[*]Corresponding author: williro@mail.utexas.edu

Biopharmaceutics Classification System (BCS) Class II drugs are classified as having poor water solubility and good permeability across biological membranes. Improving the dissolution should, therefore, increase the bioavailability of this class of drugs. The Spray-Freezing into Liquid (SFL) process has been shown to increase dissolution rates of BCS class II drugs by providing amorphous particles with high surface area and small primary particle size, stabilized using hydrophilic polymers. Modifying the physicochemical properties of the particles is accomplished by selecting the optimum feed system composition. This article reviews the influence of this composition and active pharmaceutical ingredient (API) potency on surface area, surface morphology, crystallinity, and dissolution rates of model API.

Introduction

Active pharmaceutical ingredients (API) are classified according to the Biopharmaceutics Classification System (BCS), based on their solubility in aqueous media and their intestinal permeability. Nearly half of all new API created by the pharmaceutical industry are considered poorly water-soluble and are, therefore, classified as either Class II or Class IV drugs (*1*). Class II drugs are characterized by having poor water solubility but high membrane permeability. Problems developing these API into viable drug products involve poor bioavailability because of the low solubility or erratic absorption of the API. For Class II drugs, the *in vivo* dissolution of the API is the rate limiting step for API absorption, and therefore, it is possible to predict *in vitro*/*in vivo* correlation based on the dissolution rate profile (*2,3*).

The Noyes-Whitney equation, which mathematically describes the *in vitro* dissolution rate of an API, reveals the physical factors (surface area, boundary layer thickness) that can influence the dissolution rate. As the particle size of the API is decreased, the surface area increases, increasing the dissolution rate. Likewise, decreasing the boundary layer thickness between the solid particle and the dissolution media will increase the dissolution rate. Lastly, Mura *et al.* has shown that decreasing the crystallinity of the API can increase the rate of dissolution (*4*). However, amorphous API are in a thermodynamically unstable state and can revert back to the more stable crystalline structure in the absence of stabilizing polymers such as poly(vinylpyrrolidone) (PVP) or poly(vinyl alcohol) (PVA). It is important to acknowledge that physiological factors such as gastrointestinal fluid pH, API-food interactions, and intestinal mixing motions can have a dramatic affect on the dissolution rate of Class II drugs. However, these physiological parameters are well documented in the literature and will not be discussed (*5*).

Various processes exist to increase API dissolution rates, including mechanical milling, spray-drying, supercritical fluid technologies, and solvent extraction/evaporation, and have recently been reviewed (*6-8*). Mechanical milling processes such as ball milling or colloid milling can impart large frictional forces and heat into the system and degrade thermolabile API. Spray-drying is an ideal process used in large scale production. However, processing limitations make it difficult to produce particles smaller than one micron in diameter, and high temperatures are necessary to evaporate aqueous solutions, which can cause degradation of thermolabile API. Supercritical fluid technologies are limited by the solubility of API and polymer in the solvent (supercritical CO_2 being the most preferred). Technologies such as Rapid Expansion of Supercritical Solutions (RESS) or Pressure Induced Phase Separation (PIPS) can lead to particle agglomeration and aggregation and nozzle blockages, making process scale-up more difficult (*9*). An improved particle engineering process known as Evaporative Precipitation into Aqueous Solution (EPAS) does not require the use of supercritical fluids. The EPAS process involves spraying a superheated API/stabilizer solution (water immiscible organic solvent) into a heated aqueous solution via a crimped nozzle. As the organic solvent rapidly evaporates, the API and stabilizer nucleate, forming a stable colloidal suspension (*10,11*).

The objective of this chapter is to review recently published work from this laboratory regarding the use of the Spray-Freezing into Liquid (SFL) particle engineering process to enhance dissolution and wetting of model BCS class II drugs. The chapter will focus on the influence of various feed systems, as well as examine the effect of API potency on the physicochemical properties of the SFL processed powders. The feed system comprises the API, hydrophilic excipients and the solvent system (i.e., organic solvent, aqueous-organic co-solvent, organic-in-water (o/w) emulsion). SFL has been shown to dramatically increase dissolution rates of BCS class II drugs such as danazol (DAN) and carbamazepine (CBZ), theoretically increasing the bioavailability of the API (*6,12-16*). The SFL process has also been used to manufacture stable micro-particles containing peptides (*17,18*). The microparticles produced by the SFL process are characterized by high surface areas (high porosity), enhanced wettability due to the presence of hydrophilic excipients/stabilizers, and a stable amorphous structure. These characteristics are accomplished by intense atomization of the API/polymer feed solution under pressure through an insulating nozzle, which is submerged directly into a cryogenic liquid (liquid N_2), ensuring rapid freezing of the API in a random state. The frozen particles are collected and lyophilized in a temperature-controlled tray lyophilizer. Controlling the temperature of the shelf allows for maximum sublimation rates of the solvent by lyophilizing at a temperature just below the collapse temperature, T_c, of the product (*19-21*). The desirable physicochemical properties of the microparticles produced by the SFL process enable their use in pharmaceutical drug delivery systems. For example, rapidly disintegrating tablet formulations have been produced comprising SFL engineered particles of DAN (*22*).

Materials and Methods

Materials

Carbamazepine USP (CBZ), Danazol USP (DAN; micronized grade) and the surfactants/stabilizers used in the studies, poloxamer 407 (P407), poly(vinyl-pyrrolidone), (PVP), K15, and partially hydrolyzed poly(vinyl alcohol) (PVA; MW 22,000), were purchased from Spectrum Chemicals (Gardena, CA). HPLC grade tetrahydrofuran (THF), dichloromethane (DCM) and acetonitrile (ACN) were purchased from EM Industries Inc. (Gibbstown, NJ). Sodium lauryl sulfate

(SLS), tris(hydroxymethyl)aminomethane (TRIS) and 1N hydrochloric acid were purchased from Spectrum Chemicals.

Spray-Freezing into Liquid (SFL) Apparatus

The SFL apparatus used in these studies is shown in Figure 1. The feed systems were sprayed under pressure with a flow rate of 50 ml/min through a 127-μm polyether-ether ketone (PEEK) nozzle into liquid nitrogen.

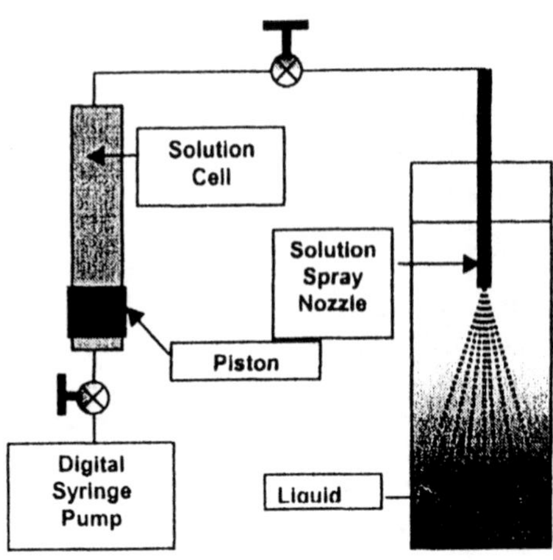

Figure 1. Schematic presentation of the SFL apparatus.

Dissolution Testing

Dissolution testing was performed on samples containing DAN, using the United States Pharmacopeia (USP) 24 apparatus II (VanKel VK6010 Dissolution testing station with Vanderkamp VK650A heater/circulator, Varian, Inc., Cary, NC). Dissolution media was composed of 900 ml of 0.75% SLS/1.21% TRIS adjusted to pH 9.0 at 37.0±0.2°C. The paddle speed was set at 50 rpm and samples were taken at 2, 5, 10, 20, 30, and 60 minutes without replacement media. Similarly, for CBZ samples the same dissolution parameters were used except the dissolution media was 900 ml of purified water. All dissolution testing was performed at sink conditions.

Methods

A Philips Model 1710 x-ray diffractometer (Philips Electronic Instruments, Inc., Mahwah, NJ) with CuKα_1 radiation and a wavelength of 1.54054 Å at 40kV and

20 mA was used to obtain the x-ray diffraction patterns of all the samples. The samples were scanned from 5 to 45° (2θ) at a rate of 0.05°/sec. A Hitachi Model S-4500 field emission scanning electron microscope (Hitachi Ltd., Rolling Meadows, IL) was used to establish the surface morphology of each sample. The samples were deposited onto the SEM stage using double-sided carbon tape and sputter coated with gold. A Nova 3000 surface area analyzer (Quantachrome Corp., Boynton Beach, FL) was used to quantify N_2 sorption at 77.40 K. The surface area was calculated using the Brunauer, Emmett, Teller (BET) equation (*23*). The sample powders were compressed into compacts using a Carver laboratory press (Fred S. Carver, Inc., Menomonee Falls, WI) with flat-faced 6 mm diameter punches. A 3-μL droplet of the respective dissolution media was placed on the face of the compact and the contact angle was determined using a goniometer (Ramè-Hart Inc., Mountain Lakes, NJ). The tangent of the curve of the droplet at the compact face was measured to calculate the angle and repeated twice for each sample.

Results and Discussion

The physicochemical properties of micronized powders produced by the SFL process can be modified by altering different processing parameters. Some important processing parameters include (i) the type of solvent used in the feed system, (ii) the concentration and type of excipient, and (iii) the rate at which the solution is atomized into the cryogenic liquid. The following sections will discuss the influence of solvent type and composition, and API concentration on the properties of the powders.

Solvent System

The composition of the liquid feed system can have a tremendous impact on the final particle size distribution, surface area, and surface morphology. For this reason, three different solvent systems (organic, aqueous-organic co-solvent, and o/w emulsion) were selected based on the SFL and lyophilization processing requirements and the solubility of the API in the solvent system. In this study, the hydrophilic excipients PVP K15 and P407 were dissolved in a 2:1:1 ratio using either ACN or THF/water co-solvent (33% w/w) systems, both containing a CBZ concentration of 0.22% w/w (*24*). The API-loaded feed systems were then processed using the SFL apparatus. API-loaded powders obtained from an o/w emulsion feed system were also produced using the SFL process. The water immiscible organic phase of the emulsion comprised the model API DAN (3.0% w/w) dissolved in DCM (30% w/w). Equal amounts of the hydrophilic stabilizer/surfactants PVA, PVP K15, and P407 were dissolved in the aqueous phase (4.5% w/w total solids). The organic phase containing the API and the aqueous phase containing the excipients were homogenized using a Polytron rotor/stator homogenizer to form a coarse emulsion. The emulsion was then transferred to an Avestin Emulsiflex-C5 high pressure homogenizer (Ottawa, Canada) and circulated through 10 cycles at 20,000 PSI (*25*). The emulsion was sprayed into liquid nitrogen by the SFL process, and the frozen particles were lyophilized to produce the powder.

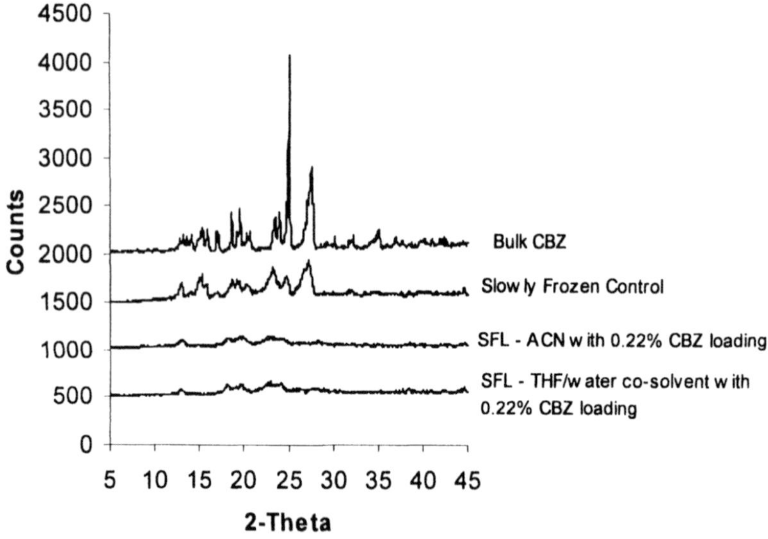

Figure 2. X-ray diffraction patterns of SFL samples from different feed systems (Reproduced with permission from ref. (24). Copyright 2003 Elsevier B.V.)

The characteristic peaks of bulk CBZ are 15.3, 25.0, and 27.6 (2θ degrees), as seen in the x-ray diffraction patterns in Figure 2. The slowly frozen control displayed the same peak pattern, indicating the presence of crystalline CBZ. The x-ray diffraction patterns for powders produced from both the ACN and the THF/water feed systems, however, lacked the characteristic peaks indicative of crystalline CBZ *(24)*. Because the powders from the two different feed solutions show similar amorphous diffraction patterns, it is concluded that the presence of crystalline CBZ is not a function of the solvent system but rather a function of the rate of freezing. A similar x-ray diffraction pattern can be seen in Figure 3 for the DAN powder prepared from the emulsion feed system *(25)*. For the slowly frozen control, the rate of freezing was sufficiently slow allowing for time for the dissolved API molecules to orient themselves in a thermodynamically stable crystalline orientation. Since the SFL feed systems are atomized directly into liquid nitrogen, there is less time for the molecules to orient themselves before the solution is frozen, thus halting molecular mobility. The API contained in the frozen particles is trapped in an amorphous state and further stabilized by the excipients in the formulation, which increase the overall glass transition temperature of the powder.

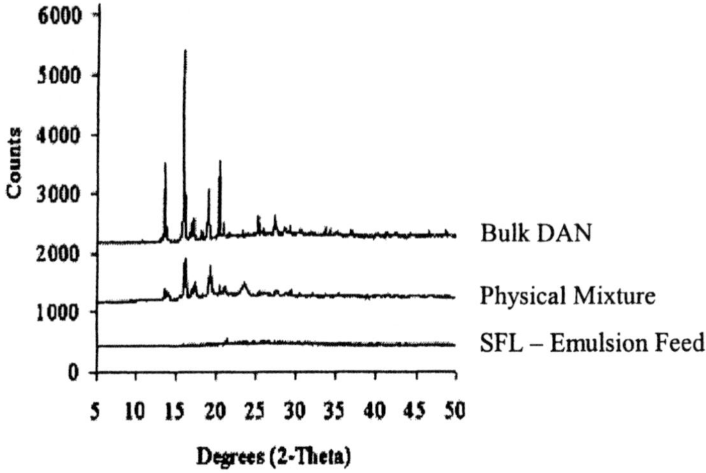

Figure 3. X-ray diffraction patterns of SFL samples from bulk DAN, a physical mixture between DAN and excipients, and the emulsion feed system. (Adapted with permission from ref. (25). Copyright 2003 Elsevier B.V.)

The particle morphologies of the SFL processed powders in Figure 4 are considerably different from each other, indicating that the type of solvent system and the total concentration have an effect on the morphology. SFL micronized powders from a one-phase system such as the organic or the aqueous-organic co-solvent feed system have an irregular particle shape with a smooth, non-porous surface as seen in Figure 4b-d. Because the emulsion feed system is a two-phase system, two distinct regions with different surface morphologies can be seen in the SEM image in Figure 4e. In addition, the total solids concentration in the feed system can influence the size of the aggregates, as seen in the SEM images. Particle size distributions of the powders determined by laser light scattering confirm these results. As the concentration of solids in the feed solution is increased, the aggregate particle size increases. As the solids content is increased, phase separation occurs, forming localized regions of solvent which leaves porous channels within the particle matrix when sublimed.

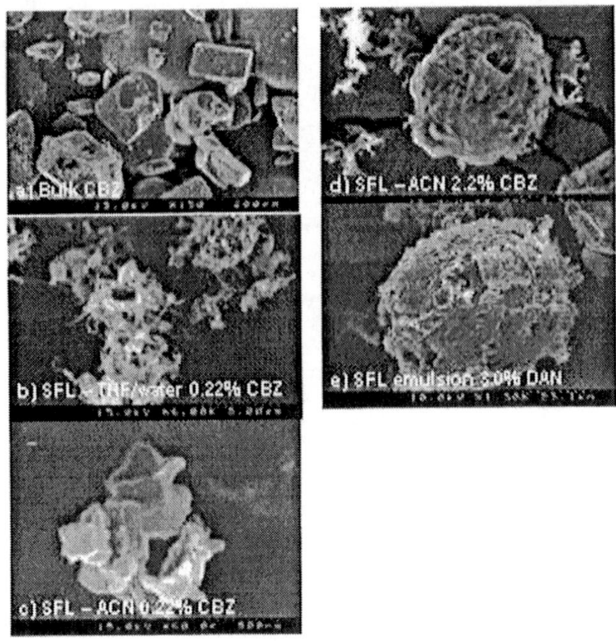

Figure 4. Unprocessed CBZ (a) with SEM micrographs of SFL processed powders from aqueous-organic co-solvent (b,c), organic (d), and organic-in-water emulsion (e). (Reproduced with permission from ref. (24). Copyright 2003 Elsevier B.V.)

Specific surface area and contact angle measurements for the SFL powders are presented in Table I. All of the powders show an increase in specific surface area compared to the control formulations and the bulk API. Both powders from the organic and aqueous-organic co-solvent feed systems gave similar surface areas of 12.9 and 13.3 m^2/g, respectively. The emulsion feed system had a higher specific surface area at 83.1 m^2/g. This increase in surface area is due in part to a reduction of the primary particle size and an increase in porosity. The contact angle measurements presented in Table I show decreased contact angles for powders produced by the SFL process, indicating an increase in wettability. Due to their hydrophobic nature, both pure API exhibit high contact angles. However, the powders contain hydrophilic surfactants, which decrease the interfacial tension between the solvent and the solute particles, resulting in lower contact angles. The type of excipient plays an important role in decreasing the contact angle of the particle since surfactant structure and orientation play a crucial role in wettability *(26)*.

Table I. Surface area and contact angle of SFL samples with various feed systems.

Sample	Surface Area (m^2/g)	Average Contact Angle
Bulk CBZ	-	45.0 (3.4)
Slow Frozen Control	1.84	43.0 (2.6)
SFL Acetonitrile w/ 0.22% CBZ	12.89	34.0 (1.5)
SFL Acetonitrile w/ 2.2% CBZ	3.88	38.0 (2.5)
SFL THF/Water w/ 0.22% CBZ	13.31	29.0 (1.5)
Bulk DAN	0.52	64.3 (1.1)
Slow Frozen Control	0.35	47.8 (2.1)
SFL Emulsion w/ 3.0% DAN	83.06	48.0 (1.4)

SOURCE: Reproduced with permission from refs. *(24)* and *(25)*. (Copyright 2003 Elsevier B.V.)

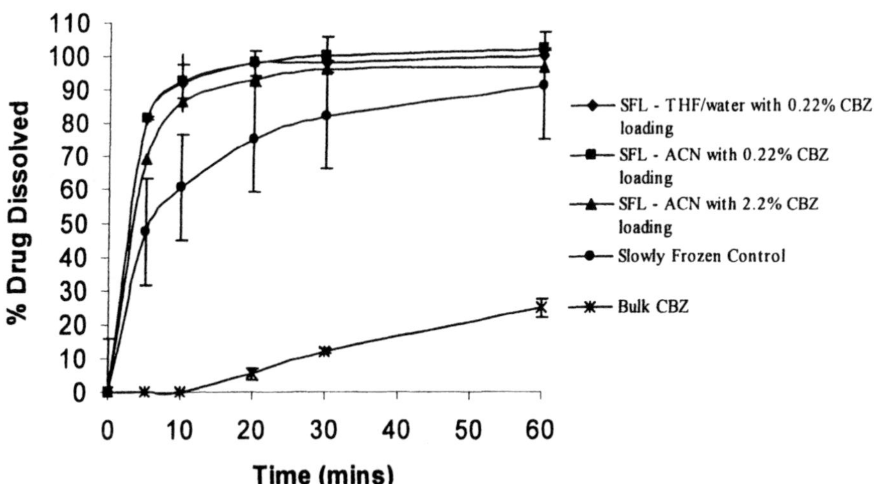

Figure 5. Dissolution of CBZ SFL powder samples. (Reproduced with permission from ref. (24). Copyright 2003 Elsevier B.V.)

The dissolution profiles for SFL powder samples containing CBZ are shown in Figure 5. The dissolution results for both powder formulations (organic and aqueous-organic co-solvent) demonstrated rapid release, with 80% of the API dissolved after 5 minutes and about 100% dissolved after 20 minutes. These dissolution rates show a dramatic improvement compared to the bulk API and the slowly frozen control. Bulk CBZ has less than 10% dissolved in 20 minutes, while the slowly frozen sample has about 70% dissolved in 20 minutes. The improved dissolution rate for the powders is attributed to the amorphous structure of the particles, an increase in specific surface area, allowing for a larger solid-liquid interface for wetting and dissolution to occur, and increased wettability of the particles due to the hydrophilic excipients in the particle matrix *(27)*.

Influence of API Concentration

Many processes designed to increase the dissolution rate of poorly water soluble API do so at the expense of API potency. Studies have shown that as the API concentration is increased, the phase behavior of the API within the polymer matrix can change, further effecting the dissolution rates (*28,29*). To form stable solid dispersions of molecularly dispersed API, formulations require large amounts of stabilizers and hydrophilic excipients. However, as the potency of the API is decreased, more processed powder is required in order to obtain the necessary amount of API required to produce a therapeutic response. Physical size restrictions of tablets and capsules limit the amount of processed powder that can be incorporated into the dosage form. Therefore, maintaining high concentrations of API in the formulation while enabling rapid dissolution rates is the challenge to researchers.

DAN/PVP K15 blends in increasing ratios from 1:2 (33%) to 10:1 (91%) were dissolved in either ACN or a DCM/ACN mixture, and sprayed by the SFL process to investigate the effect of potency on particle morphology, surface area, dissolution and stability *(16)*. Figure 6 shows the x-ray diffraction patterns of these powders (0.36-1.63% DAN loading in feed system). The results indicated an absence of crystalline DAN within the API/polymer matrix. The increase in potency had no affect on the crystallinity of the powders due to the rapid freezing rate, characteristic to the SFL process, and the ability of PVP K15 to effectively stabilize the amorphous DAN molecules even at low concentrations.

However, increasing the potency in the feed solution had a dramatic effect on the surface morphology of the resulting powders, leading to a decrease in surface area and an increase in contact angle. As seen in Figure 7, the structures of the SFL powders are composed of aggregates of nanoparticles, increasing in size from 50 nm for the 33% SFL DAN sample in Figure 7a to 750 nm to 1.5 microns for the 75% SFL DAN sample. The 91% SFL DAN sample showed a porous particle, about 3 microns in diameter (Figure 7b).

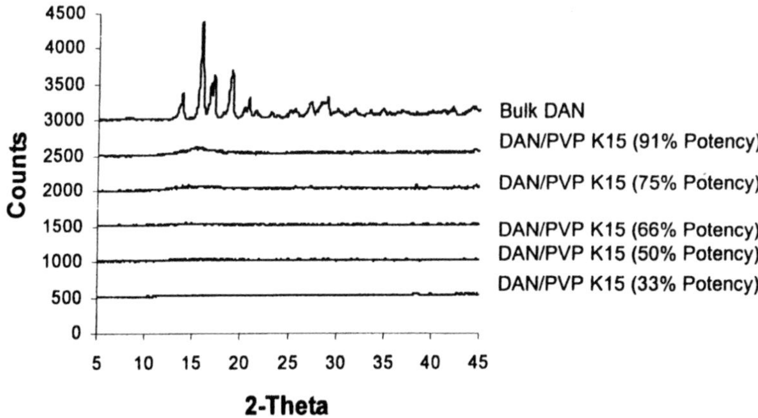

Figure 6. X-ray diffraction of SFL samples with increasing DAN potency. (Reproduced with permission from ref. (16). Copyright 2004 Elsevier B.V.)

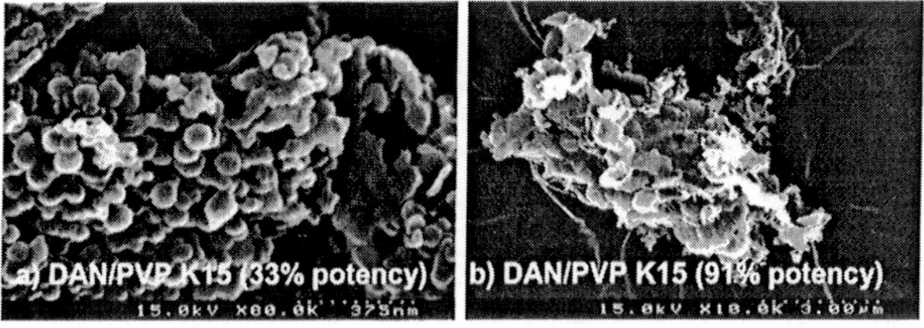

Figure 7. Surface morphology of different potency SFL DAN/PVP K15 samples. (Reproduced with permission from ref. (16). Copyright 2004 Elsevier B.V.)

According to Raoult's law, an increase in solids concentration of the feed solution leads to an overall freezing point depression of the solution, allowing more time for particle nucleation and growth. The size of the primary particles can affect the specific surface area of the sample, and was confirmed using BET surface area analysis. Smaller particle sizes result in larger surface areas for a given amount of sample. As seen in Table II, the surface areas ranged from 28.5 to 117.5 m^2/g, as potency was decreased. As the API/polymer systems are forced through the nozzle, the solution is atomized into small droplets, contributing to the increase in surface area. Within each droplet, the freezing process creates solvent channels, leading to porous regions during the freeze-drying process. Also seen in Table II are the contact angles of the powders produced by the SFL process. Bulk DAN has a contact angle of 57±2.2°, indicating the hydrophobic nature of the API. Contact angles range from 22±0.4° to 35±2.0° for the powders, indicating a correlation between the amount of hydrophobic API/hydrophilic stabilizer and wettability. As the amount of PVP K 15 is increased in the formulation, the contact angle decreased, resulting in a particle with greater ability to wet.

Table II. Surface area and contact angle of SFL DAN potency samples.

Sample	Surface Area (m^2/g)	Average Contact Angle
Bulk DAN	0.5	57.0 (2.2)
DAN/PVPK15 (91% Potency)	28.5	35.0 (2.0)
DAN/PVPK15 (75% Potency)	68.9	27.0 (1.0)
DAN /PVPK15 (66% Potency)	79.9	24.0 (0.4)
DAN/PVPK15 (50% Potency)	115.5	22.0 (0.5)
DAN/PVPK15 (33% Potency)	117.5	22.0 (0.4)

SOURCE: Reproduced with permission from ref. *(16)*. (Copyright 2003 Elsevier B.V.)

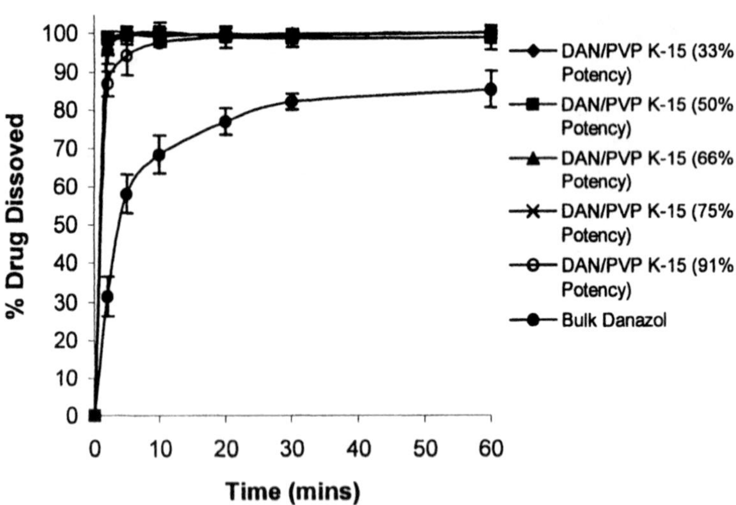

Figure 8. Dissolution of SFL DAN potency samples. (Reproduced with permission from ref. (16). Copyright 2004 Elsevier B.V.)

Dissolution profiles of the powders produced from the SFL process with increasing DAN potency can be seen in Figure 8. All of the powders had rapid dissolution rates, with complete dissolution within 2 minutes when the potency was less than 75%. Only the 91% DAN sample had a slightly slower dissolution rate with about 85% dissolving after 2 minutes and complete dissolution after 20 minutes. The slower dissolution rate was attributed to a decrease in surface area and increase in particle size compared to the other powders produced from the SFL process. However, by creating readily wettable, high surface area, amorphous DAN stabilized with PVP K15, SFL processed samples with high potencies (up to 91% DAN) had considerably faster dissolution rates compared to the bulk API.

Conclusion

Recent work in this laboratory has shown that the choice of solvent system containing the dissolved API and polymer can result in a variety of different particle morphologies, while maintaining the properties characteristic to particles produced by the SFL process. High potency formulations in organic and emulsion feed systems can offer increased API loading compared to aqueous-organic co-solvent systems, while maintaining rapid dissolution rates.

References

1. Lipinski, C. Poor aqueous solubility - an industry wide problem in drug discovery. *Am. Pharm. Rev.* **2002**, *5*, 82-85.
2. Amidon, G.L.; Lennernas, H.; Shah, V.P.; Crison, J.R. A theoretical basis for a biopharmaceutic drug classification - the correlation of in-vitro drug product dissolution and in-vivo bioavailability. *Pharm. Res.* **1995**, *12*, 413-420.
3. Dressman, J.B.; Reppas, C. In vitro-in vivo correlations for lipophilic, poorly water-soluble drugs. *Eur. J. Pharm. Sci.* **2000**, *11*, S73-S80.
4. Mura, P.; Cirri, M.; Faucci, M.T.; Gines-Dorado, J.M.; Bettinetti, G.P. Investigation of the effects of grinding and co-grinding on physicochemical properties of glisentide. *J. Pharm. Biomed. Anal.* **2002**, *30*, 227-237.
5. Mayersohn, M. In *Modern Pharmaceutics*; Banker, G.S. and Rhodes, C.T. Eds.; Drugs and the Pharmaceutical Sciences; Marcel Dekker, Inc.: New York, NY, 2002; Vol. 121, pp 23-66.
6. Rogers, T.L.; Johnston, K.P.; Williams III, R.O. Solution-based particle formation of pharmaceutical powders by supercritical or compressed fluid

CO_2 and cryogenic spray-freezing technologies. *Drug Dev. Ind. Pharm.* **2001**, *27*, 1003-1015.
7. Hu, J.H.; Johnston, K.P.; Williams III, R.O. Nanoparticle engineering processes for enhancing the dissolution rates of poorly water soluble drugs - a review. *Drug Dev. Ind. Pharm.* **2004**, *30*, 247-258.
8. Rasenack, N.; Muller, B.W. Micron-size drug particles: Common and novel micronization techniques. *Pharm. Dev. Technol.* **2004**, *9*, 1-13.
9. York, P. Strategies for particle design using supercritical fluid technologies. *Pharm. Sci. Tech. Today* **1999**, *2*, 430-440.
10. Chen, X.; Young, T.J.; Sarkari, M.; Williams III, R.O.; Johnston, K.P. Preparation of cyclosporine A nanoparticles by evaporative precipitation into aqueous solution. *Int. J. Pharm.* **2002**, *242*, 3-14.
11. Sarkari, M.; Brown, J.; Chen, X.; Swinnea, S.; Williams III, R.O.; Johnston, K.P. Enhanced drug dissolution using evaporative precipitation into aqueous solution. *Int. J. Pharm.* **2002**, *243*, 17-31.
12. Hu, J.H.; Rogers, T.L.; Brown, J.; Young, T.J; Johnston, K.P.; Williams III, R.O. Improvement of dissolution rates of poorly water soluble APIs using novel spray freezing into liquid technology. *Pharm. Res.* **2002**, *19*, 1278-1284.
13. Rogers, T.L.; Nelsen, A.C.; Hu, J.; Brown, J.N.; Sarkari, M.; Young, T.J.; Johnston, K.P.; Williams III, R.O. A novel particle engineering technology to enhance dissolution of poorly water soluble drugs: spray-freezing into liquid. *Eur. J. Pharm. Biopharm.* **2002**, *54*, 271-280.
14. Barron, M.K.; Young, T.J.; Johnston, K.P.; Williams III, R.O. Investigation of processing parameters of spray freezing into liquid to prepare polyethylene glycol polymeric particles for drug delivery. *AAPS PharmSciTech* **2003**, *4*, 90-102.
15. Rogers, T.L.; Nelsen, A.C.; Sarkari, M.; Young, T.J.; Johnston, K.P.; Williams III, R.O. Enhanced aqueous dissolution of a poorly water soluble drug by novel particle engineering technology: Spray-freezing into liquid with atmospheric freeze-drying. *Pharm. Res.* **2003**, *20*, 485-493.
16. Hu, J.; Johnston, K.P.; Williams III, R.O. Rapid dissolving high potency danazol powders produced by spray freezing into liquid process. *Int. J. Pharm.* **2004**, *271*, 145-154.
17. Rogers, T.L.; Hu, J.; Yu, Z.; Johnston, K.P.; Williams III, R.O. A novel particle engineering technology: spray-freezing into liquid. *Int. J. Pharm.* **2002**, *242*, 93-100.
18. Yu, Z.; Rogers, T.L; Hu, J.; Johnston, K.P.; Williams III, R.O. Preparation and characterization of microparticles containing peptide produced by a novel process: spray freezing into liquid. *Eur. J. Pharm. Biopharm.* **2002**, *54*, 221-228.

19. Pikal, M.J.; Shah, S. The Collapse Temperature in Freeze-Drying - Dependence on Measurement Methodology and Rate of Water Removal from the Glassy Phase. *Int. J. Pharm.* **1990**, *62*, 165-186.
20. Pikal, M.J.; Shah, S.; Roy, M.L.; Putman, R. The Secondary Drying Stage of Freeze-Drying - Drying Kinetics as a Function of Temperature and Chamber Pressure. *Int. J. Pharm.* **1990**, *60*, 203-217.
21. Searles, J.A.; Carpenter, J.F.; Randolph, T.W. The ice nucleation temperature determines the primary drying rate of lyophilization for samples frozen on a temperature-controlled shelf. *J. Pharm. Sci.* **2001**, *90*, 860-871.
22. Hu, J.H.; Johnston, K.P.; Williams III, R.O. Rapid release tablet formulation of micronized danazol powder produced by spray freezing into liquid (SFL). *J. Drug Del. Sci. Technol.* **2004**, *14*, 299-304.
23. Brunauer, S.; Emmett, P.H.; Teller, E. Adsorption of gases in multimolecular layers. *J. Am. Chem. Soc.* **1938**, *60*, 309-319.
24. Hu, J.; Johnston, K.P.; Williams III, R.O. Spray freezing into liquid (SFL) particle engineering technology to enhance dissolution of poorly water soluble drugs: organic solvent versus organic/aqueous co-solvent systems. *Eur. J. Pharm. Sci.* **2003**, *20*, 295-303.
25. Rogers, T.L.; Overhoff, K.A.; Shah, P.; Santiago, P.; Yacaman, M.J.; Johnston, K.P.; Williams III, R.O. Micronized powders of a poorly water soluble drug produced by a spray-freezing into liquid-emulsion process. *Eur. J. Pharm. Biopharm.* **2003**, *55*, 161-172.
26. Myers, D. *Surface Science and Technology*; VCH Publishers, Inc.: New York, NY, 1992; pp 292-298.
27. Horter, D.; Dressman, J.B. Influence of physicochemical properties on dissolution of drugs in the gastrointestinal tract. *Adv. Drug Del. Rev.* **2001**, *46*, 75-87.
28. Cilurzo, F.; Minghetti, P.; Casiraghi, A.; Montanari, L. Characterization of nifedipine solid dispersions. *Int. J. Pharm.* **2002**, *242*, 313-317.
29. Wang, X.; Michoel, A.; Van den Mooter, G. Study of the phase behavior of polyethylene glycol 6000-itraconazole solid dispersions using DSC. *Int. J. Pharm.* **2004**, *272*, 181-187.

Chapter 21

Preparation of Nanostructured Particles of Poorly Water Soluble Drugs via a Novel Ultrarapid Freezing Technology

J. C. Evans[1], B. D. Scherzer[1], C. D. Tocco[1], G. B. Kupperblatt[1,3], J. N. Becker[1], D. L. Wilson[1], S. Saghir[2], and E. J. Elder[1]

[1]Dowpharma and [2]Toxicology and Environmental Research and Consulting, The Dow Chemical Company, Midland, MI 48674
[3]Current address: Mylan Technologies, 110 Lake Street, St. Albans, VT 05478

Many drugs are rejected or have less-than-optimal performance because of poor water solubility. This research evaluated the dissolution rate and relative bioavailability of the poorly water soluble drug ketoconazole following processing via an ultra-rapid-freezing, particle-engineering technology. Dissolution of ketoconazole USP powder reached approximately 62% dissolved in 30 minutes. By comparison, the modified powders were 96% dissolved within 2 minutes. Improved bioavailability was demonstrated by a 4- to 7-fold increase in AUC with a corresponding 2- to 3.5-fold increase in Cmax.

Introduction

Over 40% of potential drugs are rejected because of poor water solubility (*1*). Of the drugs currently on the market, approximately 17% have less-than-optimal performance because of poor solubility and low bioavailability. Increasing the rate of dissolution may favorably impact the performance of poorly water soluble drugs (*2*).

An ultra-rapid-freezing, particle-engineering technology is under development for preparation of nanostructured particles that enhance the aqueous dissolution rate and bioavailability of poorly water soluble drugs. This process uses pharmaceutically acceptable solvents and excipients and conventional process equipment and thus is fast and scalable. The nanostructured particles have the following features:

- Particles have varying degrees of crystallinity depending on excipients and solvent.
- Particle surface area is greatly enhanced.
- Polymer absorption on the crystal surface upon freezing aids reduction of Ostwald ripening.
- Particles are isolated as agglomerates.

The purpose of the work discussed in this paper was to determine the effect that the ultra-rapid-freezing process has on the *in vitro* dissolution and *in vivo* bioavailability of the model drug ketoconazole.

Materials and Methods

Materials

Ketoconazole USP is an antifungal drug with a melt point of 150°C and particle size of ~25 µm. Aqueous solubility is 0.017 mg/mL at 25°C. Dosage form is 200-mg tablets. Poloxamer 407 and 338 are GRAS-listed stabilizers used as a crystal growth inhibitor. *t*-Butanol was used as the solvent.

Process Description

Figure 1 is an illustration of the ultra-rapid freezing process. The model drug is dissolved in a water-miscible or anhydrous solvent along with a stabilizer acting as a crystal growth inhibitor. The drug/stabilizer solution is then applied to a cryogenic substrate. The solvent is removed by lyophilization or atmospheric freeze-drying, resulting in highly porous, agglomerated particles.

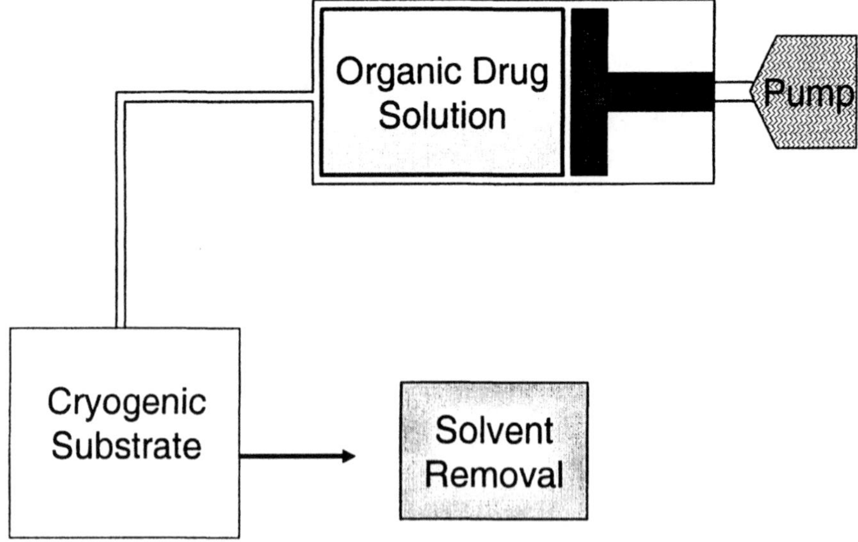

Figure 1. Process diagram for ultra-rapid freezing technology.

Characterization

The dried nanostructured particles were characterized by BET surface area analysis, scanning electron microscopy (SEM), x-ray diffraction (XRD), and particle size analysis. Dissolution rates were determined on the as-received ketoconazole USP powder and the modified ketoconazole using USP apparatus II (paddles) and media containing 0.5% sodium lauryl sulfate in water.

Bioavailability

For bioavailability determinations, a modified ketoconazole drug powder consisting of 50% ketoconazole and 50% poloxamer 407 was prepared by ultra-rapid freezing and isolated with atmospheric freeze-drying. Capsules were prepared by hand-filling size 0 gelatin capsules with the modified ketoconazole powder. Tablets were prepared from a 47.5/47.5/5 blend of drug powder, microcrystalline cellulose, and Na carboxymethylcellulose using direct compression techniques. Standard concave tablets (5/16 inch) were prepared on a hydraulic lab press. The fill weight was 150 mg, the compaction force was 200 lb, and dwell time was 10 seconds. Magnesium stearate suspended in acetone

(1% w/w) was used to lubricate tooling surfaces. Tablets were characterized by disintegration time, and dissolution.

The relative bioavailability in beagle dogs of both the capsules and tablets containing modified ketoconazole was compared to 000 gelatin capsules containing a 5% hypromellose (HPMC) suspension (1.3 mL) of ketoconazole USP (78 mg/mL, as-received). The dogs were dosed with two 000 capsules containing the 5% hypromellose/ketoconazole suspension, for a total dosage of 203 mg (78x2x1.3).

Results and Discussion

Surface Morphology

The SEMs of freeze-modified ketoconazole powder revealed a greatly enhanced surface area in comparison to that of the bulk ketoconazole (Figure 2). Particle size was reduced from about 25 μm to several microns. BET surface area increased from 2.6 m^2/g for bulk ketoconazole to 8.95 m^2/g for atmospheric freeze-dried particles of modified ketoconazole and 9.23 m^2/g for lyophilized particles.

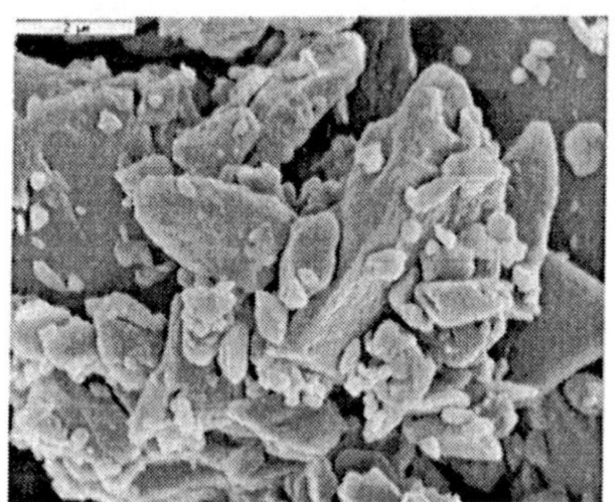

Bulk ketoconazole — Freeze-modified ketoconazole

Figure 2. Scanning electron micrographs of bulk ketoconazole and ketoconazole prepared by ultra-rapid freezing. (Adapted with permission from reference 3. Copyright 2004Drug Delivery Technology LLC.)

X-ray Diffraction

The high peak intensities indicated a high degree of crystallinity for the freeze-modified ketoconazole powder (Figure 3). Because crystallinity is dependent on the solvent and stabilizer used, ultra-rapid freezing technology has the capability of modifying crystallinity by changing solvent and/or stabilizer. This, in turn, can affect drug dissolution.

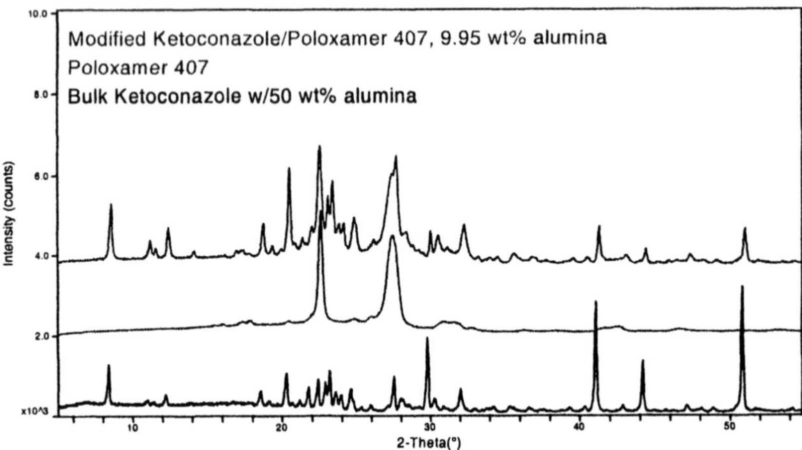

Figure 3. X-ray diffraction analysis of bulk ketoconazole, poloxamer 407, and freeze-modified ketoconazole. (Reproduced with permission from reference 3. Copyright 2004Drug Delivery Technology LLC.)

Dissolution

Dissolution of ketoconazole USP powder and a physical blend of powder and excipients reached approximately 62% in 30 minutes (Figure 4). By comparison, the freeze-modified ketoconazole powder was 96% dissolved within 2 minutes.

Effect of Stabilizer Level on Dissolution

Figure 5 shows the effect of decreasing stabilizer (poloxamer 338) on the dissolution of freeze-modified ketoconazole. Once the level of poloxamer 338 is below about 30% there is not enough stabilizer to prevent particle growth.

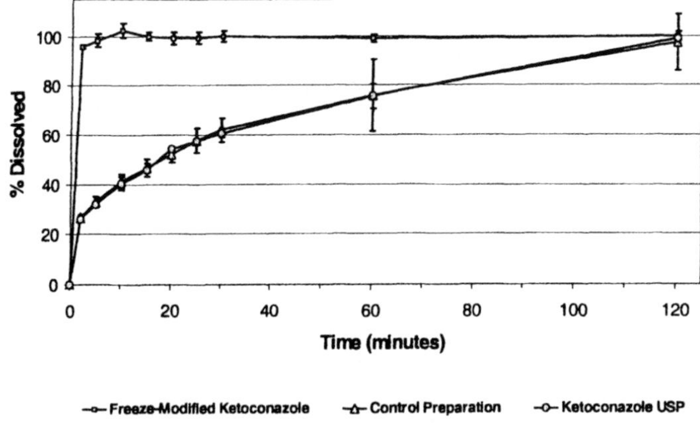

Figure 4. Dissolution profiles for freeze-modified ketoconazole, bulk ketoconazole, and a physical blend of bulk ketoconazole and excipients.

Destabilization of the nanoparticles was observed when the drug loading was increased from 70 to 80%, resulting in a decrease in particle size, decrease in surface area, and decrease in dissolution rate.

Bioavailability

The data shown in Figure 6 indicate that freeze-modified ketoconazole could lead to a dosage reduction in comparison to HPMC suspension. The freeze-modified ketoconazole delivered in capsules showed improved bioavailability with an almost 7-fold increase in AUC and correspondingly an almost 3.5-fold increase in Cmax compared to the HPMC suspension of drug as-received. The freeze-modified ketoconazole delivered in tablets showed improved bioavailability with an almost 4-fold increase in AUC and correspondingly an almost 2-fold increase in Cmax compared to the HPMC suspension of drug as-received. It should be noted that the ketoconazole tablets contained only 133 mg/mL of active in comparison to the 203 mg/mL delivered in the HPMC suspension. Characterization of the tablets showed 9 min disintegration time and dissolution of 100% in 15 min.

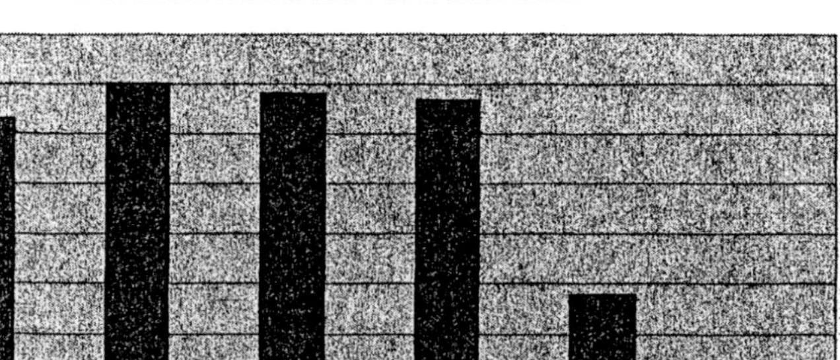

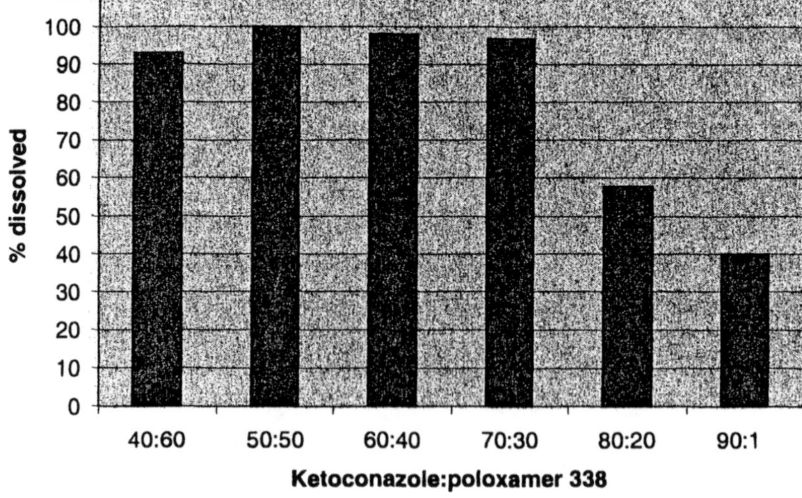

Figure 5. Effect of poloxamer 338 level on dissolution of freeze-modified ketoconazole.

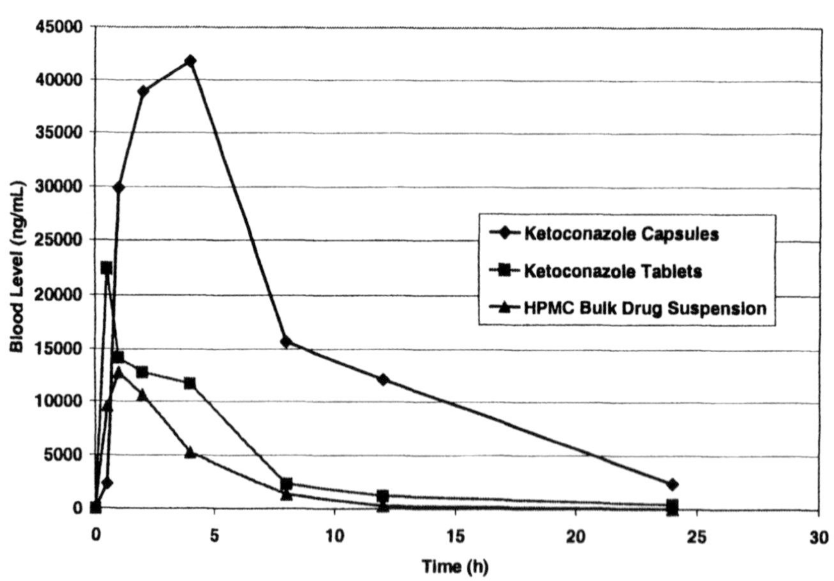

Figure 6. Blood level profiles for ketoconazole in male beagle dogs (n=4).

Effect of Process Scale

Bioavailability studies were conducted on freeze-modified ketoconazole at both laboratory and pilot process scales. SEMs of the resulting particles showed roughly equivalent particle development (Figure 7).

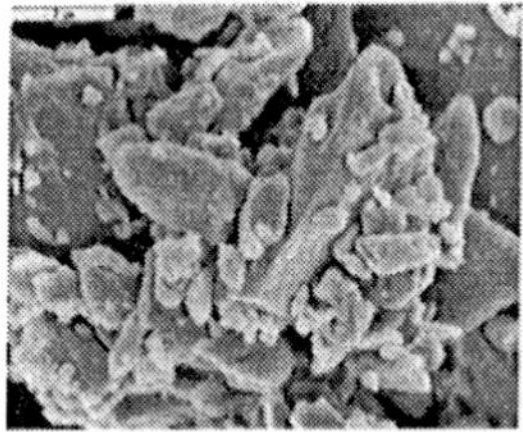

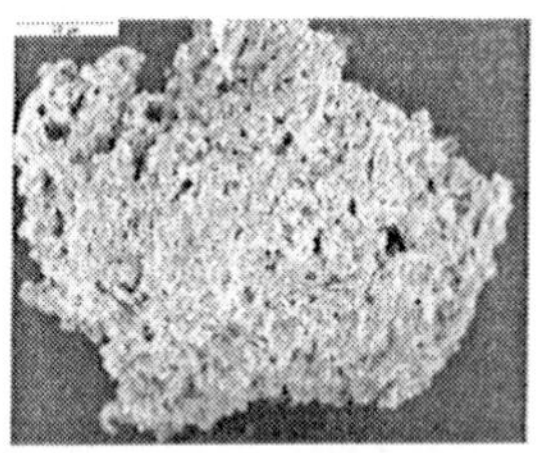

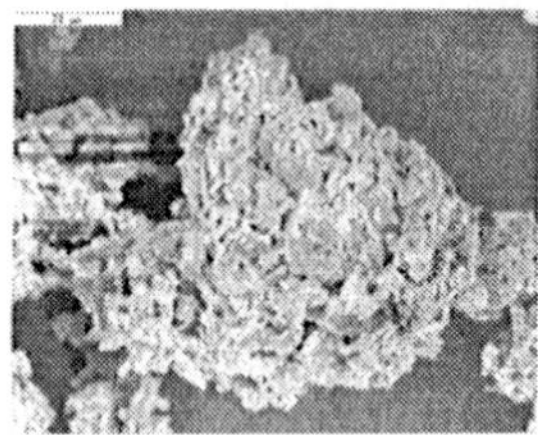

Bulk ketoconazole

Freeze-modified ketoconazole lab-scale (grams)

Freeze-modified ketoconazole pilot-scale (kilos)

Figure 7. Scanning electron micrographs of bulk ketoconazole and freeze-modified ketoconazole at lab and pilot scales. . (Adapted with permission from reference 3. Copyright 2004Drug Delivery Technology LLC.)

Figure 8 shows the comparison of AUC values for capsules containing either as-received ketoconazole or freeze-modified ketoconazole at either a laboratory (gram) or pilot (kilo) scale. The AUC curves are roughly equivalent for laboratory- and pilot-scale freeze-modified ketoconazole, indicating no deleterious effects of scale-up.

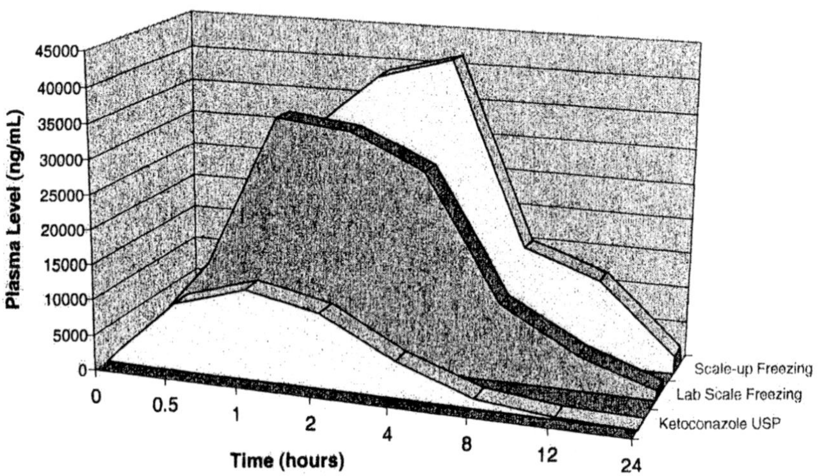

Figure 8. Mean plasma levels of ketoconazole in beagle dogs. . (Reproduced with permission from reference 3. Copyright 2004Drug Delivery Technology LLC.)

Conclusions

Ultra-rapid freezing technology was demonstrated to be a viable particle engineering tool for enhancing the bioavailability of poorly water soluble drugs. Demonstration with ketoconazole resulted in substantial improvements to *in vitro* dissolution and *in vivo* bioavailability. Scale-up of the process from gram quantities to kilos did not show any deleterious effects.

Acknowledgements

The authors gratefully acknowledge the laboratory support contributions provided by Analytical Sciences and BioAqueous^{SM} Solubility Services at Dow.

BioAqueous is a Service Mark of The Dow Chemical Company.

References

1. Bassett, P. Overcoming Formulation Challenges - Novel Techniques for Optimum Results, *D&MD Reports*, **2002**, July.
2. *Delivery of Poorly Soluble or Poorly Permeable Drugs* (Market research report), 4th ed. Falls Church, VA: Technology Catalysits International Corporation; August **2003**.
3. Connors, R. C; Elder, E. J. *Drug Delivery Technology* **2004**, 4, 78-83.

Chapter 22

Water-Soluble Polymer Formation of Monodispersed Insulin Microspheres

Larry Brown*, Yuanxi Qin, Kenneth Hogeland, John McGeehan, Kathy Mason, Travis Jeannotte, Robert Fortier, and Eric Moore

Baxter Healthcare Corporation's wholly owned subsidiary, Epic Therapeutics, Inc., 220 Norwood Park South, Norwood, MA 02062
***Corresponding author: larrybrown_baxter.com**

The PROMAXX formulation process of proteins by precipitation in the presence of water-soluble polymers is a potentially important advance for producing homogeneous, stable protein microspheres for inhalation therapy and other routes of drug delivery. We have applied this technology to produce microspheres from a large class of molecules, including low molecular weight compounds, peptides, proteins, antibodies, DNA and other oligonucleotides. The key feature of the PROMAXX process is that the microsphere formation is accomplished in aqueous solution, which eliminates the potential for protein denaturation in the presence of organic-aqueous interfaces often encountered in other microsphere technologies. Indeed, PROMAXX insulin microspheres yield a formulation with increased protein stability. Furthermore, the microspheres were produced in a size range which was shown to reach the deep lung in beagle dogs. The preservation of insulin bioactivity indicates that PROMAXX is a promising new formulation technology for protein and peptide therapeutics.

Introduction

The precipitation of proteins in the presence of nonionic water-soluble polymers has been described by many investigators in the scientific literature. The interaction of dextran with proteins such as fibrinogen and other proteins was described in 1952 *(1,2)*. For example, poly(vinylpyrolidone), (PVP), was used to precipitate Factor VIII and fibrinogen *(3)*. Poly(ethylene glycol), (PEG), has been extensively studied as a precipitating agent for proteins *(4)*. These publications focused on the conditions and mechanisms of the straightforward precipitation of proteins in aqueous solutions. However, this process of combining proteins and water-soluble polymers such as PVP or PEG typically resulted in the formation of an amorphous precipitate of no defined shape or size (Figure 1). In contrast to these amorphous precipitates, research in our laboratory has shown that we can reproducibly form discrete, monodisperse, protein microspheres by controlling the ionic strength, pH, polymer concentration, protein concentration and temperature of the aqueous protein polymer mixture.

Previously, we have shown that the PROMAXX process formed protein microspheres with very narrow particle size ranges *(5)*. These protein microspheres can be fabricated for sustained or immediate release *(5,6)*. In this article, we describe the formulation of PROMAXX insulin microspheres in the 1 to 2 micron particle size range. This particle size range was chosen in order to allow deep lung delivery of the therapeutic insulin molecule. We characterize the insulin microspheres by particle size, aerodynamic performance, long term stability and bioactivity *in vivo*.

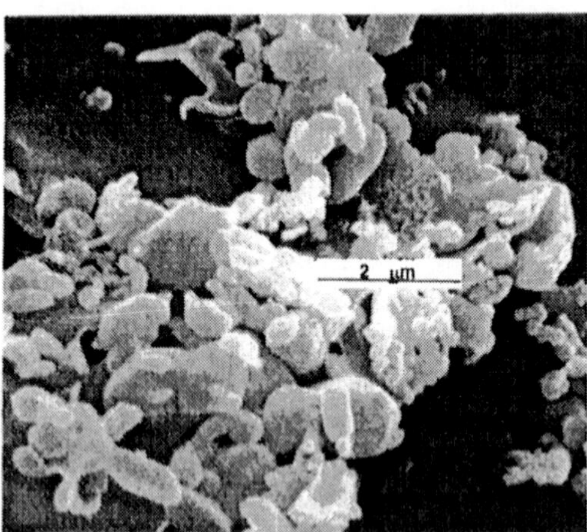

Figure 1. SEM of an amorphous mass of insulin, precipitated in the presence of PEG in aqueous solution.

Materials and Methods

Figure 2 shows crystalline recombinant human zinc insulin (Calbiochem, LaJolla, CA) being formulated into PROMAXX microspheres. The insulin crystals were suspended in deionized water. PEG 3350 polymer (Spectrum Pharmaceuticals, Gardena, CA) was dissolved at pH 5.6 in a controlled ionic strength, aqueous solution in a glass, temperature controlled, water-jacketed liquid chromatography column. The insulin suspension was added to and dissolved in the PEG solution. The insulin/polymer solution was then cooled, which subsequently resulted in the formation of a turbid suspension. The turbid suspension was diafiltered to remove the PEG and lyophilized to dryness (Virtis, Gardiner, NY).

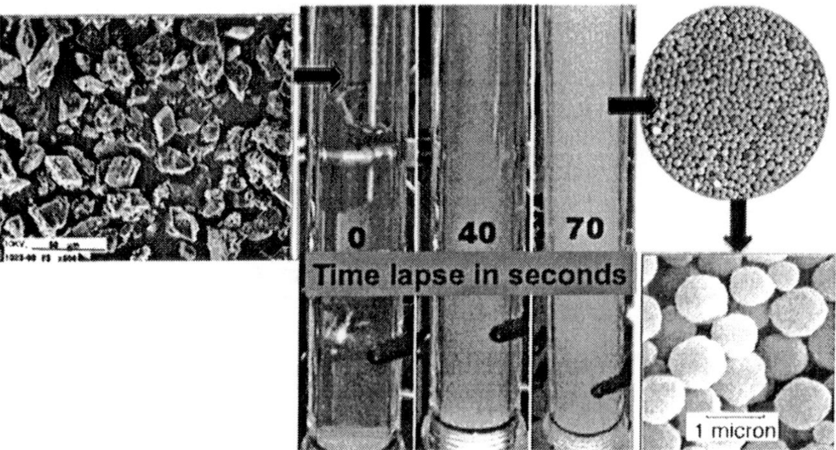

Figure 2. Transformation of zinc insulin crystals into PROMAXX microspheres is shown by first dissolving the insulin in a solution of PEG 3350 and then cooling the solution over 70 seconds. The microspheres are virtually all insulin.

The lyophilized microspheres were analyzed by scanning electron microscopy (AMRAY 1000, Bedford, MA), light scattering particle size analysis (Beckman Coulter LS230, Miami, FL), aerodynamic particle size determination (TSI Aerosizer 3225, St. Paul, MN) and the USP assay for insulin *(7)*. Insulin microsphere powder was also incorporated into Capsugel Vcaps #3CS capsules (Capsugel, Morris Plains, NJ) for testing, using a dry powder inhaler (DPI). Cascade impactor particle size distribution studies were conducted using a Thermo Andersen eight stage non-viable cascade impactor, series 20-800 Mark

II, in a 60-liter per minute configuration (Smyrna, GA). Air flow and actuation time were controlled using a Copley Instruments Critical Flow Controller (Copley Scientific Limited, Nottingham, UK). The microspheres were introduced into the cascade impactor using a Cyclohaler DPI (Pharmachemie, Haarlem, The Netherlands) as shown in Figure 3. Biological activity was shown by the administration of PROMAXX insulin powder to non-diabetic beagle dogs. The presence of deamidated insulin, insulin dimers and oligomers as well as related compounds was determined by HPLC (Waters, Milford, MA) in the lyophilized insulin starting material and compared to lyophilized insulin microspheres. The comparison was conducted at 25 and 37 °C over a 12-month period. Residual PEG was determined by size exclusion chromatography (SEC) and evaporative light scattering detection (ELSD).

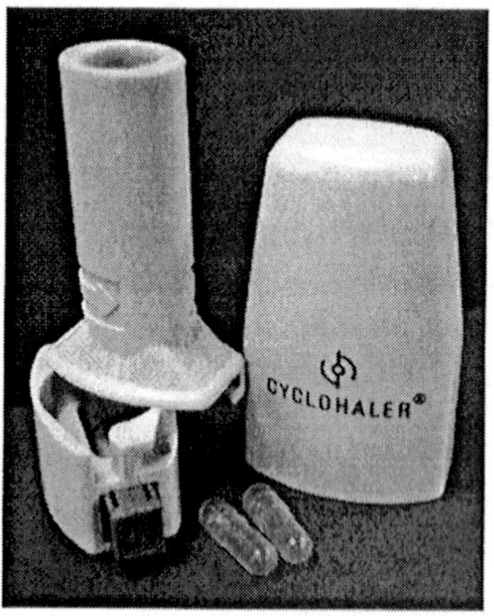

Figure 3. Cyclohaler Dry Powder Inhaler used in the Andersen Cascade Impactor Studies.

Results and Discussion

In Vitro Studies

The particle size of the insulin microspheres was determined by several methods. The SEM image in Figure 2 clearly illustrates that the particle size of the insulin microspheres was approximately 1 micron in diameter and that the insulin precipitates are spherical in shape.

Figure 4 quantifies the particle diameter by two additional analytical methods. Light scattering particle size analysis using a Beckman Coulter LS230 showed a remarkable superimposition of the percent number, percent surface area and percent volume particle size determinations for the PROMAXX microspheres. This observation is indicative of a monodisperse distribution of particles in the PROMAXX insulin microsphere preparation. These studies revealed that 95% of the microspheres were between 0.95 and 1.2 microns in diameter. Furthermore, the TSI Aerosizer device, which uses a time-of-flight method to determine particle aerodynamic diameter, showed the microspheres to be 1.47 microns in diameter.

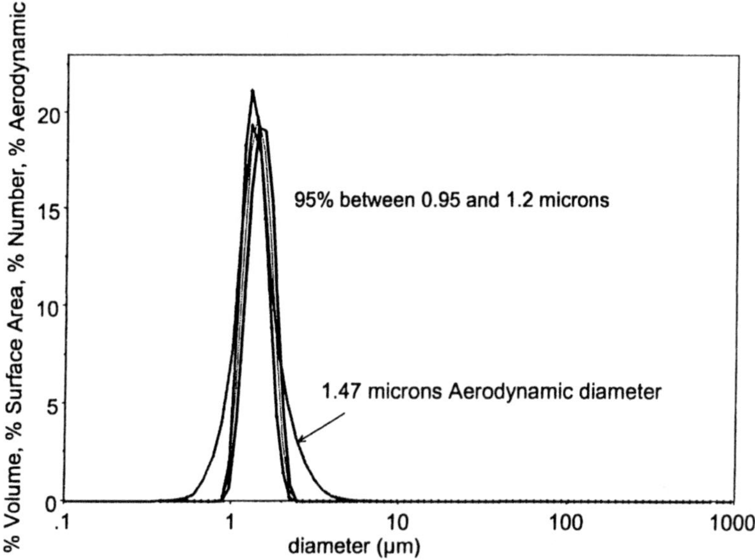

Figure 4. Light scattering and time-of-flight quantitative distribution of insulin microspheres of approximately 1 to 2 microns in size.

The cascade impactor studies were conducted to show the actual performance of the PROMAXX insulin microspheres when delivered from the Cyclohaler DPI. 10 mg of PROMAXX insulin was loaded into the capsule and then placed into the Cyclohaler DPI. A flow rate of 60 liters per minute was used in this study. The Cyclohaler was activated to release the insulin for collection onto the impactor's stages. The cascade impactor stages were disassembled, the powder on each stage was collected and submitted for HPLC analysis. Figure 5 shows that greater than 75% of the insulin microspheres deposited on the Andersen stages, corresponding to a particle size of 1.1 to 3.3 microns. Particles in this size would be predicted to deposit in the alveoli of the lung and be available for systemic absorption *(8)*.

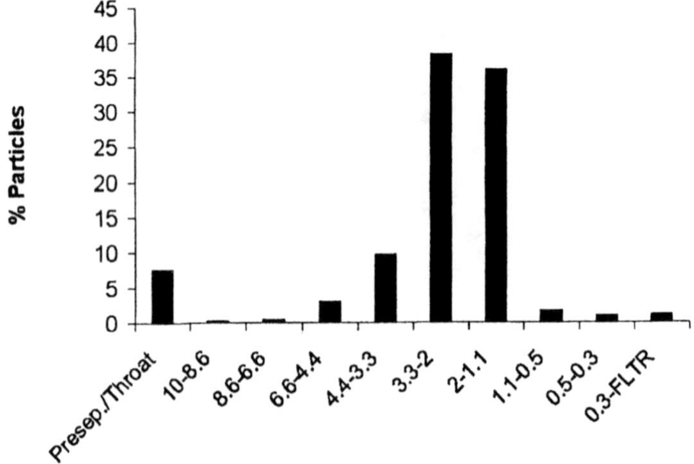

Figure 5. Andersen Cascade Impactor Results for PROMAXX insulin delivered from the Cyclohaler DPI.

In Vivo Studies

A single dose forced-maneuver inhalation and subcutaneous study in five female beagle dogs was conducted to evaluate the bioactivity of the PROMAXX insulin. The average weight of the beagle dogs was 9.5±0.5 kg. Dry powder insulin was administered via inhalation at two dose levels (1.5 and 0.5 mg) and Humulin R insulin (Eli Lilly, Indianapolis, IL) was administered via sub-cutaneous injection at a dose of 0.1 mg per animal (0.35 IU/kg body weight). All five dogs received the 3 doses of insulin in repeat dose fashion with at least a three day wash-out period between doses.

Figure 6 shows a significant decrease in blood glucose within 10 minutes after administration. Glucose levels remained depressed for approximately 200 minutes. Inhaled insulin was absorbed significantly more rapidly than the subcutaneous dose of insulin. The serum glucose concentrations of the inhaled insulin groups decreased faster than the subcutaneous group of animals. Serum glucose concentrations remained depressed below normal for 95 minutes for the

0.5 mg inhaled dose and for approximately 180 minutes for the 1.5 mg inhaled dose. The 0.1 mg subcutaneous dose remained depressed for 100 minutes after insulin administration.

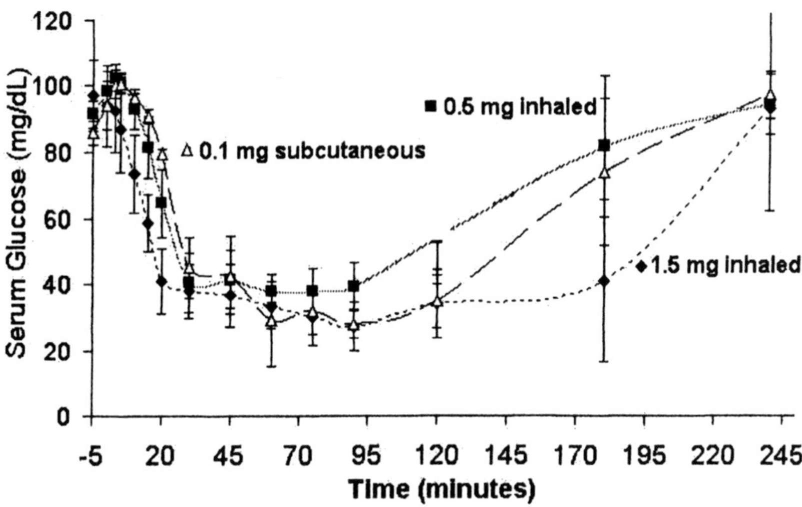

Figure 6. Efficacy of PROMAXX insulin is shown for the 0.5 (■) and 1.5 mg (♦) inhaled dosage forms. The initial serum glucose was measured 5 minutes before dosing ("-5 minutes"). The insulin dose was administered at 0 minutes. The inhaled dosage forms reduced the serum glucose more rapidly than the 0.1 mg subcutaneous dose (Δ).

Stability Indicating Studies

The stability-indicating assays for the insulin samples, designed to show the generation of insulin dimers, desamido-insulin and total insulin-related compounds, were conducted using the size exclusion chromatography or reverse phase HPLC as specified in USP 25 official monographs for human insulin *(7)*. Figure 7 shows that at 37°C, the PROMAXX microspheres demonstrated a significantly lower total percentage of dimer formation than the insulin starting material used to form these microspheres.

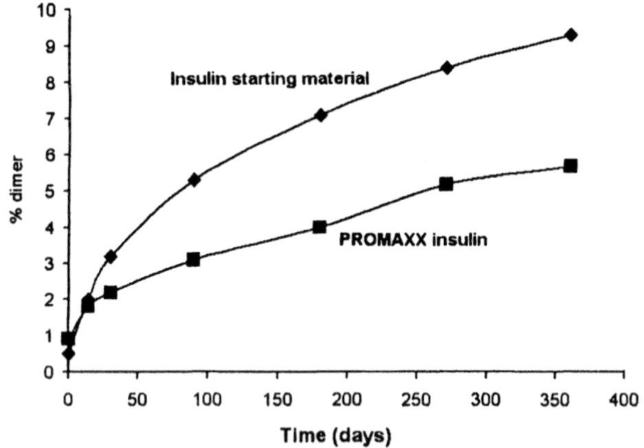

Figure 7. Dimer formation of PROMAXX insulin compared to lyophilized insulin starting material (stored at 37°C). The PROMAXX insulin showed approximately one half the dimer formation after one year.

Figure 8 shows the formation of desamido-insulin in the insulin starting material and the PROMAXX insulin microcapsules. The deamindated insulin was significantly reduced in the PROMAXX insulin microsphere group compared to the insulin starting material stored at 37°C.

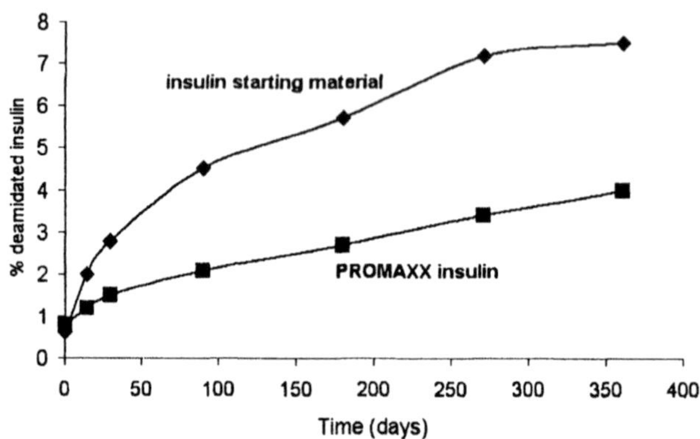

Figure 8. Desamido-insulin formation of PROMAXX insulin compared to lyophilized insulin starting material (stored at 37°C). The PROMAXX insulin showed approximately one half the desamido-insulin compound formation after one year.

The total insulin related compounds is shown in Figure 9. Again the data clearly indicates that higher molecular weight insulin related compounds are reduced in the PROMAXX insulin group compared to the lyophilized starting insulin.

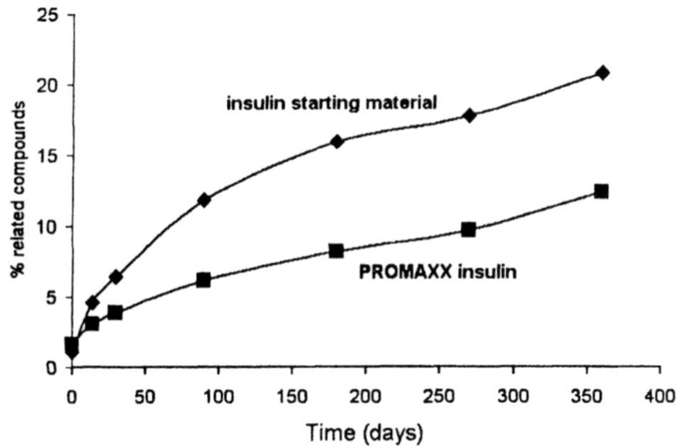

Figure 9. Total insulin related compound formation of PROMAXX insulin is compared to lyophilized insulin starting material (stored at 37°C). The PROMAXX insulin showed approximately one half the total insulin related compound formation after one year.

The higher stability of PROMAXX insulin with respect to insulin dimers, desamido-insulin and total high molecular weight insulin related compounds compared to lyophilized starting material was repeated for several lots of insulin. Corresponding results, i.e., significantly less deamidated insulin, fewer insulin dimers and less total related compounds, were found for samples stored at 25°C. Additional studies in our laboratory indicated that commercial insulin formulations generated up to 15-times more high molecular weight insulin related compounds compared to PROMAXX insulin microspheres stored at 37°C. Further analysis indicated that the weight percent insulin of the PROMAXX microspheres is higher than 96% by weight zinc insulin. Size exclusion chromatography indicated that the residual weight percent PEG in the microspheres was less than 0.2% in these early formulations. The remainder of the microsphere is residual water.

Conclusions

PROMAXX insulin microspheres were produced with a novel and important advance, which may be related to the process of protein precipitation in the presence of water-soluble polymers. However, unlike the volume exclusion process of protein precipitation, the PROMAXX process resulted in the formation of discrete spherical particles 1 to 2 microns in diameter. These microspheres were shown to be comprised of virtually all insulin. Analyses of these microspheres exhibited a narrow particle size distribution by SEM, light scattering, time of flight measurements and by the Andersen Cascade Impactor. Additional testing of the PROMAXX insulin powder delivered from the Cyclohaler DPI showed that the formulation was potentially suitable for systemic pulmonary delivery of proteins *(9)*. These lyophilized insulin microspheres also showed enhanced stability over a 12-month time period under accelerated storage conditions, compared to starting material in the absence of stabilizing agents, added to the formulation. The PROMAXX process must, therefore, be precipitating the insulin in a manner and form which inhibits the formation of insulin dimers and desamido-insulin. The preservation of the insulin's biological activity was shown in a glucose depression assay in beagle dogs.

Acknowledgements

This work was funded by Baxter Healthcare Corporation's wholly owned subsidiary, Epic Therapeutics, Inc. The authors acknowledge the assistance of William Fowle of Northeastern University Biology Department for SEM analysis. The contributions of Roxanne Lau, Karen Kuzmich and Kathleen Boutin are also very much appreciated.

References

1. Rickets, C.R. Interaction of dextran with fibrinogen. *Nature* **1952**, *169*, 970.
2. Laurent, T.C. The interaction between polysaccharides and macromolecules 5. The solubility of proteins in the presence of dextran. *Biochem. J.* **1963**, *89*, 253-257.
3. Casillas, G.; Simonetti, C. Polyvinylpyrrolidone (PVP): A new precipitating agent for human and bovine factor VIII and fibrinogen. *British J. Haemat.* **1982**, *50*, 665-672.

4. Atha D.H.; Ingham K.C. Mechanism of precipitation of proteins by polyethylene glycols. Analysis in terms of excluded volume. *J. Biol. Chem.* **1981**, *256*, 12108-12117.
5. Brown, L.; Jarpe, M.; McGeehan, J.; Qin, Y.; Moore, E.; Hogeland, K. PROMAXX microsphere characterization. *Resp. Drug. Del.* **2004**, *IX*, 477-480.
6. Rashba-Step, J.; Brown, L.; Hogeland, K.; McGeehan, J.; Proos, R.; Rulon, P.; Sullivan, A.; Scott, T. Albumin microspheres as a drug delivery vehicle for multiple routes of administration. *Proc. Int. Symp. Control. Rel. Bioact. Mater.* **2001**, *28*, 1001-1002.
7. *USP 25, Official Monographs* **2002**, p. 911.
8. Byron, P.R. Prediction of drug residence times in regions of the human respiratory tract following aerosol inhalation. *J. Pharm. Sci.* **1986**, *75*, 433-438.
9. Skyler, J.S.; Cefalu, W.T.; Kourides, I.A.; Landschulz, W.H.; Balagtas C.C.; Cheng, S.; Gelfand, R.A. Efficacy of inhaled human insulin in type 1 diabetes mellitus: A randomized proof-of-concept study. *The Lancet* **2001**, *357*, 331-335.

Chapter 23

Preparation of Fast-Dissolving Tablets Based on Mannose

Yourong Fu, Seong Hoon Jeong, Jacqueline Callihan, Jeanny Kim, and Kinam Park*

Departments of Pharmaceutics and Biomedical Engineering, Purdue University, West Lafayette, IN 47907
***Corresponding author: kpark@purdue.edu**

Recent developments in fast dissolving (or fast disintegrating) tablets have brought convenience in dosing to elderly and children, who have trouble swallowing tablets. Fast-dissolving tablets dissolve or disintegrate in the mouth without any extra fluid, and so they are highly useful for those who need to take medicine in the absence of water. The key properties of fast-dissolving tablets are fast absorption of water into the core of tablets and disintegration of associated particles into individual components for fast dissolution. The strategy of making fast-dissolving tablets presented in this study is based on using carbohydrates that have extremely high water solubility. D-Mannose is a naturally occurring sugar with aqueous solubility of 2.5 mg/ml that can be compressed at a low pressure. Therefore, mannose was chosen as the main excipient. Tablets with high porosity were made in a 3-step process. First, mannose powder was compressed into a tablet with reasonable strength. Second, this compact was exposed to relative humidity higher than the critical relative humidity of mannose to absorb water. Third, tablets were dried to gain mechanical strength. The disintegration mechanism was studied.

Introduction

Oral administration of a tablet is the most popular dosage form; however, parts of the populations such as children and the elderly have difficulties in swallowing tablets and capsules *(1)*. Fast-dissolving tablet technologies, as a new dosage form in pharmaceutical industry, have been gaining attention recently. Fast-dissolving tablets disintegrate or dissolve in the mouth without requiring extra fluid to help swallowing. Moreover, they are very useful for those who need to take medicine without immediate access to water. The name "fast-dissolving" indicates that these tablets dissolve quickly, and it implies that the tablets disintegrate into smaller particles. Fast-dissolving tablets combine the advantages of liquid dosage form, i.e., convenient drug administration, and solid dosage form, i.e., easy handling and accurate dosing. When a fast-dissolving tablet is administered, it disintegrates or dissolves in the saliva and is swallowed into the stomach. The time to reach from the mouth to the stomach is estimated to be between 5 and 10 minutes *(2-4)*. This fast passage to the stomach provides a better opportunity for the medication to be absorbed through the membrane of the buccal cavity, pharynx, and esophagus for improved bioavailability *(5,6)* and quick onset of drug action.

Currently, fast-dissolving tablets are prepared by several methods, such as freeze-drying, molding, sublimation, and direct compression. Each method has its own advantages and limitations. Common approaches to making fast-dissolving tablets are maintaining high porosity of the tablet matrix, incorporation of highly water-soluble excipients in the formulation, and addition of quick disintegrating agents *(7-10)*. Superporous hydrogel particles have been shown to be suitable disintegrating agents for fast-dissolving tablets *(11)*. Regardless of the preparation methods, the key properties for fast-dissolving tablets are fast absorption or wetting of water into the tablets, followed by disintegration of associated particles into individual components. These properties require excipients with high wettability, and a tablet structure consisting of a highly porous network. Since the strength of a tablet is directly related to the compression pressure, while porosity is inversely related to the compression pressure, it is important to find the porosity that allows fast water absorption while maintaining high mechanical strength.

The strategy of making fast-dissolving tablets presented in this study involves using a highly water-soluble carbohydrate as the main excipient. After the initial screening of a large number of candidate materials, D-mannose was chosen as the main excipient. D-Mannose is a naturally occurring carbohydrate with an aqueous solubility of about 2.5 g/ml. The compressibility of mannose at low pressure was observed to be very reasonable. For these properties, mannose was investigated as the main excipient for making fast-dissolving tablets.

Figure 1. Chemical structure of D-mannose and an SEM picture of D-mannose particles.

Figure 1 shows the molecular structure of D-mannose and a scanning electron microscopy (SEM) image of the raw material. As seen in the SEM image, mannose powder has a highly porous structure, which may help generating large surface areas when the powder is compressed into tablets. The pores between crystals are expected to allow fast absorption of water by capillary force. The particles have spherical shape and possess good flowability. To prepare tablets with high porosity, the following three steps process was applied. First, the mannose powder was compressed into a tablet at relatively low pressure. Second, the compact was placed in a humidity chamber with relative humidity higher than the critical relative humidity of the mannose. Third, the tablets gained significant mechanical strength after they were taken out from the humidity chamber and dried. The disintegration mechanism of the mannose tablet and factors affecting tablet strength were studied.

The effects of moisture on the properties of tablets, including appearance, color, hardness, disintegration, dissolution and bioavailability, were investigated in detail. It was found that the tablet hardness increased when the tablet first gained moisture and subsequently lost it during storage *(12)*. Hardness, disintegration and dissolution of the tablets did not change much on exposure to ambient room conditions. However, after equilibration under high humidity, an initial decrease in tablet hardness occurred. After drying, the moisture-treated tablets increased their hardness again, which exceeded the initial hardness level *(13,14)*. It appears that the partial loss of moisture induced recrystallization of D-mannose, which was dissolved in the absorbed moisture on the surface of the particles *(15)*. The binders used in the tablet preparation are also known to increase the tablet strength after partial loss of moisture during drying *(16)*. The objective of this research was to develop a new fast-dissolving tablet formulation based on mannose as the main ingredient.

Materials and Methods

Materials

Mannose powder was purchased from Hofman International Inc. (Calgary, Canada).

Preparation of Tablets

Tablets were compressed on a single punch Carver Laboratory Press (Carver Inc., Wabash, IN) at different compression pressures, using plane-face punches with diameter of 0.5 inch.

Moisture Treatments

The prepared tablets were placed in a Drykeeper desiccator (Sanplatec Corp., Osaka, Japan) with 75% RH at 25°C, which was created by placing a saturated sodium chloride solution in the Drykeeper desiccator. The tablets were taken out after 4 hours and air dried for 8 hours at 25 °C or at room temperature. The tablet hardness and the disintegration time were measured.

Disintegration Test

Fast-dissolving tablets are supposed to disintegrate in the mouth by saliva. The amount of saliva is limited and no simulated tablet disintegration test in the mouth was found in United States Pharmacopeia. Since it is difficult to apply a general disintegration test to reflect real conditions, a new simple testing device was designed as shown in Figure 2 to evaluate the disintegration times of various formulations.

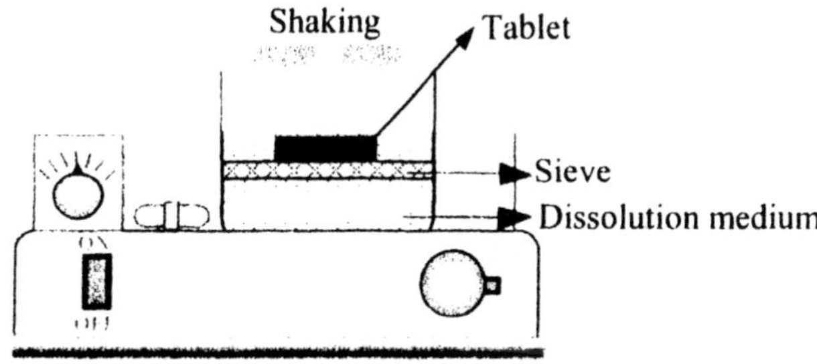

Figure 2. A device for disintegration testing of fast-dissolving tablets.

A 10-mesh sieve was placed inside a glass cylinder in such a way that 2 ml of a dissolution medium could be filled between the bottom of the cylinder and the sieve. 3 ml of water were poured into the device so that there was 2 ml of water below the sieve and 1 ml of water above the sieve. The device was placed on a reciprocal shaking bath (Precision, Winchester, VA), keeping the temperature at 37 °C. When the tablet was immersed into the disintegration medium, the shaker was run in horizontal back and forth motions at 150 rpm. The time until particles of the tablet went through the sieve was recorded as the disintegration time.

Hardness of Tablet

The hardness of the mannose tablets was determined using a VK 200 Tablet Hardness Tester (Vankel, 36 Meridan Road, Edison, NJ 08820).

Moisture Sorption Isotherms

Moisture sorption isotherms of superporous hydrogel (SPH) particles were determined using a Symmetric Gravimetric Analyzer Model 100 (SGA-100, VTI Corporation, Hialeah, FL). Sample particles (5.0 mg, 44-106 µm) were placed in the sample holder and dried at 60 °C for 3 hours. The relative humidity was then set to zero until stable mass was recorded, and the balance was zeroed. The sorption balance was programmed to generate relative humidity steps in an absorption/desorption cycle. The target relative humidity used during the absorption stage under a continuous nitrogen flow of 200 cm^3/min was in the range of 10–90% relative humidity. The relative humidity was held at each 5% relative humidity increments until equilibrium was reached.

Scanning Electron Microscope

Powder samples were adhered to the scanning electron microscope (SEM) sample holder by double-faced copper paper. The tablet samples were broken by a shock of a blade so that the exposed surface did not have contact with the blade. The samples were then mounted to a sample holder and sputter-coated for three minutes. Images of the prepared samples were then taken by a SEM.

Results and Discussion

Isothermal Moisture Absorption of Mannose

An isothermal moisture absorption test was conducted to examine moisture absorption into the mannose powders. The moisture absorption property is important for finding the optimal condition for formation of liquid bridging as well as the drying process. As shown in Figure 3, the moisture absorption into mannose powders was almost negligible up to 70% relative humidity at 25°C, and then the water absorption increased linearly with the relative humidity. Thus, 70% relative humidity is the critical value for mannose.

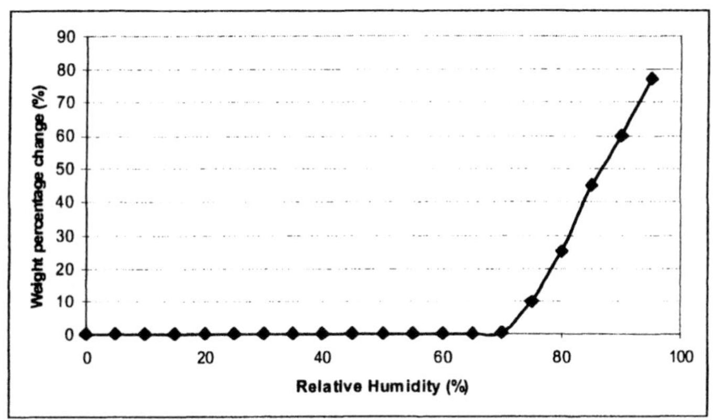

Figure 3. Isothermal moisture absorption of mannose powders.

Mechanisms of Fast Disintegration of Mannose Tablets

It was observed that water penetration into mannose tablets was instantaneous. The void space inside the tablet was quickly filled with water, unless the porosity of the tablet was too low to allow effective water penetration. Water penetration alone, however, cannot explain the fast disintegration property of mannose tablets because some tablets with intermediate pore size did not disintegrate fast. This observation can be explained by fast dissolution of mannose molecules inside the tablet due to their high aqueous solubility, which weakens the whole structure of the tablet. The tablet cannot withstand its structure and collapses, resulting in its disintegration. To test this hypothesis, disintegration tests of mannose tablets in different media were carried out. Mannose tablets compressed at 300 lbs with 0.5 inch punches were used as sample tablets for the test. The experimental conditions were kept the same except for the disintegration media, which consisted of different concentrations of mannose solutions and other solvents.

Results of the disintegration tests are summarized in Table I. It was observed that the time for the media to penetrate into the tablet was negligible compared to the whole disintegration times in all cases. The disintegration time increased as the mannose concentration in the medium increased. Solubilities of mannose in pyridine and ethanol are 0.29 g/ml and 0.004 g/ml, respectively *(17)*.

Both the concentration of the solution and the solubility of the material affect the dissolution process. Therefore, it appears that the dissolution rate of the material determines the disintegration kinetics of the tablets. Since the main mechanism of disintegration is quick dissolution of mannose, it is critical to maximize the inner surface of the tablet to improve the fast disintegration behavior of mannose tablets.

Table I. Disintegration test results of D-mannose tablets in different disintegration media.

Disintegration media	Water	Mannose soln. (0.1 g/ml)	Mannose soln. (0.2 g/ml)	Mannose soln. (0.4 g/ml)
Disintegration time (sec)	7.0 ± 0.3	11.1 ± 2.1	14.0 ± 2.3	18.7 ± 1.6

Disintegration media	Mannose soln. (0.6 g/ml)	Mannose soln. (2.5 g/ml)	Pyridine	Ethanol
Disintegration time (sec)	36.9 ± 16.3	> 7,200	532 ± 50	> 7,200

Effect of Moisture Treatment on the Hardness and Disintegration of Mannose Tablets

To test the effects of porosity and duration of moisture treatment on the properties of the resultant tablets, a series of tablets were prepared. As shown in Table II, the compression force ranged from 100 to 1000 lbs. The tablets were then placed in a 75% relative humidity chamber and sampled at 3, 4, 6 and 8 hours. The tablet strength was tested by a hardness tester. Disintegration tests of these tablets were performed as described above, and the thickness and diameter of the tablets were measured before and after the treatment to calculate the extent of volume reduction.

As summarized in Table II, the tablet strength improved after the treatment. The strength gained by tablets compressed at 100 lbs was not significant in all cases. On the other hand, the tablets compressed at 1,000 lbs gained some strength, but their disintegration times were significantly increased. Tablets compressed at 300 lbs gained strength gradually with moisture treatment time, but the disintegration times remained essentially the same. These different responses to moisture treatment may indicate differences in the pore size distribution inside the tablet, especially the lower portion of the size distribution. Tablets compressed at 100 lbs do not contain enough small pores for the moisture layers to merge together to make liquid bridges. On the other hand, tablets compressed by 1,000 lbs have many small pores. With excess amount of small pores and lack of large pores, the tablet porosity and surface area inside the tablet significantly decreased leading to the observed slow disintegration.

This point is also supported by the volume reduction data. With about the same amount of water absorbed, the volume reductions tend to increase with increase in compression. Therefore, it is very important to find the optimal pore distribution, with enough small pores to merge and gain strength and at the same time sufficient large pores necessary for disintegration.

Table II. Effects of moisture treatment on the properties of tablets made at different compression pressures.

Moisture treatment (hours)	Compression force (lb)	Hardness (KP)	Disintegration time (sec)	Volume reduction (%)
0	100	1.5 ± 0.2	5.1 ± 1.4	
	300	2.5 ± 0.3	7.7 ± 1.5	
	600	4.2 ± 0.1	10.4 ± 0.2	
	1,000	4.6 ± 0.6	15.2 ± 3.2	
3	100	2.1 ± 0.3	5.6 ± 0.7	4.1 ± 2.2
	300	4.7 ± 0.6	13.0 ± 0.2	5.7 ± 1.1
	600	5.7 ± 1.7	22.7 ± 1.4	6.1 ± 1.0
	1,000	6.1 ± 0.6	35.1 ± 1.1	5.2 ± 0.7
4	100	2.3 ± 0.6	5.4 ± 0.8	4.5 ± 0.8
	300	4.9 ± 0.8	11.4 ± 0.9	5.9 ± 1.3
	600	5.1 ± 0.7	23.5 ± 2.5	6.9 ± 1.3
	1,000	6.7 ± 0.6	47.1 ± 5.5	6.2 ± 1.3
6	100	2.8 ± 0.8	5.5 ± 0.9	5.6 ± 3.8
	300	4.1 ± 0.3	13.0 ± 0.4	5.7 ± 0.8
	600	5.1 ± 0.9	32.0 ± 6.0	7.3 ± 0.9
	1,000	6.5 ± 1.0	38.0 ± 1.6	7.4 ± 1.3
8	300	5.6 ± 0.1	15.6 ± 0.3	7.1 ± 1.5
	600	5.4 ± 0.5	35.5 ± 4.7	8.7 ± 1.7
	1,000	6.1 ± 0.6	42.8 ± 2.4	8.4 ± 1.1

Additional visual confirmation of the tablet structural changes was gained from SEM images (Figure 4). The image of a tablet compressed at 300 lbs without humidity treatment shows some individual particles that are still not merged together. However, after humidity treatment small pores were significantly decreased or had merged but larger pores were still intact, allowing fast absorption of water into the tablet core. The tablet compressed at 1000 lbs after humidity treatment, however, revealed many merged pores and few large pores, blocking the absorption of water into the tablet core. Although the images were taken at different magnifications, a comparison of the amount of pores larger than 10 µm revealed that the tablet compressed at 300 lbs after humidity treatment had many more pores than the tablet compressed at 1000 lbs after humidity treatment. This visual examination supports the results from the hardness and disintegration tests.

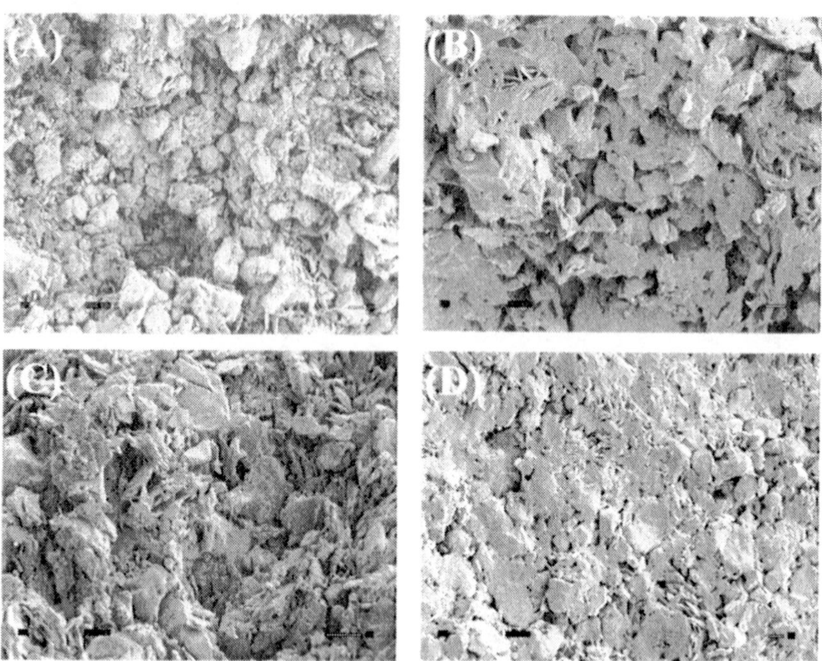

Figure 4. SEM pictures of the cross-sectional (horizontal) views of mannose tablets compressed at different pressures. (A) Tablet compressed at 300 lbs and before humidity treatment (magnification of 500). (B) Tablet compressed at 300 lbs and after humidity treatment at 75% relative humidity for 4 hours (magnification of 700). (C) Tablet compressed at 1000 lbs and before humidity treatment (magnification of 1200). (D) Tablet compressed at 1000 lbs and after humidity treatment at 75% relative humidity for 4 hours (magnification of 500).

The importance of the pore size distribution became evident when tablets made from different particle sizes were compared. Mannose particles of three different particle size ranges, i.e., 75-90 µm, 212-250 µm, and 1000-1400 µm, were collected to make tablets. 300-mg samples of each size range were compressed to make tablets of the same volume, and therefore, the same total pore volumes. The only difference among those three types of tablets was the pore size distribution. Those tablets went through the same moisture treatment, i.e., 75% relative humidity at 25°C for 4 hours. The results are listed in Table III. Tablet made from the smallest mannose particles showed the highest hardness after vapor sorption. Although there are some fluctuations, the disintegration time of those tablets remained in an acceptable range. The applied compression pressure was such that there was enough total pore volume to ensure fast disintegration. The disintegration time was not very sensitive to the pore distribution as long as the total pore volume was kept constant. However, the tablet strength was dependent on the pore distribution and could be varied by using different initial particle sizes. Smaller particles tend to have more intimate contacts than larger particles, and this may contribute to the observed higher tablet strength. Fracture of the initial particles was limited to the compression stage because of the low compression pressure used in making these mannose tablets.

Table III. Effects of different initial particle size on tablet properties.

		Hardness (KP)		
Particle Size Size (μm)	Initial Volume (mm^3)	Before Vapor Sorption	After Vapor Sorption	Volume Reduction (%)
75-90	345.5 ± 5.3	0.2 ± 0.3	4.3 ± 0.5	8.1±0.3
212-250	347.5 ± 2.4	0.1 ± 0.2	3.7 ± 0.1	7.0±0.7
1000-1400	344.2 ± 3.3	0.3 ± 0.3	3.4 ± 0.4	7.0±0.8

	Disintegration Time (Sec)	
Particle Size Size (μm)	Before Vapor Sorption	After Vapor Sorption
75-90	7.5±0.7	12.1±0.5
212-250	8.4±1.5	10.6±1.2
1000-1400	7.4±1.2	14.9±1.1

Conclusions

The strength of tablets containing mannose was enhanced by the sequential process of moisture absorption and drying. Mannose has a critical relative humidity of 70% at 25 °C. When the relative humidity is kept above this level, mannose particles within the tablet absorb water from the environment, leading to the formation of a liquid layer on the particle surfaces. As the liquid layers on the particles grow, different layers on adjacent particles merge together to form liquid bridges between these particles. Upon drying, these liquid bridges become solid bridges to form stable bonds, which significantly increase the strength of the whole tablet. By optimizing the amount of water absorbed, and therefore, the formation of liquid bridges, the strength of the tablets can be controlled. This method does not significantly reduce the tablet volume, and thus, the pores inside the tablet are maintained for fast disintegration and dissolution.

Acknowledgments

This study was supported in part by Samyang Corporation and NSF Industry/University Center for Pharmaceutical Processing Research.

References

1. Lindgren, S.; Janzon, L. Dysphagia: Prevalence of swallowing complaints and clinical finding. *Medical Clinics of North America* **1993**, *77*, 3-5.
2. Washington, N.; Wilson, C.G.; Greaves, J.L.; Norman, S.; Peach, J.M.; Pugh, K. A gamma scintigraphic study of gastric coating by Expidet, tablet and liquid formulations. *Int. J. Pharm.* **1989**, *57*, 17-22.
3. Wilson, C.G.; Washington, N.; Norman, S.; Greaves, J.L.; Peach, J.M.; Pugh, K. A gamma scintigraphic study to compare esophageal clearance of Expidet formulations, tablets and capsules in supine volunteers. *Int. J. Pharm.* **1988**, *46*, 241-246.
4. Wilson, C.G.; Washington, N.; Peach, J.; Murray, G.R.; Kennerley, J. The behavior of a fast-dissolving dosage form (Expidet) followed by g-scintigraphy. *Int. J. Pharm.* **1987**, *40*, 119-123.
5. Ishikawa, T.; Koizumi, N.; Mukai, B.; Utoguchi, N.; Fujii, M.; Matsumoto, M.; Endo, H.; Shirotake, S.; Watanabe, Y. Pharmacokinetics of acetaminophen from rapidly disintegrating compressed tablet prepared using microcrystalline cellulose (PH-M-06) and spherical sugar granules. *Chem. Pharm. Bull.* **2001**, *49*, 230-232.

6. Seager, H. Drug-delivery products and the Zydis fast-dissolving dosage form. *J. Pharm. Pharmacol.* **1998**, *50*, 375-382.
7. Dobetti, L. Fast-melting tablets: Developments and technologies. *Pharm. Technol. North Am.* **2001**, (Suppl.), *44-46*, 48-50.
8. Chang, R.-K.; Guo, X.; Burnside, B.A.; Couch, R.A. Fast-dissolving tablets. *Pharm. Technol.* **2000**, *24*, 52, 54, 56, 58.
9. Habib, W.; Khankari, R.; Hontz, J. Fast-dissolve drug delivery systems. *Crit. Rev. Ther. Drug Carrier Syst.* **2000**, *17*, 61-72.
10. Sastry, S.V.; Nyshadham, J.R.; Fix, J.A. Recent technological advances in oral drug delivery. A review. *Pharm. Sci. Technol. Today* **2000**, *3*, 138-145.
11. Yang, S.; Fu, Y.; Jeong, S.H.; Park, K. Application of poly(acrylic acid) superporous hydrogel microparticles as a super-disintegrant in fast-disintegrating tablets. *J. Pharm. Pharmacol.* **2004**, *56*, 429-436.
12. Chowhan, Z.T., Moisture, hardness, disintegration and dissolution interrelationships in compressed tablets prepared by the wet granulation process. *Drug Dev. Ind. Pharm.* **1979**, *5*, 41-62.
13. Chowhan, Z.T., The effect of low- and high-humidity ageing on the hardness, disintegration time and dissolution rate of dibasic calcium phosphate-based tablets. *J. Pharm. Pharmacol.* **1980**, *32*, 10-14.
14. Chowhan, Z.T.; Amaro, A.A. The effect of low- and high-humidity aging on the hardness, disintegration time and dissolution rate of tribasic calcium phosphate-based tablets. *Drug Dev. Ind. Pharm.* **1979**, *5*, 545-562.
15. Chowhan, Z.T.; Palagyi, L. Hardness increase induced by partial moisture loss in compressed tablets and its effect on in vitro dissolution. *J. Pharm. Sci.* **1978**, *67*, 1385-1389.
16. Chowhan, Z.T., Role of binders in moisture-induced hardness increase in compressed tablets and its effect on in vitro disintegration and dissolution. *J. Pharm. Sci.* **1980**, *69*, 1-4.
17. Stecher, P.G. The Merk Index of Chemicals and Drugs, 638 pages, 1960.

Author Index

Subject Index

A

B

C

D

E

F

K

L

M

O

P

R

S

T

U

V

W

X

Z